COMPLIANT MECHANISMS

COMPLIANT MECHANISMS

LARRY L. HOWELL
Mechanical Engineering Department
Brigham Young University

A Wiley-Interscience Publication
JOHN WILEY & SONS, INC.

New York / Chichester / Weinheim / Brisbane / Singapore / Toronto

Copyright © 2001 by John Wiley & Sons, Inc. All rights reserved.

Published simultaneously in Canada.

No part of this publication may be reproduced, stored in a retrieval system or transmitted in any form or by any means, electronic, mechanical, photocopying, recording, scanning or otherwise, except as permitted under Section 107 or 108 of the 1976 United States Copyright Act, without either the prior written permission of the Publisher, or authorization through payment of the appropriate per-copy fee to the Copyright Clearance Center, 222 Rosewood Drive, Danvers, MA 01923, (978) 750-8400, fax (978) 750-4744. Requests to the Publisher for permission should be addressed to the Permissions Department, John Wiley & Sons, Inc., 605 Third Avenue, New York, NY 10158-0012, (212) 850-6011, fax (212) 850-6008, E-Mail: PERMREQ @ WILEY.COM.

This publication is designed to provide accurate and authoritative information in regard to the subject matter covered. It is sold with the understanding that the publisher is not engaged in rendering professional services. If professional advice or other expert assistance is required, the services of a competent professional person should be sought.

Library of Congress Cataloging-in-Publication Data:

Howell, Larry L.
  Compliant mechanisms/Larry L. Howell
    p. cm.
  ISBN 0-471-38478-X (cloth : alk. paper)
      1. Mechanical movements. 2. Machinery, Kinematics of. I. Title

TJ181 .H89 2002
621.8'11—dc21   2001026196

10 9 8 7 6 5 4 3 2 1

To my wife

Peggy

and my children

Angela, Travis, Nathan, and Matthew

CONTENTS

PREFACE .... **XV**

1 INTRODUCTION **1**
   1.1 Advantages of Compliant Mechanisms 2
   1.2 Challenges of Compliant Mechanisms 6
   1.3 Historical Background 8
   1.4 Compliant Mechanisms and Nature 10
   1.5 Nomenclature and Diagrams 11
         1.5.1 Compliant Mechanisms versus Compliant Structures 12
         1.5.2 Nomenclature 12
         1.5.3 Diagrams 15
   1.6 Compliant MEMS 15
      Problems 18

2 FLEXIBILITY AND DEFLECTION **21**
   2.1 Linear versus Nonlinear Deflections 21
   2.2 Stiffness and Strength 22
   2.3 Flexibility 23
   2.4 Displacement versus Force Loads 26
   2.5 Material Considerations 28
         2.5.1 Maximum Deflection for a Flexible Beam 28

		2.5.2	Ratio of Strength to Young's Modulus 29
		2.5.3	Other Material Selection Criteria 30
		2.5.4	Creep and Stress Relaxation 32
	2.6	Linear Elastic Deflections 34	
	2.7	Energy Storage 38	
	2.8	Stress Stiffening 41	
	2.9	Large-Deflection Analysis 42	
		2.9.1	Beam with Moment End Load 43
		2.9.2	Elliptic-Integral Solutions 45
		2.9.3	Numerical Methods 55
		Problems 55	

3 FAILURE PREVENTION 61

	3.1	Stress 61	
		3.1.1	Principal Stresses 62
		3.1.2	Stress Concentrations 67
	3.2	Static Failure 67	
		3.2.1	Ductile Materials 68
		3.2.2	Brittle Materials 73
	3.3	Fatigue Failure 77	
		3.3.1	Fatigue Basics 78
		3.3.2	Fatigue Failure Prediction 79
		3.3.3	Estimating Endurance Limit and Fatigue Strength 82
		3.3.4	Endurance Limit and Fatigue Strength Modification Factors . . . 83
		3.3.5	Surface Factor 84
		3.3.6	Size Factor 84
		3.3.7	Load Factor 85
		3.3.8	Reliability 86
		3.3.9	Miscellaneous Effects 86
		3.3.10	Completely Reversed Loading 88
		3.3.11	Fluctuating Stresses 93
		3.3.12	Fatigue of Polymers 98
		3.3.13	Testing 102
		Problems 104	

4 RIGID-LINK MECHANISMS 111

	4.1	Introduction 111	
		4.1.1	Mobility 111
		4.1.2	Kinematic Chains and Inversions 112
		4.1.3	Classification of Four-Bar Mechanisms 113
		4.1.4	Mechanical Advantage 113
	4.2	Position Analysis 115	
		4.2.1	Four-Bar Mechanism: Closed-Form Equations 116
		4.2.2	Slider–Crank Mechanism: Closed-Form Equations 117
		4.2.3	Complex Number Method 118

Contents

	4.3	Velocity and Acceleration	123
	4.4	Kinematic Coefficients	125
		4.4.1 Four-Bar Kinematic Coefficients	125
		4.4.2 Slider–Crank Kinematic Coefficients	126
	4.5	Mechanism Synthesis	126
		4.5.1 Function Generation	127
		4.5.2 Path Generation	129
		4.5.3 Motion Generation	130
		Problems	131

5 PSEUDO-RIGID-BODY MODEL 135

	5.1	Small-Length Flexural Pivots	136
		5.1.1 Active and Passive Forces	140
		5.1.2 Stress	141
		5.1.3 Living Hinges	144
	5.2	Cantilever Beam with a Force at the Free End (Fixed–Pinned)	145
		5.2.1 Parametric Approximation of the Beam's Deflection Path	147
		5.2.2 Characteristic Radius Factor	148
		5.2.3 Coordinates of Beam End	150
		5.2.4 Rule of Thumb for Characteristic Radius Factor	151
		5.2.5 Angular Deflection Approximation	152
		5.2.6 Stiffness Coefficient	152
		5.2.7 Torsional Spring Constant	156
		5.2.8 Stress	157
		5.2.9 Practical Implementation of Fixed–Pinned Segments	160
	5.3	Fixed–Guided Flexible Segment	162
	5.4	End-Moment Loading	165
	5.5	Initially Curved Cantilever Beam	166
		5.5.1 Stiffness Coefficient for Initially Curved Beams	169
		5.5.2 Stress for Initially Curved Beams	170
	5.6	Pinned–Pinned Segment	170
		5.6.1 Initially Curved Pinned–Pinned Segments	172
	5.7	Segment with Force and Moment (Fixed–Fixed)	175
		5.7.1 Loading Cases	175
	5.8	Other Methods of Pin Joint Simulation	180
		5.8.1 Living Hinges	181
		5.8.2 Passive Joints	183
		5.8.3 Q-Joints	185
		5.8.4 Cross-Axis Flexural Pivots	189
		5.8.5 Torsional Hinges	190
		5.8.6 Split-Tube Flexures	193
	5.9	Modeling of Mechanisms	194
		5.9.1 Examples	195
	5.10	Use of Commercial Mechanism Analysis Software	205
		Problems	209

6 FORCE–DEFLECTION RELATIONSHIPS 219

- 6.1 Free-Body Diagram Approach 220
- 6.2 Generalized Coordinates 225
- 6.3 Work and Energy 226
- 6.4 Virtual Displacements and Virtual Work 228
- 6.5 Principle of Virtual Work 230
- 6.6 Application of the Principle of Virtual Work 231
- 6.7 Spring Function for Fixed–Pinned Members 237
- 6.8 Pseudo-Rigid-Body Four-Bar Mechanism 239
- 6.9 Pseudo-Rigid-Body Slider Mechanism 248
- 6.10 Multi-Degree-of-Freedom Mechanisms 254
- 6.11 Conclusions 256
  - Problems 256

7 NUMERICAL METHODS 259

- 7.1 Finite Element Analysis 260
- 7.2 Chain Algorithm 261
  - 7.2.1 Shooting Method 268

8 COMPLIANT MECHANISM SYNTHESIS 275

- 8.1 Rigid-Body Replacement (Kinematic) Synthesis 275
  - 8.1.1 Loop Closure Equations 280
- 8.2 Synthesis with Compliance: Kinetostatic Synthesis 286
  - 8.2.1 Additional Equations and Unknowns 287
  - 8.2.2 Coupling of Equations 288
  - 8.2.3 Design Constraints 290
  - 8.2.4 Special Case of $\theta_o = \theta_j$ 292
- 8.3 Other Synthesis Methods 297
  - 8.3.1 Burmester Theory for Finite Displacements 297
  - 8.3.2 Infinitesimal Displacements 298
  - 8.3.3 Optimization of Pseudo-Rigid-Body Model 298
  - 8.3.4 Optimization 299
- 8.4 Problems 299

9 OPTIMAL SYNTHESIS WITH CONTINUUM MODELS ... 301

Ananthasuresh, G. K., and Frecker, M. I.

- 9.1 Introduction 301
  - 9.1.1 Distributed Compliance 303
  - 9.1.2 Continuum Models 303

Contents

	9.1.3	Elastostatic Analysis Using the Finite Element Method 304
	9.1.4	Structural Optimization 305
9.2	Formulation of the Optimization Problem..................... 306
	9.2.1	Objective Function, Constraints, and Design Variables 306
	9.2.2	Measures of Stiffness and Flexibility..................... 308
	9.2.3	Multicriteria Formulations 310
9.3	Size, Shape, and Topology Optimization..................... 312
	9.3.1	Size Optimization 312
	9.3.2	Shape Optimization 319
	9.3.3	Topology Optimization 319
9.4	Computational Aspects 323
	9.4.1	Optimization Algorithms 324
	9.4.2	Sensitivity Analysis................................... 325
9.5	Optimality Criteria Methods 327
	9.5.1	Derivation of the Optimality Criterion 327
	9.5.2	Solution Procedure 329
	9.5.3	Examples 329
9.6	Conclusion 332
9.7	Acknowledgments 332
	Problems... 332

10 SPECIAL-PURPOSE MECHANISMS 337

10.1 Compliant Constant-Force Mechanisms 337
    10.1.1 Pseudo-Rigid-Body Model of Compliant Slider Mechanisms .. 338
    10.1.2 Dimensional Synthesis 339
    10.1.3 Determination of Force Magnitude 342
    10.1.4 Examples 343
    10.1.5 Estimation of Flexural Pivot Stress 344
    10.1.6 Examples 345

10.2 Parallel Mechanisms 346
    10.2.1 Compliant Parallel-Guiding Mechanisms 347
    10.2.2 Applications 347
    10.2.3 Pseudo-Rigid-Body Model 350
    10.2.4 Additional Design Considerations....................... 352

Problems... 353

11 BISTABLE MECHANISMS 355

11.1 Stability 355

11.2 Compliant Bistable Mechanisms 357

11.3 Four-Link Mechanisms 359
    11.3.1 Energy Equations 360
    11.3.2 Requirements for Bistable Behavior 362
    11.3.3 Young Bistable Mechanisms............................ 367

11.4 Slider–Crank or Slider–Rocker Mechanisms 372
    11.4.1 Energy Equations 373
    11.4.2 Requirements for Bistable Behavior 374
    11.4.3 Examples for Various Spring Positions 374

11.5 Double-Slider Mechanisms 377
    11.5.1 Double-Slider Mechanisms with a Pin Joining the Sliders 377
    11.5.2 Double-Slider Mechanisms with a Link Joining the Sliders . . . 379
    11.5.3 Requirements for Bistable Behavior 381

11.6 Snap-Through Buckled Beams 382

11.7 Bistable Cam Mechanisms 382

    Problems 383

A REFERENCES 385

B PROPERTIES OF SECTIONS 399

B.1 Rectangle 399

B.2 Circle 399

B.3 Hollow Circle 400

B.4 Solid Semicircle 400

B.5 Right Triangle 400

B.6 I Beam with Equal Flanges 400

C MATERIAL PROPERTIES 401

D LINEAR ELASTIC BEAM DEFLECTIONS 407

D.1 Cantilever Beam with a Force at the Free End 407

D.2 Cantilever Beam with a Force Along the Length 407

D.3 Cantilever Beam with a Uniformly Distributed Load 408

D.4 Cantilever Beam with a Moment at the Free End 408

D.5 Simply Supported Beam with a Force at the Center 408

D.6 Simply Supported Beam with a Force Along the Length 409

D.7 Simply Supported Beam with a Uniformly Distributed Load 409

D.8 Beam with One End Fixed and the Other End Simply Supported . . . 409

D.9 Beam with Fixed Ends and a Center Load 410

D.10 Beam with Fixed Ends and a Uniformly Distributed Load 410

D.11 Beam with One End Fixed and the Other End Guided 410

E PSEUDO-RIGID-BODY MODELS 411

- E.1 Small-Length Flexural Pivot 411
- E.2 Vertical Force at the Free End of a Cantilever Beam. 412
- E.3 Cantilever Beam with a Force at the Free End. 413
- E.4 Fixed–Guided Beam 415
- E.5 Cantilever Beam with an Applied Moment at the Free End 416
- E.6 Initially Curved Cantilever Beam 417
- E.7 Pinned–Pinned Segments. 418
  - E.7.1 Initially Curved Pinned–Pinned Segments 418
- E.8 Combined Force–Moment End Loading 420

F EVALUATION OF ELLIPTIC INTEGRALS. 421

G TYPE SYNTHESIS OF COMPLIANT MECHANISMS . . 425

Murphy, M. D.

- G.1 Matrix Representation for Rigid-Link Mechanisms 425
- G.2 Compliant Mechanism Matrices 426
  - G.2.1 Segment-Type Designation. 428
  - G.2.2 Connection-Type Designation. 428
  - G.2.3 Examples 429
- G.3 Determination of Isomorphic Mechanisms 429
  - G.3.1 Rigid-Body Isomorphic Detection Techniques 431
  - G.3.2 Isomorphism Detection for Compliant Mechanisms 431
- G.4 Type Synthesis 433
- G.5 Determination of Design Requirements. 434
- G.6 Topological Synthesis of Compliant Mechanisms. 435
  - G.6.1 Segment-Type Enumeration 436
  - G.6.2 Connection-Type Enumeration 437
  - G.6.3 Combined Segment and Connection-Type Results 438
  - G.6.4 Formation of Compliant Mechanisms 441
- G.7 Examples 442
  - G.7.1 Discussion 449

INDEX. 451

PREFACE

Compliant mechanisms offer great promise in providing new and better solutions to many mechanical-design problems. Since much research in the theory of compliant mechanisms has been done in the last few years, it is important that the abundant information be presented to the engineering community in a concise, understandable, and useful form. The purpose of this book is to fulfill this need for students, practicing engineers, and researchers.

The book presents methods for the analysis and design of compliant mechanisms and illustrates them with examples. The materials in the book provide ideas for engineers to employ the advantages of compliant mechanisms in ways that otherwise may not be possible. The analysis of small deflection devices is addressed, but emphasis is given to compliant mechanisms that undergo large, nonlinear deflections. The pseudo-rigid-body model is introduced as a method which simplifies the analysis of compliant mechanisms that undergo large deflections by modeling them with elements common to traditional mechanisms. This simplification makes it possible to design compliant mechanisms for many types of tasks. The advantages of compliant mechanisms in the emerging area of microelectromechanical systems (MEMS) are also addressed, and several MEMS examples are provided throughout the book.

The chapters are organized to flow from simple to more complex concepts; the book then concludes with the application of the previous materials to specific types of devices. This is done by organizing the chapters into major sections of introduction, fundamentals, analysis, design, and special-purpose mechanisms. In a similar way, simple examples facilitate understanding, followed by more complicated examples that demonstrate how the material can be used in applications.

Review of essential topics in strength of materials, machine design, and kinematics is provided to create a self-contained book that does not require a lot of additional references to solve compliant-mechanism problems. These reviews can help emphasize important topics the reader has studied previously, or they can be used as a resource for those from other disciplines who are working in the area of MEMS or related areas. The appendixes provide a resource for quick reference to important equations presented in the book.

The area of compliant mechanisms exists thanks to the vision and insight of Professor Ashok Midha. Many have contributed to the knowledge of compliant mechanisms, but Professor Midha may be considered the father of modern compliant mechanisms. His insight and vision have had a profound effect on the field and on those with whom he has associated. I have greatly benefited from both his work in compliant mechanisms and his example and mentorship, and I am grateful for his influence.

The earlier versions of this book were used as notes in compliant mechanisms courses offered at Brigham Young University, Purdue University, and the University of Missouri, Rolla. Students made many helpful comments to improve the quality of the notes.

Several colleagues have graciously volunteered their time and expertise by contributing parts of the book. Professor G. K. Ananthasuresh at the University of Pennsylvania and Professor Mary I. Frecker at Pennsylvania State University wrote Chapter 9. Dr. Morgan D. Murphy of Delphi Automotive Systems contributed Appendix G. Chapter 11 relies heavily on graduate work completed by Brian Jensen when he was at Brigham Young University.

Some of the text and figures in this book are summarized from previous writings, including a number of papers coauthored with graduate students and colleagues and published by the American Society of Mechanical Engineers (ASME) in various conference proceedings and in the *Journal of Mechanical Design*. Work from a number of graduate student theses has also been included. Grateful thanks is extended to all those who have participated in this work: James Derderian, Patrick Opdahl, Brian Edwards, John Parise, and Brian Jensen have generously contributed sections of this book. The contributions of Scott Lyon, Brent Weight, and Greg Roach are also greatly appreciated, as are the efforts of many other students that have made this possible. The valuable assistance of Megan Poppitz is also gratefully acknowledged.

The Mechanical Engineering Department at Brigham Young University has been very supportive of this project and has provided many resources to assist in its completion. The College of Engineering and the administration of Brigham Young University have also supported the author's efforts in many ways.

In addition to the many students who have provided recommendations and encouragement for this work, others are thanked for their helpful reviews and comments to improve the manuscript. Special thanks to Professor G. K. Ananthasuresh, Dr. Morgan D. Murphy, Professor Kenneth W. Chase, and Professor Don Norton of Brigham Young University's English Department, and the university editing service for valuable reviews and comments on the manuscript.

Much of the fundamental work in compliant mechanisms has been funded by the National Science Foundation (NSF). The resources provided were a wise investment and will have a far-reaching impact for many years to come. The following NSF grants have supported the author's work in the area of compliant mechanisms: DMI-9624574 (CAREER Award), CMS-9978737, ECS-9528238, and DMI-9980835. The Utah Center of Excellence Program is also acknowledged for support of commercialization of compliant mechanism theory through funding of the Center of Excellence in Compliant Mechanisms.

I express my love and gratitude to my wife and children for their continued love, support, and companionship. And my eternal thanks to my parents, for their love and sacrifice. Finally, I humbly acknowledge the gifts from God, for which no words could ever adequately express my gratitude.

LARRY L. HOWELL

Provo, Utah

CHAPTER 1

INTRODUCTION

A mechanism is a mechanical device used to transfer or transform motion, force, or energy [1, 2]. Traditional rigid-body mechanisms consist of rigid links connected at movable joints; the section of a reciprocating engine shown in Figure 1.1a is an example. The linear input is transformed to an output rotation, and the input force is transformed to an output torque. As another example, consider the Vise Grip pliers shown in Figure 1.1b. This mechanism transfers energy from the input to the output. Since energy is conserved between the input and output (neglecting friction losses), the output force may be much larger than the input force, but the output displacement is much smaller than the input displacement. Like mechanisms, structures may also consist of rigid links connected at joints, but relative rigid-body motion is not allowed between the links.

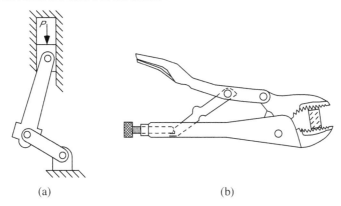

Figure 1.1. Examples of rigid-link mechanisms: (a) part of a reciprocating engine, and (b) Vise Grip.

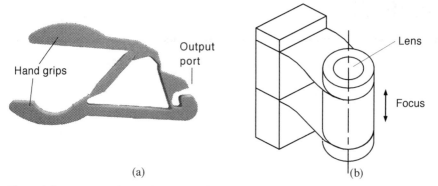

Figure 1.2. Examples of compliant mechanisms: (a) crimping mechanism (from [3]), and (b) parallel-guiding mechanism.

A compliant mechanism also transfers or transforms motion, force, or energy. Unlike rigid-link mechanisms, however, compliant mechanisms gain at least some of their mobility from the deflection of flexible members rather than from movable joints only. An example of a compliant crimping mechanism is shown in Figure 1.2a. The input force is transferred to the output port, much like the Vise Grip, only now some energy is stored in the form of strain energy in the flexible members. Note that if the entire device were rigid, it would have no mobility and would therefore be a structure. Figure 1.2b shows a device that also requires compliant members to focus a lens [4, 5].

1.1 ADVANTAGES OF COMPLIANT MECHANISMS

Compliant mechanisms may be considered for use in a particular application for a variety of reasons. The advantages of compliant mechanisms are considered in two categories: cost reduction (part-count reduction, reduced assembly time, and simplified manufacturing processes) and increased performance (increased precision, increased reliability, reduced wear, reduced weight, and reduced maintenance).

An advantage of compliant mechanisms is the potential for a dramatic reduction in the total number of parts required to accomplish a specified task. Some mechanisms may be manufactured from an injection-moldable material and be constructed of one piece. For example, consider the compliant overrunning clutch shown in Figure 1.3a [6, 7] and its rigid-body counterpart shown in Figure 1.3b. Considerably fewer components are required for the compliant mechanism than for the rigid mechanism. The reduction in part count may reduce manufacturing and assembly time and cost. The compliant crimping mechanism and its rigid-body counterpart illustrated in Figure 1.4 are other examples of part reduction.

Compliant mechanisms also have fewer movable joints, such as pin (turning) and sliding joints. This results in reduced wear and need for lubrication. These are valuable characteristics for applications in which the mechanism is not easily accessible, or for operation in harsh environments that may adversely affect joints.

Advantages of Compliant Mechanisms

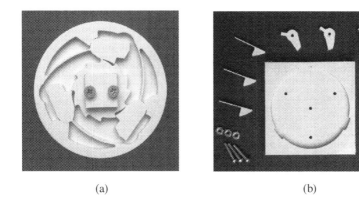

Figure 1.3. (a) Compliant overrunning clutch, and (b) its rigid-body counterpart shown disassembled. (From [6] and [7].)

Reducing the number of joints can also increase mechanism precision, because backlash may be reduced or eliminated. This has been a factor in the design of high-precision instrumentation [8, 9]. An example of a high-precision compliant mechanism is shown in Figure 1.5. Because the motion is obtained from deflection rather than by adjoining parts rubbing against each other, vibration and noise may also be reduced.

An example of a compliant mechanism designed for harsh environments is shown in Figure 1.6. This simple gripping device holds a die (such as a computer chip) during processing. The die must be transported between several different chemicals without becoming damaged. Made of Teflon—inert to the chemicals in which it is placed—the gripper holds the die without external force.

Because compliant mechanisms rely on the deflection of flexible members, energy is stored in the form of strain energy in the flexible members. This stored energy is similar to the strain energy in a deflected spring, and the effects of springs may be integrated into a compliant mechanism's design. In this manner, energy can easily be stored or transformed, to be released at a later time or in a different manner. A bow-and-arrow system is a simple example. Energy is stored in the limbs as the archer draws the bow; strain energy is then transformed to the kinetic energy of

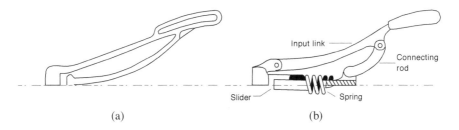

Figure 1.4. (a) Compliant crimping mechanism developed by AMP Inc., and (b) its rigid-body counterpart. Because of symmetry, only half the mechanism is shown. (From [4].)

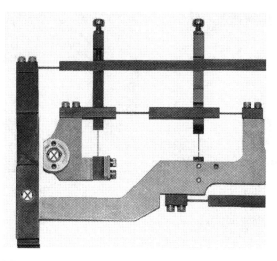

Figure 1.5. Example of a high-precision compliant mechanism.

the arrow. These energy storage characteristics may also be used to design mechanisms that have specific force–deflection properties, or to cause a mechanism to tend to particular positions. For example, the mechanism shown in Figure 1.7 is a robot end effector that was designed to have a constant output force regardless of the input displacement.

It is possible to realize a significant reduction in weight by using compliant mechanisms rather than their rigid-body counterparts. This may be a significant factor in aerospace and other applications. Compliant mechanisms have also benefited companies by reducing the weight and shipping costs of consumer products.

Figure 1.6. Compliant die grippers used to hold a die during process in several different harsh chemicals.

Figure 1.7. Compliant constant-force robot end effector.

Another advantage of compliant mechanisms is the ease with which they are miniaturized [10–19]. Simple microstructures, actuators, and sensors are seeing wide usage, and many other microelectromechanical systems (MEMS) show great promise. The reduction in the total number of parts and joints offered by compliant mechanisms is a significant advantage in the fabrication of micromechanisms. Compliant micromechanisms may be fabricated using technology and materials similar to those used in the fabrication of integrated circuits. MEMS are discussed in more detail in Section 1.6.

The compliant fishhook pliers (Compliers) illustrated in Figure 1.8 demonstrate several of the advantages discussed above. Part-count reduction is evident in that they are injection molded as a single piece. They are in a fairly harsh environment

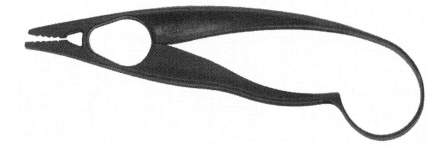

Figure 1.8. Compliant pliers, or Compliers, fishhook removal pliers.

Figure 1.9. Compliant parallel motion bicycle brakes.

where pin joints may rust and require more maintenance. They are also lightweight and are not only easy to carry, but will float if the angler drops them in water.

The high-performance bicycle brakes shown in Figure 1.9 are another example of a compliant mechanism that demonstrates several advantages of compliant mechanisms. Unlike traditional cantilever-type brakes, these brake pads do not rotate in their motion. The first brakes that had such motion used a traditional parallelogram four-bar linkage to achieve the desired motion, and a return spring was necessary to ensure that the brakes would disengage when the rider let go of the handle. The compliant brakes shown have a reduced part count because two pin joints and the return spring are integrated into a single flexible strip of titanium or stainless steel. The manufacturer claims that the manufacturing cost is approximately one-third of that for the other style of parallel motion brakes. The reliability is increased because of the reduced number of parts to fail, and the spring that was causing consumer complaints was eliminated. The reduced joints also make it more reliable in dirty environments, such as mountain biking, because foreign material has a lower probability of getting in joints. A number of common devices that are compliant mechanisms are shown in Figure 1.10.

1.2 CHALLENGES OF COMPLIANT MECHANISMS

Although offering a number of advantages, compliant mechanisms present several challenges and disadvantages in some applications. Perhaps the largest challenge is the relative difficulty in analyzing and designing compliant mechanisms. Knowledge of mechanism analysis and synthesis methods and the deflection of flexible members is required. The combination of the two bodies of knowledge in compliant mechanisms requires not only an understanding of both, but also an understanding of their interactions in a complex situation. Since many of the flexible members undergo large deflections, linearized beam equations are no longer valid. Nonlinear

Challenges of Compliant Mechanisms

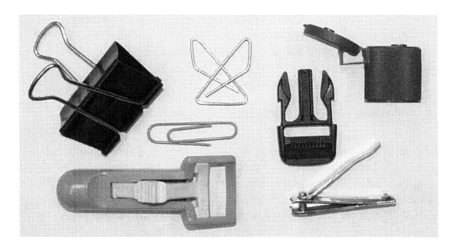

Figure 1.10. Common compliant devices. A binder clip, paper clips, backpack latch, lid eyelash curler, and nail clippers are shown.

equations that account for the geometric nonlinearities caused by large deflections must be used. Because of these difficulties, many compliant mechanisms in the past were designed by trial and error. Such methods apply only to very simple systems that perform relatively simple tasks and are often not cost-efficient for many potential applications. Theory has been developed to simplify the analysis and design of compliant mechanisms, and the limitations are not as great as they once were. Even considering these advances, however, analysis and design of compliant mechanisms are typically more difficult than for rigid-body mechanisms.

Energy stored in flexible elements has been discussed as an advantage, since it can be used to simplify mechanisms that incorporate springs, obtain specified force–deflection relationships, and store energy that is transferred or transformed by the mechanism. However, in some applications, having energy stored in flexible members is a disadvantage. For example, if a mechanism's function is to transfer energy from the input to an output, not all of the energy is transferred, but some is stored in the mechanism.

Fatigue analysis is typically a more vital issue for compliant mechanisms than for their rigid-body counterparts. Since compliant members are often loaded cyclically when a compliant mechanism is used, it is important to design those members so they have sufficient fatigue life to perform their prescribed functions.

The motion from the deflection of compliant links is also limited by the strength of the deflecting members. Furthermore, a compliant link cannot produce a continuous rotational motion such as is possible with a pin joint.

Compliant links that remain under stress for long periods of time or at high temperatures may experience stress relaxation or creep. For example, an electric motor may use the force caused by a deflected cantilever beam to hold brushes in place. With time, this holding force may decrease and the motor will no longer function properly.

The developments in this book present ways to address or overcome these challenges, but it is important to understand the difficulties and limitations of compliant mechanisms. Such knowledge is helpful in determining which applications will benefit most by use of compliant mechanism technology.

1.3 HISTORICAL BACKGROUND

The concept of using flexible members to store energy and create motion has been used for millennia. Archaeological evidence suggests that bows have been in use since before 8000 B.C. and were the primary weapon and hunting tool in most cultures [20]. Consider the longbow illustrated in Figure 1.11. Early bows were constructed of relatively flexible materials such as wood and animal sinew. Strain energy in the bow is transformed to kinetic energy of the arrow.

Catapults are another example of the early use of compliant members used by the Greeks as early as the fourth century B.C. [21]. Early catapults were constructed of wooden members that were deflected to store energy and then release it to propel a projectile. Figure 1.12 is a sketch of a compliant catapult by Leonardo da Vinci.

Flexible members have also been used to simulate the motion of turning joints. The flexural hinges of book covers, for example, have been constructed by changing the material composition and thickness at the desired point of flexure to obtain the desired motion. Other methods were developed early in the twentieth century to obtain this type of motion for other applications. The single-axis cross-flexure pivot [23] (Figure 1.13a) and Bendix Corporation's flexural (Free-Flex) pivots (Figure

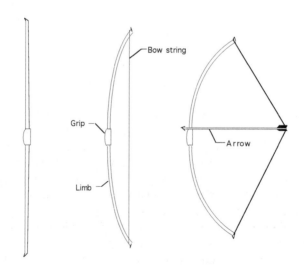

Figure 1.11. Longbow in its unstrung, strung, and drawn positions.

Historical Background

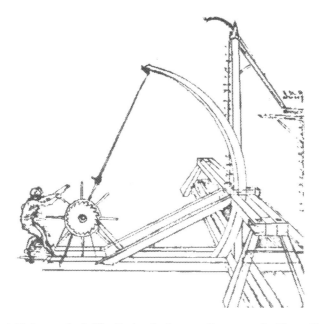

Figure 1.12. Leonardo da Vinci's sketch of a compliant catapult. (From [22].)

1.13b) are examples. The use of flexural or living hinges has been particularly important in products fabricated of injection-molded plastics [24, 25].

Flexible elements have also been used extensively in measurement instruments [8, 9]. Examples include high-accuracy load cells (Figure 1.14) for force measurement, and Bourdon tubes for pressure measurement.

The number of products that rely on flexible members to perform their functions has increased significantly over the last few decades, thanks in part to the develop-

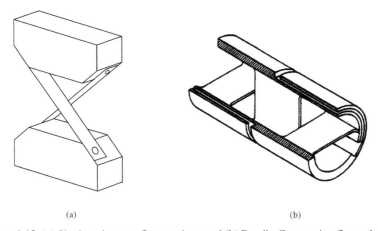

Figure 1.13. (a) Single-axis cross-flexure pivot, and (b) Bendix Corporation flexural pivot.

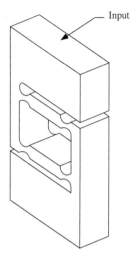

Figure 1.14. Load cell for force measurement.

ment of stronger and more reliable materials. The use of compliant mechanisms will probably continue to increase with time as materials and design methodologies are improved. The demand for increased product quality and decreased cost also pressures manufacturers to implement compliant mechanisms.

University and industry research has played an important role in the development of compliant mechanism theory and application [26–85]. Appendix A lists a number of important publications in the area.

1.4 COMPLIANT MECHANISMS AND NATURE*

Humans and nature often have differing philosophies on mechanical design. Stiff structures are usually preferred by humans because for many, stiffness means strength. Devices that must be capable of motion are constructed of multiple stiff structures assembled in such a manner as to allow motion (e.g., door hinges, linkages, and roller bearings). However, stiffness and strength cannot be equated—stiffness is a measure of how much something deflects under load, whereas strength is how much load can be endured before failure. Despite human tendencies, it is possible to make things that are flexible *and* strong. Nature uses stiff structures where needed—tree trunks, bones, teeth, and claws—but in living organisms, it more often relies on flexibility in living organisms. Bee wings (Figure 1.15), bird wings,

---

*See [26], [81], and [83].

Nomenclature and Diagrams

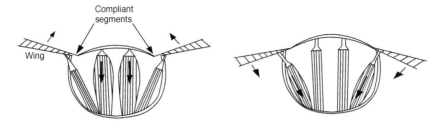

Figure 1.15. Bee wings demonstrate the use of compliance in nature.

tree branches, leaf stems, fish (Figure 1.16), and single-celled organisms are only a few examples of creations that use compliance to their advantage. Nature also has the advantage of *growing* living things, and no assembly is required [81].

The contrast between nature and human design is easily identifiable when humans try to replace one of nature's products. For example, a human heart valve is a compliant one-way valve that is capable of sustaining billions of cycles without failure. However, most current artificial heart valves use a number of assembled stiff parts with pin joints to obtain motion. They also have a comparatively short life, cause difficulty in blood flow, and often damage blood cells.

In many mechanical systems, stiff structures are a necessity. It would be quite disconcerting to feel a floor deflect as one walked across a room or to see a building sway in the wind. However, much can be learned from nature, and many devices may be made flexible and strong in order to gain some of the same advantages seen in living organisms, including increased life of some components and reduced assembly. Compliant mechanisms are one way that flexibility may be used to obtain some of these advantages in mechanical design.

1.5 NOMENCLATURE AND DIAGRAMS*

Nomenclature and skeleton diagrams are important tools in communicating mechanism design information. The following nomenclature for compliant mechanisms is

Figure 1.16. An eel uses its compliance to swim.

---

*The text and figures of this section are summarized from [38], [58], and [70].

consistent with rigid-body mechanism nomenclature, but allows for further description and identification of compliant mechanisms.

1.5.1 Compliant Mechanisms versus Compliant Structures

To illustrate the difference between a compliant mechanism and a compliant structure, consider the flexible cantilever beams shown in Figure 1.17. The diving board shown in Figure 1.17a transforms the kinetic energy of the falling diver to strain energy in the beam, and then transforms it again to kinetic energy as the diver springs off the board. Using the definition of a mechanism as a device that transfers or transforms motion or energy, the diving board qualifies as a mechanism. The same function could be performed by using a more complicated rigid-body mechanism consisting of rigid links and a spring.

The compliant cantilever shown in Figure 1.17b is used in an electric motor to maintain contact between the brush and commutator. Since this device performs its function without transferring or transforming motion or energy, it is classified as a compliant structure.

Note that the two examples in Figure 1.17 both consist of compliant cantilever beams; however, their classification as a structure or mechanism depends on their function. Many of the analysis and design methods presented here apply to both structures and mechanisms.

1.5.2 Nomenclature

Rigid-body mechanisms are constructed of rigid links joined with kinematic pairs, such as pin joints and sliders. These components are easily identified and characterized. Since compliant mechanisms gain at least some of their motion from the deflection of flexible members, components such as links and joints are not as easily distinguished. Identification of such components is useful to allow the accurate communication of design and analysis information.

Link Identification. A link is defined as the continuum connecting the mating surfaces of one or more kinematic pairs. Revolute (pin or turning) joints and

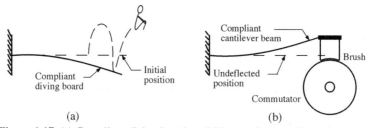

Figure 1.17. (a) Compliant diving board, and (b) compliant cantilever beam.

Nomenclature and Diagrams

prismatic (sliding) joints are examples of kinematic pairs. Links can be identified by disassembling the mechanism at the joints and counting the resulting links. For example, if the Vise Grip shown in Figure 1.1b is disassembled at the four pin joints, then there are four links. Consider the compliant device with one pin joint shown in Figure 1.18a. It is shown disassembled in Figure 1.18b. Note that it consists of one link. The mechanisms shown in Figures 1.2a and 1.3a each consist of one link and one kinematic pair.

Note that the mechanism illustrated in Figure 1.2b has no traditional joints, and therefore has zero links. Such mechanisms are termed *fully compliant mechanisms*, since all their motion is obtained from the deflection of compliant members. Compliant mechanisms that contain one or more traditional kinematic pairs along with compliant members are called *partially compliant mechanisms*.

For a rigid link, the distances between joints are fixed, and the shape of the link is kinematically unimportant regardless of the forces applied. The motion of a compliant link, however, is dependent on link geometry and the location and magnitude of applied forces. Because of this difference, a compliant link is described by its *structural type* and its *functional type*.

The structural type is determined when no external forces are applied and is similar to the identification of rigid links. A rigid link that has two pin joints is termed a *binary link*. A rigid link with three or four pin joints is a *ternary* or a *quaternary link*, respectively. A compliant link with two pin joints has the same structure as a binary link and is called a *structurally binary link*, and so on for other types of links.

A link's functional type takes into account the structural type and the number of *pseudo joints*. Pseudo joints occur where a load is applied to a compliant segment, as shown in Figure 1.19. If a force is applied on a compliant link somewhere other than at the joints, its behavior may change dramatically. A structurally binary link with force or moment loads only at the joints is termed *functionally binary*. A compliant link with three pin joints is *structurally ternary*, and if loads are applied only at the joints, it is also *functionally ternary*. The same applies for quaternary links. If a link has two pin joint connections and also has a force on a compliant segment, it is structurally binary and functionally ternary, due to the added pseudo joint caused by the force. This is illustrated for binary and ternary links in Figure 1.19.

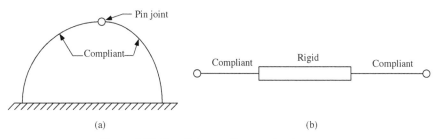

Figure 1.18. One-link compliant mechanism.

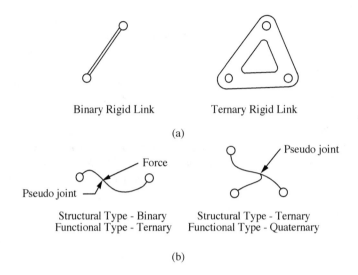

Figure 1.19. Examples of link types.

Segment Identification. While the definition of a link used above is consistent with that for rigid-body kinematics, it is not very descriptive of a compliant link. The application of a force or moment to a compliant link affects the deformation of the link, and therefore, its contribution to the mechanism's motion. Link characteristics that influence its deformation include cross-sectional properties, material properties, and magnitude and placement of applied loads and displacements. Thus a compliant link is further characterized into segments.

A link may be composed of one or more segments. The distinction between segments is a matter of judgment and may depend on the structure, function, or loading of the mechanism. Discontinuities in material or geometric properties often represent the endpoints of segments. The link shown in Figure 1.18 consists of three segments, one of which is rigid and two compliant. Since the distance between the endpoints of a rigid segment remains constant, it is considered a single segment, regardless of its size or shape.

Segment and Link Characteristics. The characteristics of individual segments and links may also be described, as shown in Figures 1.20a and b. A segment may be either rigid or compliant. This is referred to as a segment's *kind*. A compliant segment may be further classified by its *category* of either simple or compound. A simple segment is one that is initially straight, has constant material properties, and a constant cross section. All other segments are compound.

A link may be either rigid or compliant (its kind) and may consist of one or more segments. A rigid link needs no more characterization. A compliant link may

Compliant MEMS 15

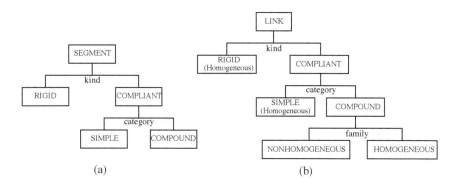

Figure 1.20. Component characteristics of (a) segments, and (b) links.

be either simple or compound (its category). A simple compliant link consists of one simple compliant segment; all others are compound links. A compound link may be either homogeneous or nonhomogeneous. This is its *family*. A homogeneous link is one that consists of all rigid segments or all compliant segments. Therefore, rigid links and simple compliant links are special cases of homogeneous links. Nonhomogeneous links contain both rigid and flexible segments.

1.5.3 Diagrams

Skeleton diagrams are often used to facilitate the description of the structure of rigid-body mechanisms. Similar diagrams for compliant mechanisms must distinguish between rigid and compliant links and segments. Symbols that represent individual joints or segments are shown in Figure 1.21. A compliant segment is shown as a single line and a rigid segment as two parallel lines. Additional information may be included in the diagram to show its categorization as a simple or compound segment. An axially compliant segment is one that is allowed to compress or extend. An extension spring is an example of an axially compliant segment.

Diagrams for the mechanisms illustrated in Figures 1.2a, b, 1.3a, and 1.11 are shown in Figures 1.22a, b, c, and d, respectively.

1.6 COMPLIANT MEMS

Microelectromechanical systems (MEMS) integrate mechanical and electrical components with feature sizes ranging from micrometers to millimeters. They may be fabricated using methods similar to those used to construct integrated circuits.

16 **Introduction**

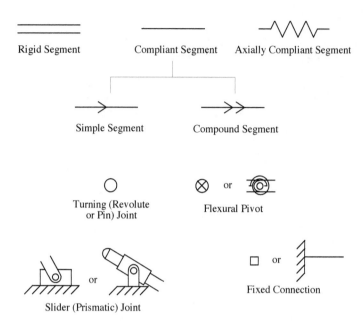

Figure 1.21. Symbol convention for compliant mechanism diagrams.

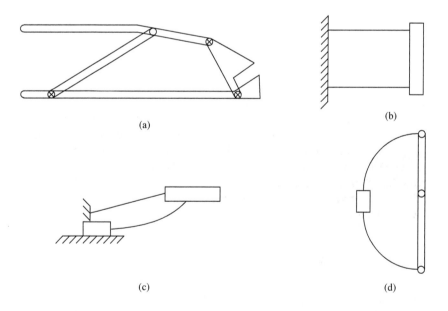

Figure 1.22. Diagrams representing the compliant mechanisms in (a) Figure 1.2a, (b) Figure 1.2b, (c) Figure 1.4a, and (d) Figure 1.11.

Compliant MEMS 17

MEMS have the great potential of providing significant cost advantages when batch fabricated. Their size also makes it possible to integrate them into a wide range of systems. Microsensors (e.g., accelerometers for automobile crash detection and pressure sensors for biomedical applications) and microactuators (e.g., for moving arrays of micromirrors in projection systems) are examples of commercial applications of MEMS. This field is expected to grow dramatically over time.

The most common methods for MEMS fabrication use planar layers of material. Surface micromachining uses multiple layers of material that are deposited, then patterned using planar lithography. Possibly the most common material for surface micromachining is polycrystalline silicon, or polysilicon, on a silicon wafer. Bulk micromachining involves selectively etching the wafer substrate to create cavities and structures. LIGA (a German acronym which stands for lithography, electroforming, injection molding) uses x-rays to construct high-aspect-ratio structures in a single layer of material. More detail on the fabrication methods used in MEMS can be found in a number of sources, including [86].

The constraints introduced by the planar nature of MEMS fabrication and the scale (which makes assembly of parts difficult) introduce a number of challenges in constructing mechanical devices at the micro level. Compliant mechanisms present solutions to many of these problems. The advantages at the micro level are that compliant mechanisms [87]:

- Can be fabricated in a plane
- Require no assembly
- Require less space and are less complex
- Have less need for lubrication
- Have reduced friction and wear
- Have less clearance due to pin joints, resulting in higher precision
- Integrate energy storage elements (springs) with the other components

There are also some challenges associated with designing compliant MEMS. The performance is highly dependent on the material properties, yet the design is limited to a few materials that are compatible with the fabrication methods. The material properties are not always well known at this scale, and there can often be significant scatter in material property data because device sizes are often on the order of the material grain size. Tests have demonstrated that compliant components can be very robust at the micro level and often last longer than do components that use pin joints or other elements that induce wear.

An example of a compliant micro device, a micro compliant bistable mechanism, is shown in the scanning electron microscope (SEM) photo in Figure 1.23. An example of a compliant member as part of a Sandia National Laboratories microengine is shown in the SEM photo in Figure 1.24a. This type of microengine uses a number of flexible components, and a view of the overall mechanism is shown in Figure 1.24b.

The compliant mechanism theory presented in this book applies for the analysis and design of micro and macro devices. Examples of micro compliant mechanisms

Figure 1.23. Scanning electron micrograph of a microcompliant bistable mechanism.

are provided throughout the following chapters. The cgs (centimeter–gram–second) system of units is often used to ensure a consistent set of units. The unit of force in this system of units is the dyne (g-cm/s^2). The length unit of micrometer, (μm, 1×10^{-6} m) is often used to describe the dimensions of various components. When performing calculations, this unit is converted to meters when using the meter–kilogram–second (mks) system or to centimeters for the cgs system.

PROBLEMS

1.1 List several possible advantages of compliant mechanisms.
1.2 List possible challenges or disadvantages associated with compliant mechanisms.

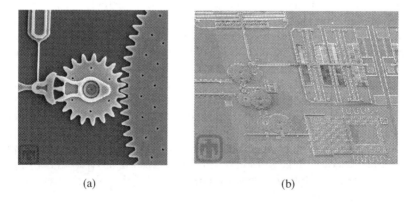

(a)   (b)

Figure 1.24. (a) Compliant member in a microengine, and (b) microengine that uses several compliant members. (Courtesy of Sandia National Laboratories, www.sandia.gov.)

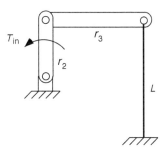

Figure P1.4. Figure for Problem 1.4.

1.3 What is the difference between a compliant mechanism and a compliant structure?
1.4 Consider the flapper mechanism shown in its undeflected position in Figure P1.4. How many links does this mechanism have? Identify the segments for the compliant link.
1.5 Consider the compliant mechanism shown in Figure P1.5.
   (a) How many links does this mechanism have?
   (b) How many compliant segments are in this mechanism? Rigid segments? (Include ground.)
   (c) Is this a partially compliant mechanism, a fully compliant mechanism, or a rigid-body mechanism?
1.6 Consider the compliant mechanism illustrated in Figure P1.6. How many links does it have? How many rigid segments does this mechanism have? Compliant segments? Identify and label the segments of each link. Is this a fully compliant or a partially compliant mechanism?

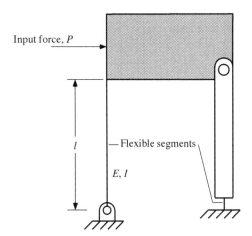

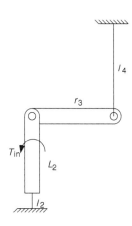

Figure P1.5. Figure for Problem 1.5.   **Figure P1.6.** Figure for Problem 1.6.

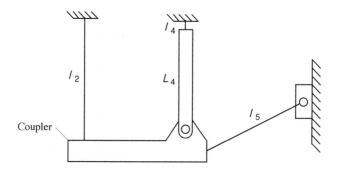

Figure P1.7. Figure for Problem 1.7.

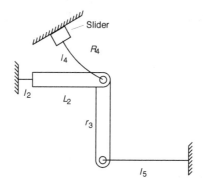

Figure P1.8. Figure for Problem 1.8.

1.7 Consider the mechanism illustrated in Figure P1.7. How many links does it have? How many rigid segments does this mechanism have? Compliant segments? Identify and label the segments of each link.

1.8 Consider the mechanism illustrated in Figure P1.8. How many links does it have? How many rigid segments does this mechanism have? Compliant segments? Identify and label the segments of each link.

CHAPTER 2

FLEXIBILITY AND DEFLECTION

Traditional kinematic analysis assumes that the mechanism links are infinitely rigid and do not deflect. Although this assumption is adequate when the deflections are very small, real materials will have some deflection when a load is applied. Since compliant mechanisms rely on deflection for their motion, their deflections cannot be ignored.

Linear and nonlinear deflection analysis are reviewed in this chapter. A basic understanding of mechanics of materials is assumed, and concepts particular to compliant mechanisms are emphasized.

2.1 LINEAR VERSUS NONLINEAR DEFLECTIONS

In most deflection analyses, it is assumed that the deflection is small compared to the dimensions of the structure, that the material is elastic, and that strain is proportional to stress. These assumptions are used to simplify the analysis by linearizing the equations. In most structural applications, the deflections are small, the stress is below the elastic limit, and linear equations provide accurate results. However, there are cases for which structural nonlinearities occur, and these assumptions are not valid. In such cases a nonlinear analysis must be performed.

Structural nonlinearities are divided into two categories: material nonlinearities and geometric nonlinearities. *Material nonlinearities* arise when Hooke's law, which states that stress is proportional to strain, does not apply. Examples of material nonlinearities include plasticity, nonlinear elasticity, hyperelasticity, and creep.

Geometric nonlinearities occur when deflections are such that they alter the nature of the problem. Examples of geometric nonlinearities are large deflections, stress stiffening, and large strains. Nonlinearities that result from large strains must be taken into account if the strain is large enough to cause significant changes in geometry, such as area or thickness. Stress stiffening occurs when the stiffness of the structure is a function of the deflection.

Geometric nonlinearities due to large deflections are commonly encountered in compliant mechanisms. In large-deflection analysis, the strain is assumed to be small, but the overall deflection is large. The analysis of large-deflection beams is discussed in detail later in this chapter.

In nonlinear analysis it is important to determine if the load is conservative or nonconservative. A *conservative* problem is one in which the final deflection is independent of the order that the loads are applied and the number of increments over which they are applied. The potential energy of a conservative system depends only on the final deflection, not on the path traversed to obtain that deflection. This knowledge may be used to simplify the analysis of conservative systems by incrementing the load in a manner that improves the convergence characteristics of the solution method. Geometric nonlinearities and nonlinear elasticity are examples of conservative problems. If the energy of the system is path dependent, such as with plasticity and creep, then the problem is *nonconservative* and the analysis must conform to the actual load path. Since compliant mechanisms most often include members that undergo large, conservative, elastic deflections, the analysis related to these kinds of deflections is summarized in this chapter.

2.2 STIFFNESS AND STRENGTH

The relationships between stiffness, strength, and deflection are often confused. If a small load causes a relatively large deflection, then the part has a low stiffness, but nothing is implied about its strength. The amount of deflection that a load will cause is related to a structure's stiffness or rigidity. Strength is a property of the material that specifies the stress it can withstand before failure. In other words, a structure's stiffness determines how much deflection will occur due to a load, while the strength determines how much stress can occur before failure.

The rigidity of a structure is a function of both material properties and geometry. In bending, the flexural rigidity is EI, where E is Young's modulus (the modulus of elasticity), and I is the cross-sectional moment of inertia. For an axial load, the axial stiffness is EA, where A is the cross-sectional area.

Consider the cantilever beam shown in Figure 2.1. The material is isotropic, with equal Young's modulus and strength in all directions. The beam geometry is such that the moment of inertia is much larger about one axis than the other. If a force, F_x, is applied in the x direction until the maximum stress is equal to the yield strength, S, the deflection in the x direction is

$$\delta_x = \frac{2SL^2}{3Eb} \tag{2.1}$$

Flexibility

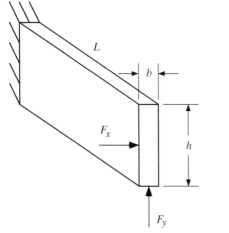

δ_x is the deflection in direction of F_x

δ_y is the deflection in direction of F_y

for the same maximum stress due to F_x or F_y: $\delta_x = \dfrac{h}{b}\delta_y$

Figure 2.1. Cantilever beam with equal strength but different stiffnesses in the x and y directions.

A force in the y direction, F_y, that causes a maximum stress equal to the yield strength produces a deflection in the y direction of

$$\delta_y = \frac{2SL^2}{3Eh} \tag{2.2}$$

The part can have a larger deflection in the x direction than in the y direction before failure, by a factor of h/b, that is,

$$\delta_x = \frac{h}{b}\delta_y \tag{2.3}$$

This example shows that the beam has equal strength in both directions but has much different stiffness in the y direction than in the x direction. Stiffness and strength are not necessarily related, and in some cases failure is avoided by making a member less stiff. This is usually the case with compliant mechanisms, as will be shown.

2.3 FLEXIBILITY

Beam flexibility and deflections are undesirable characteristics for most structural and mechanical systems. For example, in high-speed mechanisms, the flexibility of moving members causes vibrations in the system that may result in undesirable mechanism performance or failure. Deflections in most structures are also minimized to avoid the appearance of being weak, even if the strength of a structure is

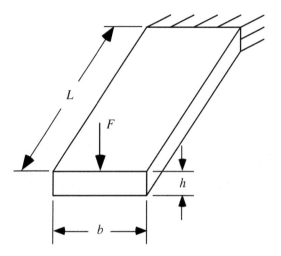

Figure 2.2. Flexible cantilever beam.

more than adequate to sustain the applied loads. Also, when deflections are small, simple linearized beam equations may be used to analyze the resulting deflections. For these reasons it is appropriate that most work in structural and mechanical design has centered on methods of increasing the rigidity of members to decrease their deflections under loads.

However, since compliant mechanisms depend on deflection for their motion, flexibility of their components is essential. It is desirable that these deflections be achieved with as small an applied load and member stress as possible.

Flexibility, the ability of a member to deflect under a load, is the inverse of rigidity. A beam's flexibility may be changed by modifying either the material properties or the geometry. This is illustrated by the example beam shown in Figure 2.2. Assuming that the deflection is in the linear range, the deflection, δ, is

$$\delta = \frac{FL^3}{3EI} \tag{2.4}$$

where the moment of inertia is $I = bh^3/12$, so

$$\delta = 4F\frac{1}{E}\frac{L^3}{bh^3} \tag{2.5}$$

The magnitude of the deflection of the beam is influenced by the magnitude of the force (F), the material properties ($1/E$), and the geometry (L^3/bh^3). For a given load, the flexibility of the beam may be altered by modifying the material or the beam geometry.

Flexibility

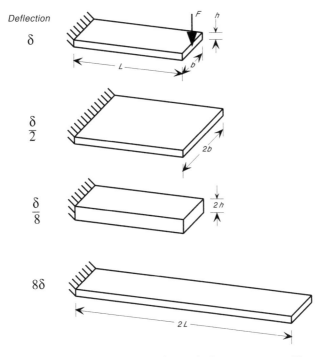

Figure 2.3. Changes in deflection due to change in beam geometry. The same force is applied to each beam.

The effect of material properties on flexibility is supported by our intuition gained from everyday experiences with materials. For example, suppose that several beams are constructed with the same geometry, have the same load, but are made of different materials. A steel ($E = 30$ Mpsi, 207 GPa) beam will have a deflection a third as large as the deflection of an aluminum ($E = 10$ Mpsi, 70 GPa) beam, while a polypropylene ($E \approx 0.2$ Mpsi, 1.4 GPa) beam's deflection will be 50 times greater than the aluminum beam's deflection.

Intuition also supports the fact that geometry affects flexibility. If you make something thinner, it becomes more flexible. It is useful to quantify this to provide a better understanding of these effects and find which is the most efficient approach to obtain flexibility.

Continuing with the example of Figure 2.2, it is noted that the flexibility is highly dependent on beam thickness. If the load, material, and geometry of the beam remain the same, but the thickness is halved, the deflection increases by a factor of 8. Note also that the length of the beam is cubed. If the length of the beam is doubled, the beam deflection is increased by a factor of 8. These simple relationships show that modifying the thickness and length can be efficient methods of changing flexibility. This is demonstrated visually for several beams in Figure 2.3. Similar relationships may be found for beams with other cross-sectional shapes and loading conditions.

The relationship between flexibility and ductility is often misunderstood. Many people would consider a brittle material such as glass not to be flexible. However, flexibility and brittleness are not necessarily related. A glass beam may be made very flexible by modifying the geometry as discussed above. This distinction is particularly important in microelectromechanical systems, where compliant mechanisms are often fabricated of polysilicon, a very brittle material. One source of the confusion comes from the behavior of different materials under high stresses. A brittle material can fail catastrophically when its failure stress is reached. However, a ductile material will yield when the stress goes beyond the yield strength. This yielding provides warning of overstress; the part does not fracture, but has some permanent deformation. Some ductile materials (such as polypropylene) may be stressed beyond their yield strength thousands of times without fracturing. In summary, brittle materials may be made to be flexible, but they will fail catastrophically when overstressed.

2.4 DISPLACEMENT VERSUS FORCE LOADS

One of the challenges of compliant mechanisms is to allow deflections large enough for the mechanism to perform its function, while maintaining stresses below an allowable maximum stress. Once the deflected position of a compliant mechanism is known, the stress analysis is relatively straightforward. Most mechanisms operate in two dimensions, and the stresses are modeled as planar. Bending and axial stresses are the predominate loading modes in compliant mechanisms. The axial stress associated with an axial force, F, is

$$\sigma = \frac{F}{A} \quad \text{(axial)} \tag{2.6}$$

where F is the axial force and A is the area of the cross section. The stress due to bending is

$$\sigma = \frac{My}{I} \quad \text{(bending)} \tag{2.7}$$

M is the moment load, y the distance from the neutral axis to the point of interest, and I the moment of inertia of the cross section. The maximum stress occurs at the location farthest from the neutral axis, c, and is

$$\sigma_{max} = \frac{Mc}{I} \quad \text{(maximum bending)} \tag{2.8}$$

Refer to texts on mechanical design and stress analysis for details on the effects of stress concentrations, temperature, environment, and other loading conditions that may also come into play.

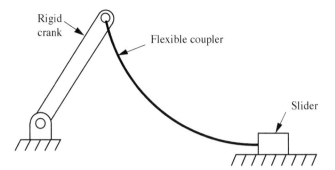

Figure 2.4. Example of a compliant mechanism with known displacement and unknown force.

Most familiar structures are loaded by applied forces. Examples of such structures include a bridge with traffic and wind loads, a high-speed linkage with dynamics loads, an aircraft wing with aerodynamic loads, and so forth. Compliant mechanisms, however, are typically loaded in a different way. Rather than calculating the deflection and stress from a known load, the displacement is known, and the resulting stress and reaction forces are calculated. A simple example is illustrated in Figure 2.4. The deflection of the flexible beam is defined by the rest of the system, and the load is unknown. The approach used to design a system with a displacement load is significantly different than for structures with a force load.

To illustrate this concept, consider the two beams shown in Figure 2.5. The first beam has a known force. The moment at the fixed end is $M = FL$, and the maximum stress is found from equation (2.8) as

$$\sigma = \frac{FLc}{I} \tag{2.9}$$

The displacement of the second beam is determined from the cam profile, and the reaction force is unknown. The maximum moment, M_{max}, is at the fixed end, and the deflection equation is

$$\delta = \frac{FL^3}{3EI} = \frac{M_{max}L^2}{3EI} \tag{2.10}$$

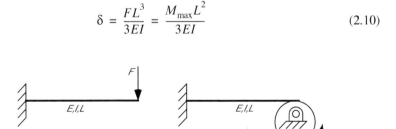

Figure 2.5. (a) Beam with known force load, and (b) beam with known displacement.

Solving this equation for M_{max} yields

$$M_{max} = \frac{3\delta EI}{L^2} \tag{2.11}$$

Substituting into equation (2.7) results in a stress of

$$\sigma = \frac{3\delta Ec}{L^2} \tag{2.12}$$

For the case of a beam with a force load (Figure 2.5a), the stress is reduced by increasing the moment of inertia. This represents the common and intuitive approach of adding material to a structure to strengthen it. However, this approach would be disastrous in the case with an applied displacement (Figure 2.5b). For such a beam, the stress is decreased by *decreasing* the stiffness. A structure that must undergo a given displacement should be made more flexible. This can be accomplished by using a material with a lower Young's modulus, or by decreasing the moment of inertia. This often seems odd to those with much experience with force-loaded structures, but it is an essential concept to understand in structures with displacement loads. An example is provided in Section 2.6.

2.5 MATERIAL CONSIDERATIONS

Many different types of materials may be used in compliant mechanism design. Although each application has its own criteria for material selection, the selection process can be guided by principles that can be used in numerous situations.

Some of the most important principles were discussed in earlier sections of this chapter. Perhaps the most important thing to remember is that stiffness and strength are not the same and it is possible to make something both flexible *and* strong. Ductility and flexibility are also not equivalent, and brittle materials may be used to construct compliant mechanisms if their geometry is made such that they are not overstressed. Another important point already made is that something can be made more flexible by modifying either its geometry or its material properties.

Unlike most other mechanical devices or structures, materials for compliant mechanisms are chosen to maximize flexibility rather than stiffness. The maximum deflection that a cantilever beam can undergo until it fails is considered in the next section.

2.5.1 Maximum Deflection for a Flexible Beam

Consider the rectangular-cross-section cantilever beam illustrated in Figure 2.2. As mentioned before, the deflection of the free end, δ, is $\delta = FL^3/3EI$, or for a rectangular cross section where $I = bh^3/12$:

Material Considerations

$$\delta = \frac{4FL^3}{Ebh^3} \quad (2.13)$$

The maximum stress occurs at the fixed end where the moment is $M_{max} = FL$ and $c = h/2$. Substituting these terms and the moment of inertia into equation (2.7) results in a maximum stress, σ_{max}, of

$$\sigma_{max} = \frac{6FL}{bh^2} \quad (2.14)$$

Considering the failure of the beam to occur when the maximum stress equals the yield strength, S_y, then

$$S_y = \frac{6FL}{bh^2} \quad \text{at failure} \quad (2.15)$$

Rearranging to solve for the force, F, results in

$$F = \frac{S_y bh^2}{6L} \quad (2.16)$$

Substituting equation (2.16) into equation (2.13) results in the maximum deflection that the beam will undergo before failure, δ_{max}, or

$$\delta_{max} = \frac{2}{3}\frac{S_y}{E}\frac{L^2}{h} \quad (2.17)$$

Equation (2.17) shows that the maximum deflection depends on both geometry, L^2/h, and material properties, S_y/E.

2.5.2 Ratio of Strength to Young's Modulus

For the beam discussed in the preceding section, the material that will result in the largest deflection is that with the highest ratio of strength to Young's modulus. Table 2.1 lists yield strength, Young's modulus, and the ratio of yield strength to Young's modulus for several materials. A low- and a high-strength alloy are listed for steel, aluminum, and titanium to provide a range for these materials. It is important to note that Young's modulus remains nearly the same for most metals regardless of the addition of alloying elements or heat treatment. For example, the low- and high-strength steels listed in Table 2.1 have the same modulus. While the yield strength and tensile strength may be increased with heat treating or cold working, the material will also become more brittle. Defects become increasingly important, including cracks, notches, voids, and impurities. Failure mechanisms and sensitivity to stress concentrations should also be considered.

The materials with the highest strength-to-modulus ratio will allow a larger deflection before failure. This ratio is one of the most important parameters available in selecting materials for compliant mechanism applications.

2.5.3 Other Material Selection Criteria

While the strength-to-modulus ratio of materials is important when choosing a material for a compliant mechanism, there are many other important criteria that should be considered. These criteria will vary depending on the application.

Metals. Metals, such as steel, stainless steel, aluminum, titanium, and so on, should be considered for use in a compliant mechanism for the following reasons:

- Predictable material properties needed (needed in high-precision instruments)
- Good performance in high-temperature environment
- Biocompatibility of some metals
- Low susceptibility to creep
- More predictable fatigue life
- Ability to operate in many harsh environments
- Electrical conductivity (such as is needed for flexible electrical connectors)

TABLE 2.1. Ratio of yield strength to Young's modulus for several materials

Material	E [Mpsi (GPa)]	S_y [kpsi (MPa)]	$(S_y/E) \times 1000$
Steel (1010 hot rolled)	30 (207)	26 (179)	0.87
Steel (4140 Q&T@400)	30 (207)	238 (1641)	7.9
Aluminum (1100 annealed)	10.4 (71.7)	5 (34)	0.48
Aluminum (7075 heat treated)	10.4 (71.7)	73 (503)	7.0
Titanium (Ti-35A annealed)	16.5 (114)	30 (207)	1.8
Titanium (Ti-13 heat treated)	16.5 (114)	170 (1170)	10
Beryllium copper (CA170)	18.5 (128)	170 (1170)	9.2
Polycrystalline silicon	24.5 (169)	135 (930)	5.5
Polyethylene (HDPE)	0.2 (1.4)	4 (28)	20
Nylon (type 66)	0.4 (2.8)	8 (55)	20
Polypropylene	0.2 (1.4)	5 (34)	25
Kevlar (82 vol %) in epoxy	12 (86)	220 (1517)	18
E-glass (73.3 vol %) in epoxy	8.1 (56)	238 (1640)	29

Material Considerations 31

Some disadvantages of metals in compliant mechanisms include:

- Cost of the material
- Cost of forming or machining the materials
- Greater need to assemble components
- High density
- Low strength-to-modulus ratio compared to many polymers

Plastics. Although the cost per pound of some plastics is comparable to that of steel, plastic processing methods such as injection molding make final plastic part costs considerably lower than for other materials in high volume. This and their high strength-to-modulus ratio are the main reasons for their use in high-volume compliant mechanism applications. Advantages of plastics in compliant mechanisms include:

- Low manufacturing cost in high volume
- High strength-to-modulus ratio
- Machinability
- Increased possibility of eliminating assembly of parts
- Low density
- Biocompatibility of some plastics
- Electrically insulating

Disadvantages of using plastics include:

- Variability in mechanical properties make plastics less predictable than many materials
- Low melting temperatures
- Material degradation in some environments
- Creep and stress relaxation
- Consumer perception of being cheap or weak
- Nonlinear elastic material properties for some materials

Glass is often used to reinforce plastics. Such reinforcement increases strength but also increases stiffness, resulting in a decrease in the strength-to-modulus ratio.

Polypropylene is a commonly used polymer in compliant mechanisms for a number of reasons. First, it has a very high ratio of strength to modulus, as shown in Table 2.1. It also has a number of advantages compared to other polymers with similar strength-to-modulus ratios. Polypropylene is readily available, inexpensive, easy to process, and has a low density. It is also very ductile and is much less likely to result in catastrophic failure when yielded. This makes it well suited for living hinges because the material must undergo large strains for millions of cycles, as discussed in later chapters. Although it is an outstanding material for many compliant mechanisms, it is not appropriate in some instances. Creep, limited temperature range, and chemical resistance are a few limitations of polypropylene that make it unsuitable for some applications.

Other Materials. Parts made of brittle materials may be made compliant, but they are very unforgiving when overstressed. When brittle materials are used as a compliant member it is usually because the design is constrained by other factors such that only certain materials may be used. This is the case when micro mechanisms are constructed with the technology used to fabricate integrated circuits. The materials are limited to those that are appropriate for the technology, most of which are very brittle. A fiber optic strand is another example of a flexible part that is made from a brittle material because of essential constraints associated with the application.

Wood has been used for compliant devices for millennia. Bows and catapults are two examples of such devices. The wide variation in mechanical properties of wood is a disadvantage for its use in modern engineering systems. Wood may be considered for use when the size of the device is comparatively large, cost is a driving constraint in the design, and a sizable factor of safety is possible.

Composite materials (such as graphite fiber composites, fiberglass, and glass-reinforced plastics), elastomers, fabrics, and other materials may be considered for use in compliant mechanisms when applications warrant their use. It is important to keep an open mind in material selection because many things that are traditionally used in stiff structures can be made flexible by properly designing the geometry. For example, graphite composites are often used to make stiff aerospace structures, but they may also be used to make something flexible, such as a fly fishing rod.

2.5.4 Creep and Stress Relaxation

Creep is the deformation that occurs in a material under load over time. For metals, creep is usually significant only at elevated temperatures, but many polymers experience creep at room temperature. A typical creep model is shown in Figure 2.6a. The springs represent the elastic component of the material that results in an immediate deflection when the load is applied. The dashpots allow continued deformation with time. The creep rate depends on the viscosity associated with the dashpots and is associated with the viscoelastic properties of the material. [88]

Stress relaxation is closely related to creep but probably occurs more often than creep in compliant mechanisms. Stress relaxation occurs when a constant deflection is applied to a structure and the resulting stress decreases with time. A stress relaxation model is shown in Figure 2.6b. Here an initial deflection and resulting strain is applied to the device. The reaction load has its largest magnitude when the strain is first applied. In the model the initial force is the spring constant times the deflection, which corresponds to the initial elastic deflection of the material. The dashpot moves with time and the spring relaxes, resulting in a lower reaction force. As with

Material Considerations

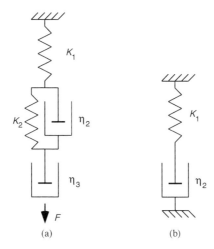

Figure 2.6. Models for (a) creep and (b) stress relaxation.

creep, the dashpot characteristics depend on the viscoelastic properties of the material. [88]

Stress relaxation can be a serious problem for compliant components that are meant to act as springs to supply a reaction force for long periods of time. Figure 2.7 is an example of how stress relaxation can be a problem in a compliant device. It shows a computer mouse with a compliant member acting as a leaf spring to hold the ball of the mouse next to the rollers. The ball rotates as the mouse is moved across a surface. Because the ball is in contact with the rollers, they also turn as the ball rotates. A signal corresponding to the roller rotation is sent to the computer to identify the mouse motion. Because the compliant member is subject to a constant deflection for long periods of time, it is important that stress relaxation be considered. If the reaction force from the spring reduces due to stress relaxation and becomes too low, it is possible for the ball to rotate without causing the rollers to rotate. This in turn sends a false signal to the computer and the mouse no longer functions properly. A similar example is the use of compliant members that act as springs that push brushes against the rotor in electric motors. In such cases it is important to consider metals as the compliant member or plastics that have an acceptable creep rate under the conditions.

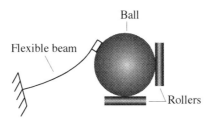

Figure 2.7. A computer mouse is an example of where stress relaxation can be a problem in a compliant device.

Adding glass fiber reinforcement reduces creep and stress relaxation in plastics. Care should be taken to ensure that other characteristics are still satisfied. Glass reinforcement increases the stiffness of the material and also makes it more brittle. In applications such as the mouse or the motor brush springs, increased stiffness and reduced ductility may not be a problem because the stress is very controlled due to the constant deflection.

2.6 LINEAR ELASTIC DEFLECTIONS

The Bernoulli–Euler equation states that the bending moment is proportional to the beam curvature, that is,

$$M = EI\frac{d\theta}{ds} \tag{2.18}$$

where M is the moment, $d\theta/ds$ the rate of change in angular deflection along the beam (curvature), E the Young's modulus of the material, and I the beam moment of inertia. The curvature may be written as

$$\frac{d\theta}{ds} = \frac{d^2y/dx^2}{[1+(dy/dx)^2]^{3/2}} \tag{2.19}$$

where y is the transverse deflection and x is the coordinate along the undeflected beam axis. When deflections are small, the square of the slope, $(dy/dx)^2$, can be assumed to be small compared to unity in the denominator of equation (2.19). This assumption leads to the classical beam moment–curvature equation:

$$M = EI\frac{d^2y}{dx^2} \tag{2.20}$$

As an example, consider the cantilever beam with a vertical end force shown in Figure 2.8. The moment may be written as

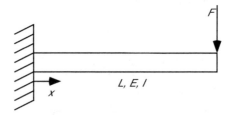

Figure 2.8. Cantilever beam with an end force and small deflections.

Linear Elastic Deflections

$$M = F(L - x) \tag{2.21}$$

where F is the value of the vertical force and L the beam length. The beam angle, or slope, is

$$\theta = \frac{dy}{dx} \tag{2.22}$$

and equation (2.20) may be written as

$$M = EI\frac{d\theta}{dx} \tag{2.23}$$

Substituting equation (2.21) into equation (2.23) results in

$$F(L - x) = EI\frac{d\theta}{dx} \tag{2.24}$$

Separating variables and integrating yields

$$\int d\theta = \frac{F}{EI}\int (L - x)\, dx \tag{2.25}$$

or

$$\theta = \frac{F}{EI}\left(Lx - \frac{x^2}{2}\right) + C_1 \tag{2.26}$$

The constant of integration, C_1, is found by applying the boundary conditions, that is, $\theta = 0$ at $x = 0$. This results in $C_1 = 0$ and

$$\theta = \frac{dy}{dx} = \frac{Fx}{2EI}(2L - x) \tag{2.27}$$

The vertical deflection is found by separating variables and integrating:

$$y = \int \frac{Fx}{2EI}(2L - x)\, dx = \frac{F}{2EI}\left(Lx^2 - \frac{x^3}{3}\right) + C_2 \tag{2.28}$$

The boundary conditions of $y = 0$ at $x = 0$ yield $C_2 = 0$ and

$$y = \frac{Fx^2}{6EI}(3L - x) \tag{2.29}$$

Results for many different loading and boundary conditions have been tabulated in various forms, one of the most comprehensive of which is found in [89]. Appendix D also lists equations for a number of beam types.

In most structural applications, the load in terms of force, moment, pressure, and so on, is known, and the deflections and stresses are found from equations similar to equation (2.29). In compliant mechanisms analysis, however, the required deflection is known, and the force necessary to produce the deflection and the associated stress must be calculated.

Example: Compliant Connector. A snap connector is shown in Figure 2.9. To install the connector, the beam must deflect a distance of $\delta = 2$ mm. The beam has a length of $L = 9$ mm, width $b = 6$ mm, and thickness $h = 2$ mm. It is made of a polymer with a modulus of $E = 1.3 \times 10^9$ Pa. Determine the maximum stress due to bending.

Solution: In this example, and in most compliant structures and mechanisms, the deflection is specified and the force is unknown. The maximum stress occurs at the fixed end of the beam, where

$$M_{max} = FL \qquad (2.30)$$

where M_{max} is the moment and F is the applied force at the beam end. Substituting equation (2.30) and $x = L$ into equation (2.29) yields

$$\delta = \frac{M_{max} L^2}{3EI} \qquad (2.31)$$

Solving for M_{max} and substituting into equation (2.7) results in

$$\sigma = \frac{3\delta E h}{2L^2} = 96 \text{ MPa} \qquad (2.32)$$

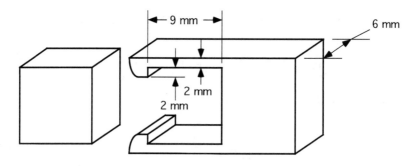

Figure 2.9. Compliant connector device.

Linear Elastic Deflections

Note that the stress is expressed in terms of the deflection, and a calculation of the force was not required.

Equation (2.32) shows that the maximum stress is not a function of the beam width because the width term, b, cancels. The force required to deflect the beam is proportional to the width, but the stress is not affected. This provides a useful way of modifying the reaction force without changing the maximum stress. If the width is increased, the reaction force is increased but the stress does not change.

It is important to note that a decrease in beam thickness decreases the stress. As discussed in Section 2.4, for most structures a force is specified as the input, and the stress is decreased by increasing the moment of inertia of the structure. Most compliant mechanisms and structures will have displacement boundary conditions as inputs, and the stress will be reduced by decreasing the stiffness, allowing the structure to deflect to the desired position.

Example: Compliant Ice-Cream Scoop. Figure 2.10 shows two ice-cream scoops—one with rigid links and the other with a compliant segment (the flexible beam). The flexible beam of the compliant version replaces the function of the pin joint, a rigid link, and the spring. The rack is easily integrated into the compliant member also. The deflection, required input forces, and resulting stresses may be calculated using linearized beam equations. The required beam deflection is a function of the pinion size and rotation requirement. The blade must rotate 180°, which requires a linear motion of the rack, δ, of

$$\delta = \pi r_p \qquad (2.33)$$

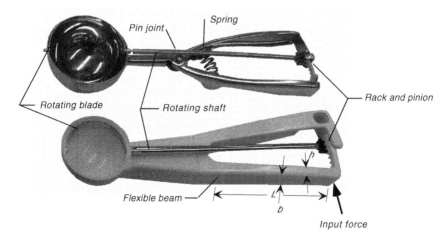

Figure 2.10. Rigid-link ice-cream scoop and a compliant ice-cream scoop.

where r_p is the radius of the pinion gear. Since the rack is part of the flexible beam, the beam end deflection is the same as the rack motion. For a fixed beam with a force at the free end,

$$\delta = \frac{FL^3}{3EI} \tag{2.34}$$

Combining equations (2.33) and (2.34) results in an input force of

$$F = \frac{3\pi r_p EI}{L^3} \tag{2.35}$$

where E is Young's modulus of the beam, L the beam length, and the moment of inertia for the rectangular beam is $I = bh^3/12$. The maximum stress, σ_{max}, occurs at the fixed connection where the moment is a maximum of FL. The magnitude of the stress is

$$\sigma_{max} = \frac{Mc}{I} = \frac{FLh}{2I} \tag{2.36}$$

Combining equations (2.35) and (2.36) results in

$$\sigma_{max} = \frac{3\pi}{2}\frac{Ehr_p}{L^2} \tag{2.37}$$

2.7 ENERGY STORAGE

Because compliant mechanisms have elastic members that deflect, energy is stored in the system in the form of strain energy. Thus the energy available at the output may be considerably less than was provided at the input. This energy is not lost, but is stored and released later. For many compliant mechanism applications, the stored energy is a desirable characteristic of the mechanism and is often a reason to select a compliant mechanism over other types of mechanisms. Examples of such applications include bistable mechanisms and mechanisms where the effects of rigid links and springs may be integrated into flexible components. However, there are other applications where the energy available at the output should be maximized, such as with a gripper device. Whether energy storage is desirable or undesirable in an application, it is often important to understand what factors affect the amount of energy storage.

The energy storage may be minimized by designing a compliant mechanism such that the constraints on a deflecting member allow it to take on a shape that minimizes the work required to obtain the shape. To illustrate, consider a number of different beam loadings and the work required to obtain a specified deflection. Suppose that each beam has a length of L, its left end is fixed, and its free end must

Energy Storage

obtain a deflection of δ. Assume that the deflections are small enough that linearized beam equations of Appendix D may be used.

For a beam with a vertical force at the free end,

$$F = \frac{3EI\delta}{L^3} \tag{2.38}$$

The work required to obtain the required deflection is calculated as

$$W = \int_0^\delta F \, d\delta \tag{2.39}$$

Combining equations (2.38) and (2.39) and integrating gives

$$W = \int_0^\delta \frac{3EI\delta}{L^3} \, d\delta = \frac{3}{2} \frac{EI\delta^2}{L^3} \tag{2.40}$$

For a beam with a moment (M_o) at the free end,

$$M_o = \frac{EI\theta}{L} \tag{2.41}$$

where θ is the angle of the free end, and

$$\theta = \frac{2\delta}{L} \tag{2.42}$$

The work done by the moment is

$$W = \int_0^\theta M \, d\theta = \int_0^\theta \frac{EI\theta}{L} \, d\theta = \frac{EI\theta^2}{2L} = 2\frac{EI\delta^2}{L^3} \tag{2.43}$$

Using a similar approach for a fixed–guided beam yields

$$W = 6\frac{EI\delta^2}{L^3} \tag{2.44}$$

If the work required to obtain a deflection δ is written as

$$W = C\frac{EI\delta^2}{L^3} \tag{2.45}$$

the value of C may be used to compare the required work for various types of load conditions. The values of C for the loading conditions discussed above are summarized in Table 2.2. Note that it requires four times more work to deflect a beam with fixed–guided boundary conditions than it would to deflect a fixed–free beam with a vertical force at the end. This also means that four times as much energy is stored in the fixed–guided beam.

Work is done to deflect the beam, and the energy is stored in the beam as strain energy. This is similar to deflecting a spring where the energy is stored as potential energy in the spring. If a spring is twice as stiff as another, it will store twice as much energy for the same deflection. The knowledge of the efficiency of the loading may be used to design mechanisms that obtain the desired deflections while storing little energy.

Now consider a structure with an axial load. The work required to obtain an axial deflection of δ is

$$W = \frac{EA\delta^2}{2L} \tag{2.46}$$

where A is the cross-sectional area. If the beam has a rectangular cross section where h is the thickness, then

$$W = 6\left(\frac{L}{h}\right)^2 \frac{EI\delta^2}{L^3} \tag{2.47}$$

or $C = 6(L/h)^2$. For a typical flexible beam, h is much smaller than L, and C is very large compared to the values for other loading conditions shown in Table 2.2. For the axial load to be comparable to a fixed–guided beam, the thickness would have to be as large as the length. This means that axial loading is a very inefficient way to obtain deflections, and a lot of energy is required to obtain a relatively small deflection. This is one reason why axial loads are often avoided in compliant mechanisms. Tension causes stress stiffening, which can be another problem caused by axial loads that reduce deflection. Stress stiffening is discussed next.

TABLE 2.2. Values of C for various loading conditions (work, $W = CEI\delta^2/L^3$)

Type of loading	C
Vertical force	1.5
Moment	2
Guided	6
Axial	$AL^2/2I$
Axial (rectangular)	$6(L/h)^2$

2.8 STRESS STIFFENING

Stress stiffening occurs when the stiffness of a structure changes as its deflection changes. Since the stiffness must be known before the deflection may be calculated, but the deflection must be known before the stiffness can be calculated, the solution is nonlinear. An example of stress stiffening is shown in Figure 2.11. The beam is fixed at both ends with a vertical load, F, in the middle. For extremely small deflections, the equation for the maximum deflection may be expressed as

$$y = \frac{Fl^3}{192EI} \quad (2.48)$$

where E is Young's modulus and I is the moment of inertia. However, if the vertical deflection is not extremely small, the length must increase to allow the structure to have a vertical deflection. The increase in length comes from axial strain caused by stretching the beam. The stress associated with the axial strain stiffens the structure. To illustrate, imagine a taut rope tied at both ends. If a vertical force is placed on the rope and the ends are fixed, the tension in the rope increases. As the vertical force increases, the rope becomes stiffer, since an incremental increase of force causes a smaller increase in deflection than was the case for a smaller deflection. Since the stiffness depends on the deflection, the analysis is nonlinear, and an iterative solution is required.

Consider the beam illustrated in Figure 2.12a. The small, linear deflection of this beam may be calculated using equation (2.29). Now suppose a large tension axial force is applied, as shown in Figure 2.12b. This force tends to oppose a horizontal deflection, and the deflection is smaller than for the beam with no axial load. This is called *axial stiffening*. Next, consider the beam illustrated in Figure 2.12c, which has a large compressive force. The deflection caused by the horizontal force provides a moment arm for the compressive force, resulting in a moment in the beam from the compressive force. The addition of the moment results in a deflection larger than if no axial force were present. This has the same effect as making the beam more flexible.

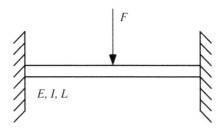

Figure 2.11. Example of geometric nonlinearities caused by stress stiffening.

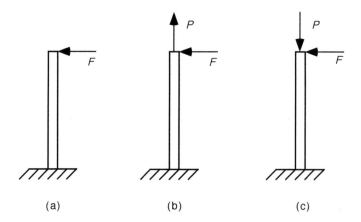

Figure 2.12. Beam with various combinations of transverse and axial loads.

These concepts may be used to determine how efficiently a load produces deflections. If a compliant member is loaded in a manner such that it experiences considerable tension, the load is relatively inefficient and the deflections will be small. In general, tension loads should be reduced or eliminated in compliant members when deflections are desired.

2.9 LARGE-DEFLECTION ANALYSIS

In general, compliant mechanism members undergo large deflections, which introduce geometric nonlinearities and require special considerations in deriving methods for their analysis. The major difference between large- and small-deflection analysis lies in the assumptions made to solve the Bernoulli–Euler equation, stated in equation (2.18). Recall that for small deflections, the slope was assumed to be small, and the curvature was approximated by the second derivative of the deflection, as expressed in equation (2.20). If the slope is not small, this assumption is not valid. If the curvature is written as

$$\frac{d\theta}{ds} = C\frac{d^2y}{dx^2} \tag{2.49}$$

then

$$C = \frac{1}{[1 + (dy/dx)^2]^{3/2}} \tag{2.50}$$

For small deflections, it is assumed that $C = 1$. As the deflection increases, C changes, and its deviation from unity represents the factor by which the small

Large-Deflection Analysis

deflection assumption is inaccurate. Table 2.3 lists the value of the slope, the corresponding beam angle, and C. Note that for very small values of the slope, C is nearly equal to 1. As the slope increases, C decreases.

2.9.1 Beam with Moment End Load

Consider the cantilever beam with a moment applied to the free end, as shown in Figure 2.13. The Bernoulli–Euler equation is

$$\frac{d\theta}{ds} = \frac{M_o}{EI} \qquad (2.51)$$

where M_o is constant along the beam length. The deflected angle of the beam end, θ_o, is found by separating variables and integrating, that is,

$$\int_0^{\theta_o} d\theta = \int_0^L \frac{M_o}{EI} ds \qquad (2.52)$$

$$\theta_o = \frac{M_o L}{EI} \qquad (2.53)$$

where θ_o is in radians. No small deflection assumptions have been made in this equation, because the integration was done with respect to s, the distance along the beam rather than with respect to x, the horizontal distance.

The vertical deflection may be found by using the chain rule of differentiation to obtain

$$\frac{M_o}{EI} = \frac{d\theta}{ds} = \frac{d\theta}{dy}\frac{dy}{ds} \qquad (2.54)$$

TABLE 2.3. Small-deflection assumption assumes $C = 1$

dy/dx	θ (deg)	C
0.01	0.6	0.9999
0.05	2.9	0.9963
0.10	5.7	0.9852
0.25	14.0	0.9131
0.50	26.6	0.7155
1.00	45.0	0.3536
2.00	63.4	0.0894

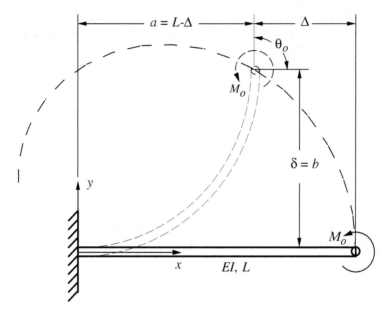

Figure 2.13. Flexible cantilever beam with a moment applied to the free end.

Because $dy/ds = \sin\theta$, equation (2.54) may be simplified as

$$\frac{M_o}{EI} = \frac{d\theta}{dy}\sin\theta \tag{2.55}$$

Separating variables and integrating yields

$$\int_0^b dy = \frac{EI}{M_o}\int_0^{\theta_o}\sin\theta\, d\theta \tag{2.56}$$

Integrating this equation results in

$$b = \frac{EI}{M_o}(-\cos\theta_o + 1) \tag{2.57}$$

and dividing both sides by L results in the nondimensionalized vertical beam deflection:

$$\frac{b}{L} = \frac{EI}{M_o L}(1 - \cos\theta_o) \tag{2.58}$$

Substituting equation (2.53) into equation (2.58) yields

Large-Deflection Analysis

$$\frac{b}{L} = \frac{1 - \cos \theta_o}{\theta_o} \quad (2.59)$$

or

$$\frac{b}{L} = \frac{1 - \cos[M_o L/(EI)]}{M_o L/(EI)} \quad (2.60)$$

The horizontal deflection is found in a similar manner. The chain rule of differentiation is used to write the Bernoulli–Euler equation as

$$\frac{M_o}{EI} = \frac{d\theta}{ds} = \frac{d\theta}{dx}\frac{dx}{ds} \quad (2.61)$$

Since $dx/ds = \cos \theta$, equation (2.54) may be simplified as

$$\frac{M_o}{EI} = \frac{d\theta}{dx} \cos \theta \quad (2.62)$$

Separating variables and integrating yields

$$\int_0^a dx = \frac{EI}{M_o} \int_0^{\theta_o} \cos \theta \, d\theta \quad (2.63)$$

Integrating and dividing both sides by L results in the nondimensionalized horizontal coordinate of the beam end:

$$\frac{a}{L} = \frac{\sin \theta_o}{\theta_o} = \frac{\sin[M_o L/(EI)]}{M_o L/(EI)} \quad (2.64)$$

The nondimensionalized horizontal deflection is $1 - a/L$. The end deflection described by equations (2.58) and (2.64) is shown by the dashed curve in Figure 2.13.

2.9.2 Elliptic-Integral Solutions

It is much more difficult to derive closed-form solutions for structures with large deflections and loading conditions more complicated than in the example above. Elliptic integrals are commonly used to solve large-deflection problems with other loading conditions [90–96]. A brief description of elliptic integrals is presented next, followed by a derivation of large-deflection solutions for a cantilever beam with a force at the free end. Simpler solution methods are presented in later chapters, but this method provides a foundation on which the approximate methods described later are based.

Elliptic Integrals. Elliptic integrals are briefly introduced in this section (for more detail, see [97–100]). The approach is very simplistic, but its purpose is to provide enough information for the reader to be comfortable with the use of elliptic integrals.

Consider the analogy of elliptic integrals to trigonometric functions. For a cosine function, for example, an angle is input into the function, and a value is returned. The value may be computed by a series expansion or by accessing a look-up table. In the same way, elliptic integrals may be viewed as functions where an input is given and a result calculated by a series expansion or a table. Just as there is more than one trigonometric function, there are multiple elliptic integrals. Just as there are sine, cosine, and tangent functions, there are elliptic integrals of the first, second, and third kinds. One difference in this analogy is that trigonometric functions have one independent variable (the angle), while some elliptic integrals may require two or three independent variables.

The definition of the elliptic integral of the first kind, $F(\phi, k)$, is

$$F(\phi, k) = \int_0^\phi \frac{d\theta}{\sqrt{1 - k^2 \sin^2 \theta}} \tag{2.65}$$

where ϕ is called the amplitude and k the modulus. The elliptic integral of the second kind, $E(\phi, k)$, is defined as

$$E(\phi, k) = \int_0^\phi \sqrt{1 - k^2 \sin^2 \theta}\, d\theta \tag{2.66}$$

There is also an elliptic integral of the third kind, $\Pi(\phi, \alpha^2, k)$, which will not be discussed here because it does not commonly occur in large-deflection analysis.

The elliptic integrals above are often called the *incomplete elliptic integrals of the first and second kind*. The *complete elliptic integrals* are special cases of the incomplete elliptic integrals for which $\phi = \pi/2$. The complete elliptic integrals of the first and second kinds commonly occur and are denoted as $F(\pi/2, k) = F(k)$ and $E(\pi/2, k) = E(k)$, respectively.

A simplified method of solving equations that involve elliptic integrals is as follows: First, derive the equations that describe the system. These equations will include integral terms that cannot be solved by usual methods. Second, manipulate these integral terms until they are in a form that is integrable using elliptic integrals. Third, choose a form available in the extensive *Handbook of Elliptic Integrals for Engineers and Physicists*, by P. F. Byrd and M. D. Friedman [97]. Once this step is completed, the solution will contain elliptic integrals. Appendix F describes an algorithm that may be used to determine numerical values for elliptic integrals.

Large-Deflection Analysis

Solution for Cantilever Beam with Force at Free End. An initially straight cantilever beam with combined horizontal and vertical end forces is shown in Figure 2.14. The combined end forces may also be treated as a single force, acting at an angle ϕ, where $\phi = \operatorname{atan}(1/-n)$ (and atan is the arctangent function), as shown. An elliptic integral solution for the end deflection of the beam may be derived, with the assumption that the beam is linearly elastic, inextensible, rigid in shear, and of constant cross section.

The internal moment at any point in the beam is given by

$$M = P(a-x) + nP(b-y) \tag{2.67}$$

where a is the horizontal distance from the fixed end to the free end, and b is the vertical distance of the free end from its undeflected position.

Using the Bernoulli–Euler equation described above ($d\theta/ds = M/EI$), the curvature, κ, may be written as

$$\kappa = \frac{d\theta}{ds} = \frac{M}{EI} \tag{2.68}$$

Substituting equation (2.67) into equation (2.68) for M results in

$$\kappa = \frac{d\theta}{ds} = \frac{P}{EI}[(a-x) + n(b-y)] \tag{2.69}$$

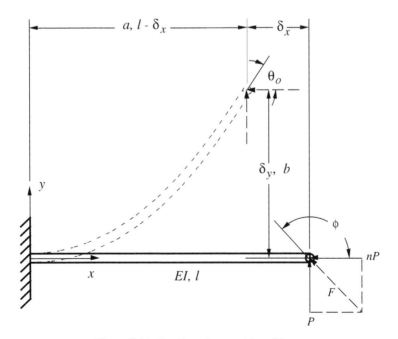

Figure 2.14. Cantilever beam with end forces.

Differentiating with respect to s yields

$$\frac{d\kappa}{ds} = \frac{d^2\theta}{ds^2} = \frac{P}{EI}\left(-\frac{dx}{ds} - n\frac{dy}{ds}\right) \tag{2.70}$$

Since dx, dy, and ds are infinitesimal, it is true that

$$\frac{dx}{ds} = \cos\theta \tag{2.71}$$

and

$$\frac{dy}{ds} = \sin\theta \tag{2.72}$$

Substituting these relations into equation (2.70) results in

$$\frac{d^2\theta}{ds^2} = \frac{-P}{EI}(n\sin\theta + \cos\theta) \tag{2.73}$$

The second derivative of θ, $d^2\theta/ds^2$, may be written as

$$\frac{d^2\theta}{ds^2} = \frac{d}{ds}\left(\frac{d\theta}{ds}\right) \tag{2.74}$$

Using the chain rule of differentiation, equation (2.74) may be written as

$$\frac{d}{ds}\left(\frac{d\theta}{ds}\right) = \frac{d}{d\theta}\left(\frac{d\theta}{ds}\right)\frac{d\theta}{ds} \tag{2.75}$$

Substituting equation (2.75) and $\kappa = d\theta/ds$ into equation (2.74) results in

$$\frac{d^2\theta}{ds^2} = \frac{d\kappa}{d\theta}\kappa \tag{2.76}$$

but this equation may also be written as

$$\frac{d^2\theta}{ds^2} = \frac{d}{d\theta}\left(\frac{\kappa^2}{2}\right) \tag{2.77}$$

Equation (2.77) may be differentiated to show that it is equivalent to equation (2.76).

Large-Deflection Analysis

Substituting equation (2.77) into equation (2.73), separating variables, and integrating yields

$$\int d\left(\frac{\kappa^2}{2}\right) = \frac{-P}{EI}\int (n\sin\theta + \cos\theta)\,d\theta \tag{2.78}$$

or

$$\frac{\kappa^2}{2} = \frac{P}{EI}(n\cos\theta - \sin\theta) + C_1 \tag{2.79}$$

where C_1 is the constant of integration.

If we use the conditions at the beam end, where $\theta = \theta_o$ and $d\theta/ds = \kappa = 0$, C_1 is found as

$$C_1 = \frac{P}{EI}(\sin\theta_o - n\cos\theta_o) \tag{2.80}$$

Substituting equation (2.80) into equation (2.79) and solving for κ yields

$$\kappa = \frac{d\theta}{ds} = \sqrt{2\frac{P}{EI}(\sin\theta_o - n\cos\theta_o - \sin\theta + n\cos\theta)} \tag{2.81}$$

or in nondimensionalized form

$$\kappa = \frac{d\theta}{ds} = \sqrt{2}\frac{\alpha}{l}\sqrt{\lambda - \sin\theta + n\cos\theta} \tag{2.82}$$

where

$$\alpha^2 = \frac{Pl^2}{EI} \tag{2.83}$$

$$\lambda = \sin\theta_o - n\cos\theta_o \tag{2.84}$$

or

$$\lambda = \eta\cos(\theta_o - \phi_1) \tag{2.85}$$

and

$$\eta = \sqrt{1 + n^2} \tag{2.86}$$

$$\phi_1 = \operatorname{atan}\frac{1}{-n} \tag{2.87}$$

Separating variables in equation (2.82) and integrating yields

$$\frac{\alpha\sqrt{2}}{l}\int_0^l ds = \int_0^{\theta_o} \frac{d\theta}{\sqrt{\lambda - \sin\theta + n\cos\theta}} \tag{2.88}$$

or

$$\alpha = \frac{1}{\sqrt{2}}\int_0^{\theta_o} \frac{d\theta}{\sqrt{\lambda - \sin\theta + n\cos\theta}} \tag{2.89}$$

This equation may be solved using numerical integration or elliptic integrals. Since many other sources exist for information on numerical integration, the elliptic integral solution is shown here. Using the elliptic integral tables of Byrd and Friedman [97] with equation (2.89) results in

$$\alpha = \frac{1}{\sqrt{\eta}}[F(t) - F(\gamma, t)] \qquad \text{for } \theta_o < \phi_1 \tag{2.90}$$

where

$$\gamma = \operatorname{asin}\sqrt{\frac{\eta - n}{\eta + \lambda}} \tag{2.91}$$

$$t = \sqrt{\frac{\eta + \lambda}{2\eta}} \tag{2.92}$$

where asin is the arcsin operator.

Equation (2.90) may be used to relate the end angle to the force. Assuming known beam geometry (L and I), Young's modulus (E), and the direction of the load (n), the force may be calculated for a given beam end angle (θ_o). This is done by substituting the value of θ_o into equation (2.84), which is then substituted into equations (2.91) and (2.92). These values are used in equation (2.90) to calculate α. The force (P) is then calculated from equation (2.83).

The equation for the final vertical position of the end of the beam may now be derived. The curvature, κ, may be rewritten as

$$\kappa = \frac{d\theta}{ds} = \frac{d\theta}{dy}\frac{dy}{ds} = \frac{d\theta}{dy}\sin\theta \tag{2.93}$$

Substituting equation (2.93) into equation (2.82), separating variables, and integrating yields

$$\int_0^b dy = \frac{l\sqrt{2}}{2\alpha}\int_0^{\theta_o} \frac{\sin\theta\, d\theta}{\sqrt{\lambda - \sin\theta + n\cos\theta}} \tag{2.94}$$

Large-Deflection Analysis

or

$$\frac{b}{l} = \frac{\sqrt{2}}{2\alpha} \int_0^{\theta_o} \frac{\sin\theta \, d\theta}{\sqrt{\lambda - \sin\theta + n\cos\theta}} \tag{2.95}$$

Integrating the equations above [97] results in

$$\frac{b}{l} = \frac{1}{\alpha \eta^{5/2}} \{\eta[F(t) - F(\gamma, t) + 2(E(\gamma, t) - E(t))] + n\sqrt{2\eta(\eta + \lambda)} \cos\gamma\} \tag{2.96}$$

The final horizontal position may be found in a similar manner, but the curvature is written as

$$\kappa = \frac{d\theta}{ds} = \frac{d\theta \, dx}{dx \, ds} = \frac{d\theta}{dx} \cos\theta \tag{2.97}$$

Substituting equation (2.97) into equation (2.82), separating variables, and integrating [97] yields

$$\frac{a}{l} = \frac{1}{\alpha \eta^{5/2}} \{-n\eta[F(t) - F(\gamma, t) + 2(E(\gamma, t) - E(t))] + \sqrt{2\eta(\eta + \lambda)} \cos\gamma\} \tag{2.98}$$

Figures 2.15 and 2.16 show vertical and horizontal deflections for various values of n and α^2. Equation (2.84) shows that for the loading specified, the nondimensionalized load parameter, λ, is not a function of the flexural rigidity, EI, or the length, l. This leaves α as the only load parameter that is a function of the geometric and material properties, that is, E, I, and l. This implies that for a beam with a given n, its deflection follows the curves specified in Figures 2.15 and 2.16, regardless of E, I, and l. In other words, the rigidity and length of the beam do not affect the shape of the deflection curve, but only dictate where on the curve a particular deflection point lies.

Numerical Example. Consider a flexible aluminum beam ($E = 71.7$ GPa) that is 100 mm long and has a circular cross section with a diameter of 2 mm. The beam end is to undergo an angular deflection of 1 rad ($\theta_o = 1$ rad). (a) Find the vertical load that is required to cause this deflection. (b) Calculate the resulting coordinates of the beam end (a and b).

Solution: (a) The moment of inertia, I, is

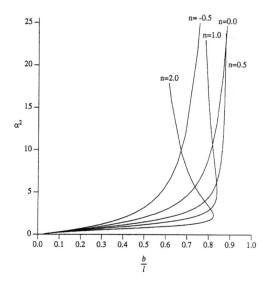

Figure 2.15. Vertical deflections of a beam for various end forces.

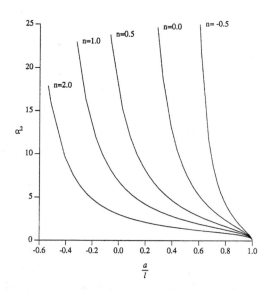

Figure 2.16. Horizontal deflections of a beam for various end forces.

Large-Deflection Analysis

$$I = \frac{\pi d^4}{64} = 7.85 \times 10^{-13} \text{ m}^4 \tag{2.99}$$

Because a vertical load is applied, $n = 0$. The values for λ and η are calculated from equations (2.84) and (2.86), respectively, as $\lambda = 0.841$ and $\eta = 1$. Equations (2.91) and (2.92) result in values for γ and t of $\gamma = 0.828$ rad and $t = 0.960$. The elliptic integral terms can now be evaluated, resulting in $F(t) = 2.69$, $F(\gamma, t) = 0.931$, $E(t) = 1.09$, and $E(\gamma, t) = 0.745$. (See Appendix F for one method of calculating the elliptic-integral values.) Substituting into equation (2.90) results in $\alpha = 1.76$. Rearranging equation (2.83) for force results in

$$P = \alpha^2 \frac{EI}{l^2} = 17.4 \text{ N} \tag{2.100}$$

(b) The end coordinates of the beam are calculated by substituting the values above into equations (2.96) and (2.98), which results in $b = 61$ mm and $a = 74$ mm.

Combined End Force, Moment, and Initial Curvature. Consider the initially curved cantilever beam with a force and moment applied at the free end (Figure 2.17). The deflection of this beam may be expressed as follows: For $|\lambda| < \eta$; $\phi - \text{acos}(-\lambda/\eta) \leq -\theta_o < \phi'$ and $\alpha \neq 0$ (where acos is the arccosine operator),

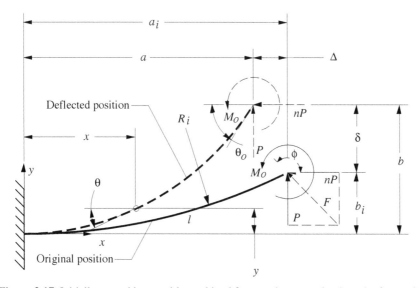

Figure 2.17. Initially curved beam with combined force and moment loads at the free end.

$$\alpha = \frac{1}{\sqrt{\eta}}[F(\gamma_2, t) - F(\gamma_1, t)] \qquad (2.101)$$

and for $\lambda > \eta$; $\phi' - \pi \leq -\theta_o < \phi'$ and $\alpha \neq 0$,

$$\alpha = \sqrt{\frac{2}{\lambda + \eta}}[F(\psi_2, r) - F(\psi_1, r)] \qquad (2.102)$$

where

$$\lambda = \frac{1}{2}\left(\frac{M_o}{EI} + \frac{1}{R_i}\right)^2 \frac{l^2}{\alpha^2} + \sin\theta_o - n\cos\theta_o \qquad (2.103)$$

$$\eta = \sqrt{1 + n^2} \qquad \phi' = \operatorname{atan}\frac{1}{n} \qquad (2.104)$$

$$\gamma_1 = \operatorname{asin}\sqrt{\frac{\eta - n}{\eta + \lambda}} \qquad \gamma_2 = \operatorname{asin}\sqrt{\frac{\eta + \sin\theta_o - n\cos\theta_o}{\eta + \lambda}} \qquad (2.105)$$

$$\psi_1 = \operatorname{asin}\sqrt{\frac{\eta - n}{2\eta}} \qquad \psi_2 = \operatorname{asin}\sqrt{\frac{\eta + \sin\theta_o - n\cos\theta_o}{2\eta}} \qquad (2.106)$$

$$t = \sqrt{\frac{\eta + \lambda}{2\eta}} \qquad r = \sqrt{\frac{2\eta}{\eta + \lambda}} \qquad (2.107)$$

The vertical deflection is: For $|\lambda| < \eta$ and $(\alpha \neq 0)$,

$$\frac{b}{l} = \frac{1}{\alpha\eta^{5/2}}\{\eta[F(\gamma_2, t) - F(\gamma_1, t) + 2(E(\gamma_1, t) - E(\gamma_2, t))] \\ + n\sqrt{2\eta(\eta + \lambda)}(\cos\gamma_1 - \cos\gamma_2)\} \qquad (2.108)$$

and for $\lambda > \eta > 0$ and $\alpha \neq 0$,

$$\frac{b}{l} = \frac{\sqrt{2(\eta + \lambda)}}{\alpha\eta^2}\left\{\frac{\lambda}{\eta + \lambda}[F(\psi_2, r) - F(\psi_1, r)] + [E(\psi_1, r) - E(\psi_2, r)] \\ + n\left[\sqrt{1 - \frac{\eta - n}{\eta + \lambda}} - \sqrt{1 - \frac{\eta + \sin\theta_o - n\cos\theta_o}{\eta + \lambda}}\right]\right\} \qquad (2.109)$$

The horizontal deflection is: For $|\lambda| < \eta$ and $\alpha \neq 0$,

Large-Deflection Analysis

$$\frac{a}{l} = \frac{1}{\alpha\eta^{5/2}}\{-n\eta[F(\gamma_2, t) - F(\gamma_1, t) + 2(E(\gamma_1, t) - E(\gamma_2, t))] \\ + \sqrt{2\eta(\eta + \lambda)}(\cos\gamma_1 - \cos\gamma_2)\}$$

(2.110)

and for $\lambda > \eta > 0$ and $\alpha \neq 0$,

$$\frac{a}{l} = \frac{\sqrt{2(\eta + \lambda)}}{\alpha\eta^2}\left\{-n\left[\frac{\lambda}{\eta + \lambda}(F(\psi_2, r) - F(\psi_1, r)) + (E(\psi_1, r) - E(\psi_2, r))\right] \\ + \left[\sqrt{1 - \frac{\eta - n}{\eta + \lambda}}\sqrt{1 - \frac{\eta + \sin\theta_o - n\cos\theta_o}{\eta + \lambda}}\right]\right\}$$

(2.111)

2.9.3 Numerical Methods

The disadvantages of the elliptic-integral solutions discussed above are that the derivations are complicated, and solutions can be found only for relatively simple geometries and loadings. The method also requires several simplifying assumptions, including linear material properties and inextensible members. An alternative method is required to analyze more general flexible members.

Nonlinear finite element methods are common numerical alternatives to elliptic-integral solutions. Because many such programs are available commercially, they are not discussed in detail here. The user must ensure that the program selected is capable of nonlinear analyses.

The chain algorithm uses the same theory as conventional finite element methods, but uses a different technique to combine and solve the resulting equations, making it computationally more efficient in many applications. The chain algorithm requires discretization of the object being modeled into beam elements and analyzes each element in succession. Each element is treated as a beam cantilevered at the end of the preceding element. Equivalent loads are found for each cantilevered element, and its deflections are calculated. The advantage of this method is that the global stiffness matrix and the resulting solution of the system of equations are not required for the problem solution. Details of the chain algorithm are included in Chapter 7.

PROBLEMS

2.1 Use the Bernoulli–Euler equation for small deflections, $M = EI(d^2y/dx^2)$, to derive the angular and vertical deflection equations for any point along the beam shown in Figure P2.1.

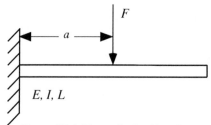

Figure P2.1. Figure for Problem 2.1.

2.2 Answer the following questions:
   (a) How does an axial force affect the flexibility of a compliant member? If flexibility is desired, is tension or compression desired?
   (b) What is the difference between stiffness and strength?
   (c) A flexible snap fastener you are designing yields during installation. What can you do to improve the design? Be specific.
   (d) State a major advantage and disadvantage of using elliptic integrals to solve large deflection beam problems.

2.3 A flexible snap fastener you are designing has a stress in the acceptable range, but the installation force is too small. What can you do to increase the installation force without causing other problems?

2.4 Determine the maximum bending stress in the telephone connector shown in Figure P2.4. Assume that the maximum deflection occurs when the tip of the rigid section touches the base.

2.5 Design a snap connector for attaching panels in automobile interiors (see Figure P2.5). The holes are square (with rounded corners), 1/4 in. each side, and the connector must fasten a 1/8-in. panel to a 1/4-in. support structure (see Figure P2.5). The vertical force required to release the fastener should exceed the force that may be applied by dynamic loading (assume about 1.5 lb) but should not be so large that it is difficult to assemble or causes damage to the panel (assume a maximum assembly force of 5 lb). Use high-density

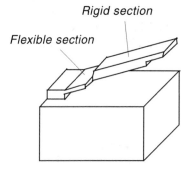

Dimensions of flexible section
width - 6 mm
thickness - 0.5 mm
length - 2 mm
Angle of beam - 15°
Length of beam - 10 mm
(flexible plus rigid)
Assume that E = 200,000 psi
Beam connection raised 1 mm

Figure P2.4. Figure for Problem 2.4.

Large-Deflection Analysis 57

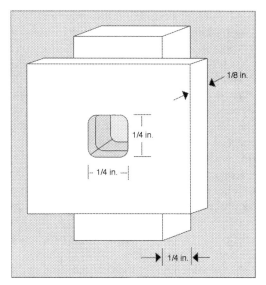

Figure P2.5. Figure for Problem 2.5.

polyethylene (HDPE) with $E = 200{,}000$ psi and $S_y = 7000$ psi. Assume the coefficient of friction between the HDPE and the panel to be approximately $v = 0.3$.

(a) Discuss different possible designs and the advantage and disadvantage of each. Choose a design for this application.
(b) Determine the length, shape, and size of the connector and its members.
(c) Calculate the vertical force required to deflect the beam and the maximum bending stress.
(d) Discuss how the end shape affects the assembly force required to install the fastener. Calculate the assembly force required to install the fastener.

2.6 Consider the fingernail clippers shown in Figure P2.6. Make the following assumptions for simplicity: F_{in} acts normal to the rigid segment; the flexible beam deflection is small and linear beam equations may be used; the contact

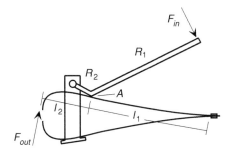

Figure P2.6. Figure for Problem 2.6.

point, point A, changes only slightly in the range of motion such that l_1 and l_2 may be considered to be constant; and the force on the flexible segment at point A is normal to the flexible segment.
  (a) Would each of the following changes increase, decrease, or have no change on the mechanical advantage of the device: (i) increase R_1; (ii) increase R_2; (iii) increase l_2; (iv) increase the moment of inertia associated with the flexible segment; (v) increase F_{in}.
  (b) Suppose the stress is acceptable but that too much force is required to actuate the mechanism. How could you modify the flexible beam to change the input force without changing the stress?
  (c) Suppose that in another case the stress was too high in the flexible segment and it yields before it closes all the way, even when not clipping nails. How could you modify the flexible beam to improve its stress characteristics?
  (d) If point A deflects $0.1 l_1$ before contacting the nail, how far does the output deflect?

2.7 A cantilevered beam has a moment load, M_o, at the free end. If the beam has a maximum stress of 50,000 psi, calculate the values of the moment (M_o), the end angle (θ_o), and the horizontal and vertical coordinates of the end point (a and b, respectively). It is a steel beam with a 1/32 in. x 1.25 in. cross section, and a length of 24 in.

2.8 Derive the large-deflection equations for a flexible cantilever beam with a vertical force at the free end. You may want to use *Handbook of Elliptic Integrals for Engineers and Physicists,* by P. F. Byrd and M. D. Friedman [97].

2.9 Use elliptic integrals to calculate the deflection (a, b, θ_o) for a steel beam with a 1/32 in. x 1.25 in. cross section, a length of 24 in., and the following load conditions:
  (a) A straight beam with a vertical force of 0.2 lb and a horizontal force (causing compression) of 0.4 lb.
  (b) A curved beam (radius of curvature of $R = 20$ in.) with a vertical force of 0.2 lb and a moment of 0.2 in.-lb.
  (c) Find the vertical load for a straight beam that will cause a vertical deflection of 12 in.

2.10 A flexible segment must undergo a particular deflection to perform its function, but it fails before achieving the entire deflection. If the material must stay the same, what modification could be made to the geometry to improve the part?

2.11 Which of the loading conditions in Figure P2.11 is more efficient if large deflections are desired? Explain why.

2.12 List two types of geometric nonlinearities.

2.13 List two things that can be changed to make a member more flexible.

2.14 What types of materials can be used to make compliant mechanisms?

2.15 State the Bernoulli–Euler assumption for beams (either in words or equation form). Describe the small deflection assumption often made with the Bernoulli–Euler assumption.

Large-Deflection Analysis

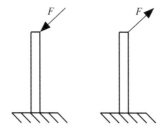

Figure P2.11. Figure for Problem 2.11.

2.16 What is the difference between a complete and an incomplete elliptic integral?

2.17 A large-deflection beam has two end forces and an end moment as shown in Figure P2.17, and a width w and a height h. The beam end has coordinates a and b.
  (a) Where is the maximum stress in this beam?
  (b) Write an equation for the stress at the top of the beam at the wall.
  (c) If the beam is made of plastic and the maximum stress exceeds the yield strength, should this be considered failure of the part? Explain.

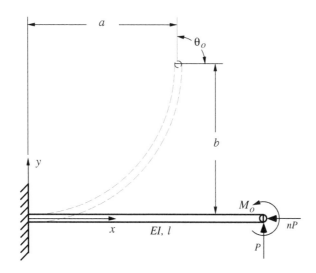

Figure P2.17. Figure for Problem 2.17.

CHAPTER 3

FAILURE PREVENTION

One of the most critical aspects of compliant mechanism design is ensuring that the mechanism will undergo its specified task without failing. Static failure and fatigue failure theories commonly used in mechanical design and examples of their use with compliant mechanisms are reviewed in this chapter. These failure theories are valuable tools that aid in the design of compliant mechanisms.

3.1 STRESS

Most mechanisms operate in two dimensions, and the stresses are modeled as planar. Bending and axial stresses are the predominate loading modes in compliant mechanisms; their associated stresses are

$$\sigma = \begin{cases} \dfrac{F}{A} & \text{(axial)} \\ \dfrac{My}{I} & \text{(bending)} \end{cases} \quad (3.1)$$

where F is the axial force, A the area of the cross section, M the moment load, y the distance from the neutral axis to the point of interest, and I the moment of inertia of the cross section. The maximum stress from bending occurs at the outer fibers, or

$$\sigma_{max} = \dfrac{Mc}{I} \quad \text{(bending)} \quad (3.2)$$

where c is the distance from the neutral axis to the outer fibers.

The bending moment in a compliant beam is usually caused by a force acting at a distance from the point considered. This implies that there will be both a moment and a shear force along the length of the beam. A shear stress will be caused by the shear force. However, this shear stress has its maximum at the center of the beam, whereas the maximum stress due to bending has its maximum stress at the outer fibers. The highest stress will be due to bending except for very short, stubby beams, such as bolts in shear loading. Because compliant mechanisms use flexible beams, this condition seldom occurs and bending is the predominate loading condition.

Torsion occurs in compliant members when an out-of-plane load is applied to a compliant mechanism. Torsion also occurs in torsion bars that are used to obtain angular deflection using torsionally compliant members. The maximum shear stress, τ_{max}, due to torsion on a circular-cross-section rod is

$$\tau_{max} = \frac{Tr}{J} \quad \text{(torsion)} \tag{3.3}$$

where T is the applied torque, r the rod radius, and J the polar moment of inertia.

Torsional stresses sometimes occur in flexible beams with rectangular cross sections. Such stresses are usually undesirable but are often the result of offset bending loads. The shear stress from torsion in a rectangular cross section can be approximated as

$$\tau_{max} = \frac{T}{bh^2} C_T \quad \text{(torsion of rectangular cross section)} \tag{3.4}$$

where b and h are the side lengths of the rectangle such that $b \geq h$, and C_T is as listed in Table 3.1. Different equations are required for other types of cross sections [89].

3.1.1 Principal Stresses

A three-dimensional stress element is shown in Figure 3.1a. The σ_i terms represent normal stresses and the τ_2 terms represent shear stresses. Loading conditions are often such that the predominate stresses may be assumed to be in a plane, as shown in Figure 3.1b. The *principal stresses* are the maximum stresses for any orientation of the stress element. The principal normal stresses are

TABLE 3.1. Values of C_T for various ratios of b to h

b/h	1.0	1.2	1.5	2.0	2.5	3	4	5	10	∞
C_T	4.81	4.56	4.34	4.07	3.88	3.74	3.55	3.43	3.2	3

Source: Data from [89].

Stress

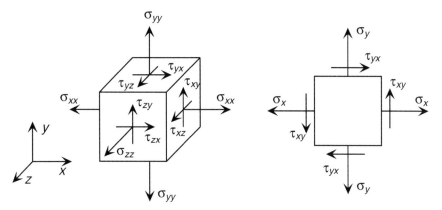

Figure 3.1. (a) Three-dimensional stress element, and (b) planar stress element.

$$\sigma_A, \sigma_B = \frac{\sigma_x + \sigma_y}{2} \pm \sqrt{\left(\frac{\sigma_x - \sigma_y}{2}\right)^2 + \tau_{xy}^2} \tag{3.5}$$

and the principal shear stress in the plane is

$$\tau_A, \tau_B = \pm\sqrt{\left(\frac{\sigma_x - \sigma_y}{2}\right)^2 + \tau_{xy}^2} \tag{3.6}$$

Although the stress is acting in a plane, when predicting failure it is important to consider the maximum stress in three dimensions. The principal stresses for planar stress are

$$\sigma_1 = \max(\sigma_A, \sigma_B, 0) \tag{3.7}$$

$$\sigma_3 = \min(\sigma_A, \sigma_B, 0) \tag{3.8}$$

and σ_2 is whichever of σ_A, σ_B, and 0 is between σ_1 and σ_3. The principal shear stress, τ_1, is $|\tau_A|$ from equation (3.6), or it can be calculated from the principal normal stresses as

$$\tau_1 = \frac{\sigma_1 - \sigma_3}{2} \tag{3.9}$$

Example: Principal Stresses. Consider the planar stress element shown in Figure 3.2. Calculate (a) the principal normal stresses (σ_1, σ_2, and σ_3) and (b) the principal shear stress (τ_1).
Solution: (a) First, use equation (3.5) to find σ_A and σ_B as

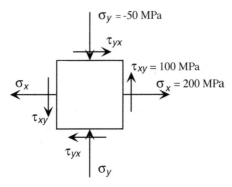

Figure 3.2. Planar stress element with normal and shear stresses for the principal stress example.

$$\sigma_A = \frac{200 \text{ MPa} - 50 \text{ MPa}}{2} + \sqrt{\left(\frac{200 \text{ MPa} + 50 \text{ MPa}}{2}\right)^2 + (100 \text{ MPa})^2} \quad (3.10)$$
$$= 235 \text{ MPa}$$

and

$$\sigma_B = \frac{200 \text{ MPa} - 50 \text{ MPa}}{2} - \sqrt{\left(\frac{200 \text{ MPa} + 50 \text{ MPa}}{2}\right)^2 + (100 \text{ MPa})^2} \quad (3.11)$$
$$= -85 \text{ MPa}$$

For this example $\sigma_A > 0 > \sigma_B$, so the principal normal stresses are

$$\sigma_1 = \sigma_A = 235 \text{ MPa}$$
$$\sigma_2 = 0 \quad (3.12)$$
$$\sigma_3 = \sigma_B = -85 \text{ MPa}$$

(b) The principal shear stress is calculated from equation (3.9) as

$$\tau_1 = \frac{235 \text{ MPa} - (-85 \text{ MPa})}{2} = 160 \text{ MPa} \quad (3.13)$$

Example: Principal Stresses for an Orthoplanar Spring. The *compliant orthoplanar spring* [101] shown in Figure 3.3 is a compact spring capable of obtaining large displacements without relative rotation between the platform and base. Each leg has the same dimensions, and the symmetry is such that each leg experiences the same stress conditions. If the platform is displaced a distance δ, determine the maximum principal stresses symbolically in terms of δ and the

Stress

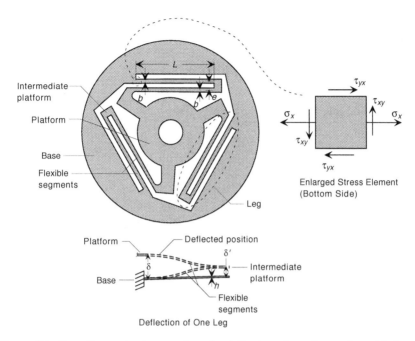

Figure 3.3. Compliant orthoplanar spring, the deflection of one leg, and a critical stress element.

geometry variables shown. For simplicity, assume that the deflections are small enough to be in the linear range and that linear beam equations are adequate for the analysis. Numerical values will be used in a later example.

Solution: Because of symmetry, each flexible segment undergoes a deflection, δ', which is half the total deflection, or

$$\delta' = \frac{\delta}{2} \tag{3.14}$$

The maximum stress in the flexible segments will occur at their ends. The flexible segment farthest from the platform will also experience some torsion, so it will have the highest stress. The end connected to the base is likely to have the highest stress concentration, so it is assumed to be the critical stress location that will be analyzed. This stress element is shown in Figure 3.3. The σ_x term comes from the bending stress induced by the deflection and the τ term is caused by the torsion due to the fact that the two flexible segments are offset.

The force–deflection relationship for the flexible segment can be found from Section D.11 in Appendix D as

$$\delta' = \frac{FL^3}{12EI} \tag{3.15}$$

where F is the force applied at the end of the flexible segment. Rearranging equation (3.15) in terms of force yields

$$F = \frac{12\delta'EI}{L^3} \qquad (3.16)$$

The bending stress is found from equation (3.2) as

$$\sigma_x = \frac{Mc}{I} = \frac{FLh}{4I} \qquad (3.17)$$

where $c = h/2$ and $M = FL/2$ for fixed–guided segments (where a fixed–guided segment has a deflection at the beam end, but the angle of the beam does not change). Substituting equations (3.14) and (3.16) into equation (3.17) yields the bending stress in terms of the deflection, or

$$\sigma_x = \frac{3\delta'Eh}{L^2} = \frac{3\delta Eh}{2L^2} \qquad (3.18)$$

There is no normal stress in the y direction, so for this example

$$\sigma_y = 0 \qquad (3.19)$$

The shear stress from torsion of the rectangular cross section is approximated from equation (3.4) as

$$\tau_{xy} = \frac{TC_T}{bh^2} = \frac{FeC_T}{bh^2} \qquad (3.20)$$

Substituting equations (3.14), and (3.16) and $I = bh^3/12$ (rectangular cross section) into equation (3.20) results in

$$\tau_{xy} = \frac{\delta'eEhC_T}{L^3} = \frac{\delta eEhC_T}{2L^3} \qquad (3.21)$$

The principal stresses are calculated by substituting the results of equations (3.18), (3.19), and (3.21) into equations (3.5) and (3.6). This results in

$$\sigma_A, \sigma_B = \frac{3\delta Eh}{4L^2} \pm \sqrt{\left(\frac{3\delta Eh}{4L^2}\right)^2 + \left(\frac{\delta eEh}{2L^3}C_T\right)^2} \qquad (3.22)$$

and

$$\tau_A, \tau_B = \pm\sqrt{\left(\frac{3\delta Eh}{4L^2}\right)^2 + \left(\frac{\delta eEh}{2L^3}C_T\right)^2} = \frac{\delta Eh}{2L^2}\sqrt{\frac{9}{4} + \left(\frac{e}{L}C_T\right)^2} \qquad (3.23)$$

Static Failure 67

3.1.2 Stress Concentrations

The stress equations discussed above assume a constant-cross-sectional geometry of the part under stress. This is called the *nominal stress*. However, geometry changes are locations of *stress concentration*, or *stress raisers*. The stress at a stress concentration can be significantly higher than the nominal stress. The amount that the stress is increased at a stress concentration is represented by the *stress-concentration factor*. The magnitude of the stress-concentration factor depends on the geometry of the stress concentration and the material. The stress concentration factor found taking into account the geometry alone is called the *theoretical stress-concentration factor*, K_t. Perhaps the most comprehensive source of stress-concentration factors can be found in *Peterson's Stress Concentration Factors* [102]. Other sources include [89], [103], and [104].

The material properties can be taken into account by finding the *notch sensitivity* of a material, q, and modifying the theoretical stress-concentration factor, K_t, to find the *fatigue stress-concentration factor*, K_f, as

$$K_f = 1 + q(K_t - 1) \tag{3.24}$$

The notch sensitivity takes into account material properties and the size of the notch. A material's sensitivity to notches tends to be less for ductile materials than for brittle materials. It also decreases for smaller notch sizes.

The stress concentration factors are used in different ways depending on whether the loading condition is static or fluctuating, and whether the material is ductile or brittle. Table 3.2 summarizes the use of the stress-concentration factors for these different conditions. Each is explained in more detail in the following sections.

3.2 STATIC FAILURE

The main reason for calculating the stress in a member is to determine if it will fail under the induced loads. This type of analysis allows devices to be designed such that they have sufficient strength to withstand operating stresses. Different failure

TABLE 3.2. Use of stress-concentration factors in design

	Material	
Loading	Ductile	Brittle
Static	$K_t = 1$	$\sigma_{max} = \sigma_{nom} K_t$
Fatigue	$c_{misc} = 1/K_f$	$\sigma_{max} = \sigma_{nom} K_f$

theories are used depending on the type of loading (static or fluctuating), the type of material (ductile or brittle), and the desired accuracy (conservative estimate or average failure). Failure for steady (static) loads is reviewed next, followed by a review of fluctuating fatigue loads.

3.2.1 Ductile Materials

If the induced stress exceeds the yield strength of a ductile material, a permanent deformation occurs, and the device may not be able to perform its intended function. The maximum shear stress (or Tresca) theory and the distortion energy (or von Mises–Hencky) theory are two of the most commonly used failure theories for ductile materials under static loads.

The *maximum shear stress theory* maintains that failure will occur when the maximum shear stress is equal to or greater than the shear stress in a tensile-test specimen at yield, or

$$\tau_{max} \geq \frac{S_y}{2} \quad \text{for failure} \tag{3.25}$$

By this theory, failure is prevented by maintaining the maximum shear stress below half the yield strength. The safety factor, SF, is then

$$SF = \frac{S_y}{2\tau_{max}} = \frac{S_y}{\sigma_1 - \sigma_3} \tag{3.26}$$

The *distortion energy theory* uses an effective stress to determine failure. The *effective stress*, also known as the *von Mises stress*, is the tensile stress that would cause the equivalent distortion energy as created by the actual stresses. The theory states that failure occurs when the effective stress, σ', is equal to the yield strength, or

$$\sigma' \geq S_y \quad \text{for failure} \tag{3.27}$$

where

$$\sigma' = \sqrt{\frac{(\sigma_1 - \sigma_2)^2 + (\sigma_2 - \sigma_3)^2 + (\sigma_3 - \sigma_1)^2}{2}} \tag{3.28}$$

For a two-dimensional stress state the effective stress can be simplified in terms of the applied stress as

$$\sigma' = \sqrt{\sigma_x^2 + \sigma_y^2 - \sigma_x\sigma_y + 3\tau_{xy}^2} \tag{3.29}$$

The safety factor, SF, is

Static Failure

$$\text{SF} = \frac{S_y}{\sigma'} \quad (3.30)$$

Both of these theories are acceptable for failure prediction of ductile materials with static loads, isotropic material properties, and equivalent tensile and compressive strengths. The two theories provide the same results for uniaxial stress, but in most other cases the maximum shear stress theory is more conservative than the distortion energy theory. Because it is more conservative and easier to calculate, the maximum shear stress theory is the most commonly used in design. The increased accuracy of the distortion energy theory makes it especially useful in estimating stresses in a part that failed.

Figure 3.4 shows the failure lines for the maximum shear stress and the distortion energy theories. If the point represented by the coordinates (σ_1, σ_3) on the plot are inside the safe region, failure will not occur. If the point is outside, failure is expected to occur. Ductile materials are commonly used in compliant mechanisms because they are more forgiving when overstressed. If the stress exceeds the material yield strength, plastic deformation will occur. The components should be designed to avoid this occurrence. However, in many cases this is only a local effect and is not catastrophic. For example, consider a cantilever beam with a force at the free end. Suppose a stress calculation determines that the maximum effective stress exceeds the yield strength. The maximum stress occurs only at the outer fibers of the beam at the fixed end. A ductile material such as polypropylene will yield at that point and redistribute the loads to adjacent material. The resulting plastic deformation may be so small as not to be noticeable. The effects of stress concentrations for ductile materials under static loads can usually be ignored because the high stress localized at a stress concentration will be redistributed as the material yields, and it can essentially cold work the material at that location. This explains why $K_t = 1$ (i.e., no stress concentration effect) is listed in Table 3.2 for ductile materials under static loads. Experienced engineers will often use these effects to

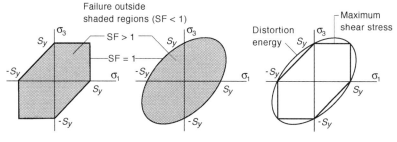

Figure 3.4. Failure lines for the maximum shear stress theory and the distortion energy theory.

their advantage in the design of ductile materials under static loads. A simple example is a plastic container that is fabricated from a flat sheet of material (dental floss containers are an example). The cross section is reduced at the edges and the sides are folded and fastened into the final shape. The stresses at the edges exceed the predicted stress for failure, but it is a one-time loading.

Example: Ductile Failure. Consider the stress element in the example described in Section 3.1.1 and shown in Figure 3.2. Recall that the principal stresses were $\sigma_1 = 235$ MPa, $\sigma_2 = 0$, $\sigma_3 = -85$ MPa, and $\tau_1 = \tau_{max} = 160$ MPa. Assuming that the material is 1050 HR steel, calculate the safety factor, SF, for ductile failure (yielding) using both (a) the maximum shear stress theory and (b) the distortion energy theory.

Solution: (a) The yield strength is listed in Table C.5 as $S_y = 345$ MPa. The safety factor using the maximum shear stress theory can be calculated using equation (3.26) as

$$SF = \frac{S_y}{2\tau_{max}} = \frac{345 \text{ MPa}}{2(160 \text{ MPa})} = 1.1 \quad \text{maximum shear stress theory} \quad (3.31)$$

(b) The effective stress for the distortion energy theory can be calculated using the principal stresses from equation (3.28) [or from the planar stresses using equation (3.29)] as

$$\sigma' = \sqrt{\frac{(235 \text{ MPa} - 0)^2 + [0 - (-85 \text{ MPa})]^2 + [235 \text{ MPa} - (-85 \text{ MPa})]^2}{2}} \quad (3.32)$$

$$= 287 \text{ MPa}$$

The safety factor is calculated from equation (3.30) as

$$SF = \frac{S_y}{\sigma'} = \frac{345 \text{ MPa}}{287 \text{ MPa}} = 1.2 \quad \text{distortion energy theory} \quad (3.33)$$

Note that the two theories give similar answers, but the maximum shear stress theory is slightly more conservative.

Example: Maximum Deflection of an Orthoplanar Spring. A compliant orthoplanar spring of the type discussed in the example in Section 3.1.1 is used as a "flapper" for a nozzle on a pneumatic valve positioner. It is made from 0.01-in. (0.25-mm)-thick type 410 annealed stainless steel and is shown in Figure 3.5. The flexible segments each have a width of $b = 0.03$ in. (0.762 mm), a length of $L = 0.646$ in. (16.4 mm), and the spacing between their centers is $e = 0.070$ in. (1.778 mm). (a) Calculate the safety factor for the spring if the maximum

Static Failure

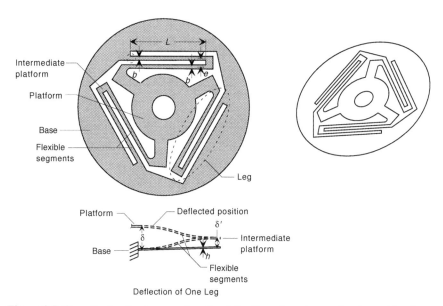

Figure 3.5. Compliant orthoplanar spring, the deflection of one leg, and an isometric view.

displacement of the platform, δ, is $\delta = 0.03$ in. (0.762 mm). (b) Calculate the maximum displacement before yielding. (c) Repeat the safety factor and maximum displacement calculations for the case where the material is heat treated rather than annealed.

Solution: (a) The Young's modulus for stainless steel is listed in Table C.1 as $E = 27.5$ Mpsi (189.6 GPa), and the yield strength for type 410 annealed stainless steel is listed in Table C.6 as $S_y = 45$ kpsi (310 MPa). The ratio of b to h is 3, and Table 3.1 gives a value of $C_T = 3.74$ for torsional stress in a rectangular cross section. The principal stresses for this configuration of the orthoplanar spring were found symbolically in the example in Section 3.1.1. The bending stress is found from equation (3.18) as

$$\sigma_x = \frac{3\delta Eh}{2L^2} = \frac{3(0.03 \text{ in.})(27.5 \text{ Mpsi})(0.01 \text{ in.})}{2(0.646 \text{ in.})^2} = 29{,}700 \text{ psi} \qquad (3.34)$$

The shear stress from torsion is calculated using equation (3.21) as

$$\tau_{xy} = \frac{\delta e E h C_T}{2L^3} = \frac{(0.03 \text{ in.})(0.07 \text{ in.})(27.5 \text{ Mpsi})(0.01 \text{ in.})(3.74)}{2(0.646 \text{ in.})^3} \qquad (3.35)$$
$$= 4000 \text{ psi}$$

Note that the bending stress is over seven times greater than the torsional shear stress.

Substituting the values for the material properties and geometry of this specific example into equations (3.22) and (3.23), or substituting the results of equations (3.34) and (3.35) directly into equations (3.5) and (3.9), results in

$$\sigma_A, \sigma_B = \frac{29{,}700 \text{ psi} + 0}{2} \pm \sqrt{\left(\frac{29{,}700 \text{ psi} - 0}{2}\right)^2 + (4000 \text{ psi})^2} \quad (3.36)$$

$$= 30{,}200 \text{ psi}, \ -529 \text{ psi}$$

Equations (3.7) and (3.8) show that the principal stresses are $\sigma_1 = 30{,}200$ psi (208 MPa), $\sigma_2 = 0$, and $\sigma_3 = -529$ psi (−3.65 MPa). The principal shear stress is

$$\tau_1 = \frac{30{,}200 \text{ psi} - (-529 \text{ psi})}{2} = 15{,}370 \text{ psi} \quad (3.37)$$

Now that the principal stresses are known, equation (3.26) can be used to calculate the safety factor using maximum shear stress theory as

$$SF = \frac{S_y}{2\tau_{max}} = \frac{45 \text{ kpsi}}{2(15.37 \text{ kpsi})} = 1.5 \quad (3.38)$$

The effective stress, σ', can be calculated from equation (3.28) as

$$\sigma' = \sqrt{\frac{(30{,}200 \text{ psi} - 0)^2 + [0 - (-529 \text{ psi})]^2 + [30{,}200 \text{ psi} - (-529 \text{ psi})]^2}{2}} \quad (3.39)$$

$$= 30{,}500 \text{ psi}$$

The safety factor using the distortion energy theory is calculated from equation (3.30) as

$$SF = \frac{S_y}{\sigma'} = \frac{45 \text{ kpsi}}{30.5 \text{ kpsi}} = 1.5 \quad (3.40)$$

Note that in this case the two theories provide the same answer. The reason for this is that the shear stress from torsion is small compared to the bending stress and σ_1 is approximately equal to the maximum bending stress. This will be the case in most well-designed compliant mechanisms, and the only significant stresses will be normal stresses, and the maximum shear stress theory and distortion energy theory will result in the same failure prediction.

(b) Because the bending stress dominates, the maximum deflection will be estimated by assuming that it occurs when the bending stress is equal to the yield strength. Substituting $\sigma_x = S_y$ into equation (3.18) and rearranging to solve for the displacement (δ) at failure (δ_{max}) results in

Static Failure

$$\delta_{max} = \frac{2S_y L^2}{3Eh} = \frac{2(45 \text{ kpsi})(0.646 \text{ in.})^2}{3(27.5 \text{ Mpsi})(0.01 \text{ in.})} = 0.046 \text{ in.} \quad (3.41)$$

(c) If the material were heat-treated type 410 stainless steel, Table C.6 lists the yield strength as $S_y = 140$ kpsi (965 MPa). The maximum shear stress theory and the distortion energy theory provide the same results, so either one may be used to calculate the safety factor as

$$SF = \frac{S_y}{\sigma'} = \frac{140 \text{ kpsi}}{30.5 \text{ kpsi}} = 4.6 \quad (3.42)$$

The maximum displacement is found using equation (3.41) as before as

$$\delta_{max} = \frac{2S_y L^2}{3Eh} = \frac{2(140 \text{ kpsi})(0.646 \text{ in.})^2}{3(27.5 \text{ Mpsi})(0.01 \text{ in.})} = 0.14 \text{ in.} \quad (3.43)$$

3.2.2 Brittle Materials

While ductile materials are more commonly used in compliant mechanisms, brittle materials may also be used. One application of brittle materials in compliant mechanisms is in microelectromechanical systems (MEMS), where materials such as polycrystalline silicon (polysilicon) are used to make mechanical devices.

Two failure theories are reviewed here, including the Coulomb–Mohr theory and the modified Mohr theory. Unlike ductile materials, brittle materials usually have different strengths in tension and compression. Thus it is important that theories for brittle fracture are based on both the ultimate tensile strength, S_{ut}, and the ultimate compressive strength, S_{uc}, of the material. The theories also assume that different failure criteria are used when all the principal stresses are tensile than if some are tensile and others are compressive.

The *Coulomb–Mohr theory* predicts the safety factor, SF (failure when SF $\leq$ 1), as

$$SF = \frac{S_{ut}}{\sigma_1} \quad \text{when } \sigma_3 \geq 0 \text{ (and } \sigma_1 > \sigma_3) \quad (3.44)$$

$$SF = \frac{S_{ut} S_{uc}}{S_{uc}\sigma_1 - S_{ut}\sigma_3} \quad \text{when } \sigma_1 \geq 0 \text{ and } \sigma_3 < 0 \quad (3.45)$$

Note that if the compressive and tensile strengths are the same, this theory is similar to the maximum shear stress theory discussed for ductile materials.

The *modified Mohr theory* predicts the following safety factor for brittle materials:

$$SF = \begin{cases} \dfrac{S_{ut}}{\sigma_1} & \text{when } \sigma_3 \geq -S_{ut} \\ \dfrac{S_{ut}S_{uc}}{S_{uc}\sigma_1 - S_{ut}(\sigma_1 + \sigma_3)} & \text{when } \sigma_3 < -S_{ut} \end{cases} \quad (3.46)$$

Both of these theories are acceptable for design and analysis of compliant mechanisms constructed in brittle materials under static loads. The Coulomb–Mohr theory is slightly more conservative than the modified Mohr theory and is commonly used in design for that reason. The modified Mohr theory is generally more accurate in its prediction of failure and therefore should be used in an analysis to determine why something has failed.

Figure 3.6 shows the failure lines for the Coulomb–Mohr and the modified Mohr theories. If the point represented by the coordinates (σ_1, σ_3) on the plot is inside the safe region, failure will not occur, and if it is outside, failure is expected to occur.

Unlike ductile materials, brittle materials are unforgiving when overstressed, and failure is catastrophic. This is particularly true for compliant mechanisms. If a brittle but flexible member is under load in a compliant mechanism, there is usually considerable elastic strain energy in the beam. If the load is high enough to cause fracture, there is usually also a relatively large magnitude of strain energy throughout the beam, not just at the location of high stress. When fracture occurs this energy is released, and it is not uncommon for the resulting dynamic loads to cause multiple fractures throughout the member. The result is that an excessively high stress in one location can cause fracture at that location, and the release of strain energy will make the device appear to explode as multiple fractures occur and pieces of broken material are catapulted away. Perhaps this is why one textbook author nates that ". . . there is no use in discussing ideal brittle materials, since no

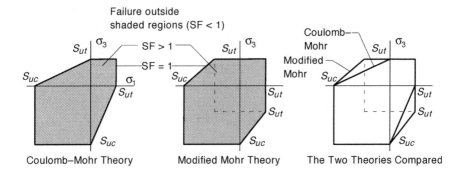

Figure 3.6. Failure lines for the Coulomb–Mohr and modified Mohr theories.

Static Failure

sane designer would ever specify one for a load-carrying member [107]." Sane or not, many modern applications involving materials, such as fiber optics and polycrystalline silicon, require the use of brittle materials.

Brittle materials are sensitive to stress concentrations, even for static loading conditions. Unlike ductile materials, the localized stress does not cause yielding and redistribution of stress, but will cause fracture. It is important to take the stress concentration factor into account when performing a stress analysis of brittle materials. Because of the energy storage and catastrophic failure considerations discussed above, it is common to be conservative and use the full stress concentration factor (K_t) for brittle materials under static loads, as listed in Table 3.2.

Example: Brittle Failure. Consider again the stress element in the example described in Section 3.1.1 and shown in Figure 3.2. But now assume that the material is class 50 gray cast iron with an ultimate tensile strength of $S_{ut} = 359$ MPa and ultimate compressive strength of $S_{uc} = 1113$ MPa. Recall that the principal stresses were calculated as $\sigma_1 = 235$ MPa, $\sigma_2 = 0$, $\sigma_3 = -85$ MPa, and $\tau_{max} = 160$ MPa. Calculate the safety factor, SF, for brittle failure using both (a) the Coulomb–Mohr theory and (b) the modified Mohr theory. Then, using the modified Mohr theory, (c) estimate the magnitude of stress concentration factor the part could tolerate before failure.

Solution: (a) For the Coulomb–Mohr theory we note that $\sigma_1 \geq 0$ and $\sigma_3 < 0$, which implies that equation (3.45) is used to calculate the safety factor as

$$SF = \frac{S_{ut}S_{uc}}{S_{uc}\sigma_1 - S_{ut}\sigma_3}$$

$$= \frac{(359 \text{ MPa})(1131 \text{ MPa})}{(1131 \text{ MPa})(235 \text{ MPa}) - (359 \text{ MPa})(-85 \text{ MPa})} = 1.4 \qquad (3.47)$$

(b) For the modified Mohr theory, $\sigma_3 \geq -S_{ut}$, so equation (3.46) is used as

$$SF = \frac{S_{ut}}{\sigma_1} = \frac{359 \text{ MPa}}{235 \text{ MPa}} = 1.5 \qquad (3.48)$$

Note that the two theories give similar answers, but the Coulomb–Mohr theory is more conservative. If a stress concentration were introduced, the modified Mohr theory predicts that the part would fail when SF = 1. Modifying equation (3.46) for this case results in

$$\frac{S_{ut}}{K_t\sigma_1} = 1 \qquad \text{at failure} \qquad (3.49)$$

(c) Solving for the stress concentration, K_t, at failure gives

$$K_t = \frac{S_{ut}}{\sigma_1} = \frac{359 \text{ MPa}}{235 \text{ MPa}} = 1.5 \qquad (3.50)$$

In other words, failure would be expected to occur if a stress concentration were introduced with a stress concentration factor of 1.5 or greater.

Example: Brittle Failure Prevention of a Polysilicon Electrostatic Comb Actuator. Electrostatic comb drives, such as the one shown in Figure 3.7, are sometimes used as actuators in micromechanical devices. The actuation force comes from electrostatic forces when a voltage difference is applied between the stationary electrode and the movable shuttle. The folded beams deflect in the plane and act as springs that suspend the shuttle. The symmetry is such that the flexible beams are fixed–guided beams. If the flexible beams have a length of $L = 250$ μm, an out-of-plane thickness of $b = 3.5$ μm, a width of $h = 3$ μm, and the shuttle undergoes a deflection of $\delta = 10$ μm, calculate the factor of safety for brittle failure using both (a) the Coulomb–Mohr theory and (b) the modified Mohr theory. The material is polycrystalline silicon (polysilicon), and assume properties of $E = 170$ GPa, $S_{ut} = 0.93$ GPa, and $S_{uc} = 1.2$ GPa. Assume a stress concentration factor of $K_t = 1.2$.

Solution: For small, linear deflections, the relations in Section D.11 can be used for fixed–guided beams. If the overall deflection of the shuttle is δ, the deflection of an individual beam, δ', is $\delta' = \delta/2$. Because both the folded beams and the

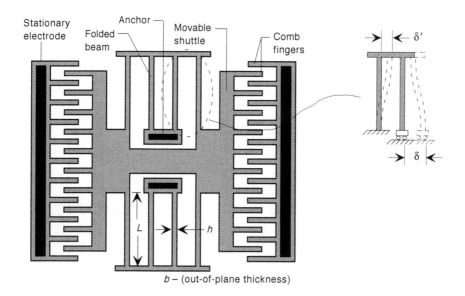

Figure 3.7. Electrostatic comb drive.

Fatigue Failure

compliant orthoplanar spring discussed earlier have fixed-guided springs in series, the equation relating stress to deflection is the same for both. Therefore, because there is negligible torsion, the maximum nominal stress (the stress calculated before taking into account the stress concentration) can be found from equation (3.18) as

$$\sigma_{o(max)} = \frac{3\delta E h}{2L^2} \tag{3.51}$$

Substituting in the values for this example results in

$$\sigma_{o(max)} = \frac{3(10 \ \mu m)(170 \ GPa)(3 \ \mu m)}{2(250 \ \mu m)^2} = 122 \ MPa \tag{3.52}$$

For a brittle material the full stress concentration factor is taken into account, so the maximum stress is

$$\sigma_{max} = K_t \sigma_{o(max)} \tag{3.53}$$

or

$$\sigma_{max} = 1.2(122 \ MPa) = 146 \ MPa \tag{3.54}$$

The shear stresses are negligible, so the principal stresses are $\sigma_1 = 146$ MPa, $\sigma_2 = 0$, and $\sigma_3 = 0$.

Because $\sigma_3 \geq 0$ and $\sigma_3 \geq -S_{ut}$, equations (3.44) and (3.46) are used to predict the safety factor for brittle failure using the Coulomb–Mohr theory and the modified Mohr theory, respectively. These two equations provide the same result:

$$SF = \frac{S_{ut}}{\sigma_1} = \frac{930 \ MPa}{146 \ MPa} = 6.4 \tag{3.55}$$

3.3 FATIGUE FAILURE

For some compliant structures the desired motion may occur only once and the static failure theories above may be enough for analysis. However, by the definition of compliant mechanisms it is known that deflection of flexible members is required for motion. Usually, it is desired that the mechanism be capable of undergoing the motion many times, and design requirements may be many millions of cycles or "infinite" life. This repeated loading causes fluctuating stresses in the members and can result in *fatigue failure*. Fatigue failure can occur at stresses that are significantly lower than those that cause static failure. Premature or unexpected failure of a device can result in unsafe design, or it may reduce consumer confidence in products that fail prematurely. For these and other reasons it is critical that the fatigue life of a compliant mechanism be analyzed and determined experimentally. This section contains a summary of fatigue failure theory used in design. It is presented as a review and it is assumed that the reader has had some exposure to fatigue analysis, such as in a machine design course. The review will emphasize concepts that are particularly important for compliant mechanisms.

3.3.1 Fatigue Basics

Fatigue failure occurs in three phases: crack initiation, crack growth, and fracture. The first phase involves a crack that may even be too small to detect at first. Cracks may result from material imperfections, manufacturing processes, or many other sources. The sharp crack causes a stress concentration that increases the local stress at the crack tip. A crack at a stress concentration of the part (such as a notch or a hole) amplifies the stress further. When a stress normal to the crack causes the crack to open (a tensile stress) the high stress at the crack tip causes the material to yield at that point and the size of the crack increases. With a fluctuating load the stress will be released (or at least decreased) and applied again. The same sequence will occur the next time the load is applied and the notch size will increase further. If the fluctuating stress is always in compression, such that it causes the crack to close rather than open, the crack will not grow. For this reason tensile stresses are much more critical in fatigue than are compressive stresses. If the crack continues to grow, it will eventually cause fracture.

Depending on the application, some devices may need to be designed to undergo only a few cycles, while others may be expected to go for millions or billions of cycles without failure. Fatigue life is often classified into the categories of *low-cycle fatigue* (fatigue failure occurs between 1 and 1000 cycles), *high-cycle fatigue* (fatigue failure at greater than 1000 cycles), and *infinite life* (no fatigue failure at the given loads). If it is known that a component will undergo only a few cycles in its life, it is not necessary to design for infinite life. However, many applications require that a device undergo many cycles. For example, a device connected to a motor operating at 1000 rpm will undergo a million cycles in 17 hours of operation.

The *S-N diagram*, or Woehler strength–life diagram, is a common way of viewing fatigue strength data. An *S–N* diagram typical for steel is shown in Figure 3.8. The failure line is made by plotting the number of cycles (N) at failure for a given stress (S). It is plotted on a log-log plot to show the knees in the data. For some materials, if the stress is below a certain point, failure will not occur regardless of the number of cycles. This is called the *endurance limit* and is discussed in more detail later.

Fatigue Failure

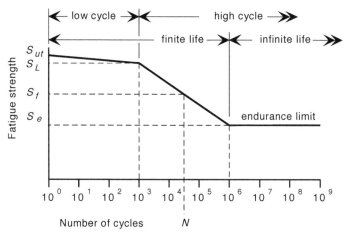

Figure 3.8. Typical *S-N* diagram for steel.

3.3.2 Fatigue Failure Prediction

Fatigue testing results in a large amount of scatter in the test results because many different factors influence fatigue failure. Some of these factors are in the engineer's control and others are not. Examples include stress concentrations, surface finish, material imperfections, initial size of cracks, temperature, corrosive elements in the environment, and so on. Even polished fatigue specimen from the same batch of material and of the same size exhibit considerable scatter. A part in service has an even wider range of failure data. Because of this, it is very difficult to predict fatigue failure with much accuracy, and testing is essential.

Although fatigue failure is difficult to predict accurately, an understanding of fatigue failure prediction and prevention is very helpful in the design of compliant mechanisms. The theory can be used to design devices that will withstand the fluctuating stresses. Testing then reveals the success of the design, but the initial analysis and design are critical for obtaining a design that is acceptable without undergoing many build-and-test iterations.

Several models are available for fatigue failure prediction. The *stress-life model* is the most straightforward and most commonly used in the design of mechanical components. This theory is most appropriate for parts that undergo consistent and predictable fluctuating stresses. Many machine components fit into this category because their motion and loads are defined by the kinematics of the mechanism. For the same reason, the stresses in most compliant mechanisms are also consistent and predictable. For example, the compliant mechanism shown in Figure 3.9 will undergo predictable deflections and the resulting strains and stresses can be calculated. Other models, such as the fracture-mechanics approach, can be used for other

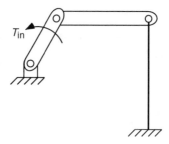

Figure 3.9. Compliant mechanism with consistent and predictable stresses on the flexible member.

types of components, such as airplane wings, that undergo randomly varying stresses.

The stress-life model is used to predict the fatigue strength for a number of cycles. The stresses are then kept well below the fatigue strength to ensure that fatigue failure will not occur. For some materials it is possible to keep the stresses below a level where cracks will not propagate at all.

Fatigue Strength. The *fatigue strength*, S_f, is the maximum totally reversed stress that a fatigue specimen can endure for N cycles. Recall that Figure 3.8 illustrates a typical *S-N* diagram for steel. This type of diagram is constructed by testing many different specimens at different values of stress and counting the number of cycles to failure. Plotting these points on a log-log scale results in the *S-N* diagram. There is a knee in the *S-N* diagram for many materials at the transition between low- and high-cycle fatigue, around 1×10^3 cycles. Some materials, such as "many low-strength carbon and alloy steels, some stainless steels, irons, molybdenum alloys, titanium alloys and some polymers" [108] have a second knee, called the *endurance limit*, S_e. If the stress is maintained below the endurance limit, fatigue failure will not occur, and the part has *infinite life*. This is very important for devices that must undergo many millions of cycles without failure.

On a log-log scale, the *S-N* diagram can be approximated as a series of straight lines. The first line goes from the strength at 1 cycle (the ultimate strength, S_{ut}) to the strength at the first knee (S_L). This first knee is assumed to occur at 1×10^3 cycles. The low-cycle fatigue regime is then approximated by a straight line on the log-log *S-N* diagram between the coordinates $(0, S_{ut})$ and $(1 \times 10^3, S_L)$. The value of S_L can be approximated as [109]

$$S_L = c_f S_{ut} \tag{3.56}$$

where

Fatigue Failure

$$c_f = \begin{cases} 0.9 & \text{bending} \\ 0.75 & \text{axial loading} \end{cases} \quad (3.57)$$

The nature of the high-cycle fatigue portion of the S-N diagram depends on whether or not the material has an endurance limit. First, consider materials with an endurance limit. The endurance limit is often taken to occur at about 1×10^6 cycles for steels and other materials. The finite life portion of the high-cycle fatigue region is then approximated on the log-log scale of the S-N diagram as a straight line from the previous knee at $(1 \times 10^3, S_L)$ to the second knee at the endurance limit of $(1 \times 10^6, S_e)$. If the stress is maintained below the endurance limit, the part will not fail regardless of the number of cycles it undergoes (infinite life). Therefore, the next line for materials with an endurance limit is a horizontal line at S_e that extends from $N = 1 \times 10^6$ cycles to infinity.

For materials without an endurance limit, the line for high-cycle fatigue starts at the same point at $(1 \times 10^3, S_L)$, but extends to a point of known fatigue strength for a known number of cycles (N, S_f). For example, for aluminum the fatigue strength at $N = 500 \times 10^6$ cycles is commonly used as the second point of the line.

The high-cycle fatigue line segment can be described by the equation

$$S_f = a_f N^{b_f} \quad (3.58)$$

where a_f and b_f are curve-fit parameters that must be found for a particular material or component. Because there are two unknowns (a_f and b_f), two equations are needed. These equations come from substituting the values of S_f and N at the two known points on the curve. The substitution and solution are simplified by writing equation (3.58) as

$$\log S_f = \log a_f + b_f \log N \quad (3.59)$$

For materials with an endurance limit $S_{f1} = c_f S_{ut}$, $N_1 = 1 \times 10^3$, $S_{f2} = S_e$, and $N_2 = 1 \times 10^6$. Substituting into equation (3.59) results in

$$\log c_f S_{ut} = \log a_f + b_f \log(1 \times 10^3) = \log a_f + 3b_f \quad (3.60)$$

and

$$\log S_e = \log a_f + b_f \log(1 \times 10^6) = \log a_f + 6b_f \quad (3.61)$$

Subtracting equation (3.61) from equation (3.60) results in

$$\log c_f S_{ut} - \log S_e = -3b_f \quad (3.62)$$

Solving for b_f and rearranging results in

$$b_f = -\frac{1}{3}\log\frac{c_f S_{ut}}{S_e} \tag{3.63}$$

Substituting equation (3.63) into equation (3.60) and solving for a_f results in

$$a_f = \frac{(c_f S_{ut})^2}{S_e} \tag{3.64}$$

A similar approach is used for materials without an endurance limit. For this case

$$b_f = \frac{1}{3 - \log N_2}\log\frac{c_f S_{ut}}{S_{f2}} \tag{3.65}$$

and

$$\log a_f = \frac{(\log N_2)(\log c_f S_{ut}) - 3 \log S_{f2}}{\log N_2 - 3} \tag{3.66}$$

For an arbitrary set of points known on the line

$$b_f = \frac{\log(S_{f2}/S_{f1})}{\log(N_2/N_1)} \tag{3.67}$$

and

$$\log a_f = \frac{(\log N_2)(\log S_{f1}) - (\log N_1)(\log S_{f2})}{\log N_2 - \log N_1} \tag{3.68}$$

Equation (3.58) can be rearranged to estimate the number of cycles to failure if the other terms are known.

$$N = \left(\frac{S_f}{a_f}\right)^{1/b_f} \tag{3.69}$$

3.3.3 Estimating Endurance Limit and Fatigue Strength

The endurance limit, or the fatigue strength at 5×10^8 cycles for materials without an endurance limit, is a critical parameter for fatigue life estimation. For some materials this information is available for tests performed on small, polished fatigue specimen under completely reversed bending or axial loads. Data obtained from such test specimen are known as the uncorrected endurance limit, $S_{e'}$, or the uncorrected fatigue strength, $S_{f'}$. Often, this information is not available and rough

Fatigue Failure

approximations must be made to obtain an estimate for design calculations. The following relationships can be used as to make these initial approximations [103, 107, 110, 111]:

$$\text{Steels: } S_{e'} \approx \begin{cases} 0.5 S_{ut} & \text{for } S_{ut} < 200 \text{ kpsi (1400 MPa)} \\ 100 \text{ kpsi (700 MPa)} & \text{for } S_{ut} \geq 200 \text{ kpsi (1400 MPa)} \end{cases} \quad (3.70)$$

$$\text{Irons: } S_{e'} \approx \begin{cases} 0.4 S_{ut} & \text{for } S_{ut} < 60 \text{ kpsi (400 MPa)} \\ 24 \text{ kpsi (160 MPa)} & \text{for } S_{ut} \geq 60 \text{ kpsi (400 MPa)} \end{cases} \quad (3.71)$$

$$\text{Aluminums: } S_{f'5 \times 10^8} \approx \begin{cases} 0.4 S_{ut} & \text{for } S_{ut} < 48 \text{ kpsi (330 MPa)} \\ 19 \text{ kpsi (130 MPa) for } S_{ut} \geq 48 \text{ kpsi (330 MPa)} \end{cases} \quad (3.72)$$

$$\text{Copper alloys: } S_{f'5 \times 10^8} \approx \begin{cases} 0.4 S_{ut} & \text{for } S_{ut} < 40 \text{ kpsi (280 MPa)} \\ 14 \text{ kpsi (100 MPa) for } S_{ut} \geq 40 \text{ kpsi (280 MPa)} \end{cases} \quad (3.73)$$

It is important to remember that these are rough estimates for polished specimens in a controlled environment. Because the component being designed will be significantly different than a test specimen, the endurance limit must be modified to be more closely related to the actual part. These estimates are valuable for the design phase to size parts but should not be substituted for testing real components under realistic conditions.

3.3.4 Endurance Limit and Fatigue Strength Modification Factors

The endurance limit or fatigue strength provided by fatigue specimen tests, or as estimated by equations (3.70) to (3.73), must be modified to reflect more accurately the actual component under consideration. This will be done using *Marin correction factors* [104, 113], c_i, as

$$S_e = c_{\text{surf}} c_{\text{size}} c_{\text{load}} c_{\text{reliab}} c_{\text{misc}} S_{e'} \quad (3.74)$$

or for materials without an endurance limit

$$S_f = c_{\text{surf}} c_{\text{size}} c_{\text{load}} c_{\text{reliab}} c_{\text{misc}} S_{f'} \quad (3.75)$$

where $S_{e'}$ and $S_{f'}$ are the theoretical endurance limit and fatigue strength, respectively, and S_e and S_f are the modified endurance limit and fatigue strength for the part. The correction factors are described below.

3.3.5 Surface Factor

The surface finish is one of the most critical things to consider in fatigue life prediction. Shigley and Mischke [114] suggest an expression for the Marin surface factor of

$$c_{surf} = \begin{cases} aS_{ut}^b & \text{if } aS_{ut}^b < 1 \\ 1 & \text{if } aS_{ut}^b \geq 1 \end{cases} \qquad (3.76)$$

where a and b are curve-fit parameters listed in Table 3.3 for various surface finishes. Note that the ultimate strength, S_{ut}, must be entered into equation (3.76) as either kpsi or MPa (not psi or Pa).

3.3.6 Size Factor

A part being analyzed is usually a different size than a standard fatigue specimen. If the cross-sectional area is larger, there is a larger probability of a surface imperfection. Shigley and Mischke recommend the following approximations for the Marin size factors [104]:

Bending and torsion of steel:

$$c_{size} = \begin{cases} 1 & \text{for } d < 0.11 \text{ in. } (2.79 \text{ mm}) \\ \left(\dfrac{d}{0.3}\right)^{-0.1133} & \text{if } d \text{ is in inches and } 0.11 \leq d \leq 2 \text{ in.} \\ \left(\dfrac{d}{7.62}\right)^{-0.1133} & \text{if } d \text{ is in millimeters and } 2.79 \leq d \leq 51 \text{ mm} \\ 0.6 & \text{for } d > 2 \text{ in. } (51 \text{ mm}) \end{cases} \qquad (3.77)$$

TABLE 3.3. Curve-fit parameters for Marin surface factor

Surface finish	a		b
	kpsi	MPa	
Ground	1.34	1.58	−0.086
Machined, cold-rolled	2.67	4.45	−0.265
Hot-rolled	14.5	58.1	−0.719
As-forged	39.8	271	−0.995

Source: Standard Handbook of Machine Design, 2nd ed., by Shigley and Mischke, 1996 [114]. Reproduced by permission of the McGraw-Hill Companies.

Fatigue Failure

Axial loading:

$$c_{size} = 1 \quad (3.78)$$

The equations above take into account the size effect assuming a circular cross section in rotating bending or torsion. An *equivalent diameter*, d_e, for other conditions may be approximated and substituted into the equations above. This approximation is made by comparing the amount of cross-sectional area that has stress above 95% of the maximum stress (A_{95}) to that of a circular test specimen in rotating bending. The equivalent diameter is then

$$d_e = 3.61\sqrt{A_{95}} \quad (3.79)$$

For a solid or hollow part with circular cross section in nonrotating bending

$$d_e = 0.37d \quad \text{nonrotating bending} \quad (3.80)$$

and for a nonrotating rectangular cross section with side dimensions b and h:

$$d_e = 0.808\sqrt{bh} \quad \text{rectangular} \quad (3.81)$$

Once calculated, the equivalent diameter, d_e, can be used in place of the diameter, d, in equations (3.77) to (3.77).

3.3.7 Load Factor

A reduction in endurance limit or fatigue strength occurs for loadings other than rotating bending. However, there is considerable scatter in test data for axial loadings and there is not agreement between various authors and researchers that have reported results in this area. The problem is most likely due to the difficulty of accurately controlling the eccentricity of the load for axial fatigue tests. Values for c_{load} have been suggested for axial loadings that range from 0.60 to 1.0. A somewhat conservative value is recommended by Norton [103] and is suggested here for use in design:

$$c_{load} = 0.70 \quad \text{axial} \quad (3.82)$$

For bending and torsion:

$$c_{load} = \begin{cases} 1.0 & \text{bending} \\ 0.577 & \text{torsion and shear} \end{cases} \quad (3.83)$$

3.3.8 Reliability

There is considerable scatter in fatigue test data. If the mean value of the endurance limit is predicted, some test values are higher and others are lower than the predicted mean. Table 3.4 lists values of c_{reliab} that can be used to account for the scatter in the data. Assuming a normal distribution with the mean equal to the median, c_{reliab} is equal to 1 for 50% reliability. The other values are for steel assuming a standard deviation of the endurance limit that is 8% of the mean value [107, 115].

3.3.9 Miscellaneous Effects

There are other factors that can influence the fatigue life of a part. Stress concentrations are associated with the geometry of the part and can be a major influence on the fatigue life. Environmental factors such as temperature and corrosion are also important influences for many applications.

Stress Concentration Effects. Most fatigue failures will occur at a stress concentration. For static loading it is common to ignore stress concentration effects for ductile materials, but in fatigue a ductile material will fail in a brittle manner and stress concentration effects are very important. To account for the fact that the stress concentration effects are important for high cycle fatigue, but not as important at low cycles for ductile materials, the stress concentration factor is used to reduce the endurance limit or fatigue strength. In this way the full value of the fatigue stress-concentration factor, K_f, is used at the high-cycle end of the curve, no stress-concentration factor is used at the low-cycle end, and a reduced value of K_f is used in between. This effect is obtained mathematically by applying a miscellaneous effect factor for ductile materials with a stress concentration as

$$c_{misc} = \frac{1}{K_f} \quad \text{for ductile materials} \quad (3.84)$$

For brittle materials the stress-concentration factor is applied for both static and fatigue loads. This is easily taken into account by using the stress-concentration factor to increase the stress as

$$\sigma_{max} = K_f \sigma_{nom} \quad \text{for brittle materials} \quad (3.85)$$

TABLE 3.4. Values for c_{reliab} **for steel (assuming a standard deviation of 8%)**

Reliability (%)	50	90	99	99.9	99.99	99.9999999
c_{reliab}	1.000	0.897	0.814	0.753	0.702	0.520

Fatigue Failure

where σ_{max} is the maximum stress that will be used in the fatigue analysis and σ_{nom} is the nominal stress that is calculated as if there were no stress concentration. This is summarized in Table 3.2.

Temperature Effects. The strength of a material can be affected dramatically by the temperature at which it operates. Materials that behave as ductile at room temperature may be brittle at very low temperatures. High temperatures are typically a more common concern because a material's strength can decrease dramatically as the temperature increases. One way of taking temperature effects into account is to determine the endurance limit or fatigue strength using a test specimen at the specified temperature. However, this approach is usually not practical. A simpler method is to use the ultimate strength at the elevated temperature in equations such as (3.70) to (3.73) to approximate the endurance limit or fatigue strength.

Corrosion. Another environmental factor that can have a significant effect on fatigue life is corrosion. For example, Norton [103] recommends that a value of $S_{e'}$ = 15 kpsi (100 MPa) be used for carbon steels in fresh water. Fontana [116] reports that aluminum bronzes and austenitic stainless steels retain only about 70 to 80% of their normal fatigue resistance in seawater (i.e., c_{misc} = 0.70 to 0.80) and high-chromium steels retain only about 30 to 40% of their normal fatigue resistance in seawater. Materials that would otherwise exhibit an endurance limit must be assumed to have a finite life in a corrosive environment.

Fretting corrosion [116] occurs at contact locations between materials that are subjected to vibration or slip. Bolted joints and press fits are examples of locations where fretting corrosion occurs. This can also dramatically reduce the fatigue life of a part, and high stresses and stress concentrations should be avoided at locations where it may be a problem.

Other. Compressive residual stresses can increase the fatigue life of a part. Residual stresses are introduced into parts by material processes such as shotpeening, cold rolling, and case hardening. The compressive residual stresses on the surface reduce the overall tension load that result from uniaxial tension or bending. It is also possible to introduce compressive stresses by preloading a part. This can be particularly important for materials that are weak in tension but will experience tensile loads from bending. Prestressed concrete may be the most common example, but the same concept can be used for other situations.

Zinc plating does not have an effect on the fatigue strength, but chromium plating or nickel plating can cause a significant reduction in endurance limit such that c_{misc} = 0.5 [104]. Anodic oxidation of light alloys can also adversely affect the endurance limit.

3.3.10 Completely Reversed Loading

Most fatigue data available are based on *completely reversed loading*. In this type of loading it is assumed that the stress alternates between tensile and compressive stress and that the magnitudes of the maximum tensile and the maximum compressive stress are equal. The stress as a function of time is shown in Figure 3.10.

Rotating bending induces completely reversed loads. In this case a rotating circular shaft is loaded with a load that is normal to the axis of rotation. At one instant in time one side of the shaft will be in compression and the opposite side will be in tension. After a 180° turn of the shaft, the side in compression will be in tension and the side that was in tension will be in compression. Rotating fatigue tests are designed to take advantage of this to collect fatigue data. Because test data are based on a specific size of polished test specimen loaded in this way is one reason why the correction factors must be made on the endurance limit or fatigue life for other cases.

Completely reversed loading is possible for loading conditions other than rotating bending, including bending (cycling between equal bending loads in opposite directions) and axial (cycling between equal magnitudes of tension and compression loads).

A simplified analysis procedure for completely reversed loading is as follows:

- Determine the maximum stress, σ_{max}. (The magnitudes of the maximum tensile and maximum compressive stresses will be equal for completely reversed loading.)
- Estimate the endurance limit, S_e, (or fatigue strength, S_f) for the component using the equations in the previous sections.
- If the completely reversed stress is less than the endurance limit (or the fatigue strength), calculate the safety factor, SF, as

$$SF = \frac{S_e}{\sigma_{max}} \qquad (3.86)$$

or

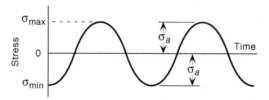

Figure 3.10. Plot of stress versus time for completely reversed loading.

Fatigue Failure

$$SF = \frac{S_f}{\sigma_{max}} \quad (3.87)$$

- If the completely reversed stress is greater than the endurance limit (or the fatigue strength), use the finite life equations to estimate the cycles to failure (Section 3.3.2).

The methods discussed here are most appropriate for determining dimensions for parts in the design phase. After a design has been specified, it may be built and fatigue tests conducted.

Example: Completely Reversed Bending of a Cantilever Beam. A cantilever beam made from hot-rolled 1095 steel has a completely reversed vertical end load of $F = 8$ lb, as shown in Figure 3.11. The beam has a length of $L = 20$ in., a width of $b = 3$ in., and a thickness of $h = 0.2$ in. The fixed end is connected such that the fatigue stress concentration is $K_f = 1.4$. (a) Estimate the factor of safety for infinite life. (b) What is the expected life if the load were $F = 16$ lb?

Solution: (a) The first step in calculating the safety factor for completely reversed bending is to calculate the maximum stress. Substituting the maximum moment at the fixed end, $M = FL$, the moment of inertia for a rectangular cross section, $I = bh^3/12$, and $c = h/2$ into the bending stress equation of equation (3.2) results in

$$\sigma_{max} = \frac{6FL}{bh^2} = \frac{6(8 \text{ lb})(20 \text{ in.})}{(3 \text{ inch})(0.2 \text{ in.})^2} = 8000 \text{ psi} \quad (3.88)$$

Next, the uncorrected endurance limit can be estimated from equation (3.70) as

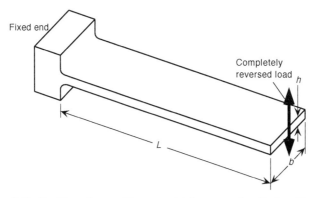

Figure 3.11. Cantilever beam with a completely reversed vertical end load.

$$S_{e'} = 0.5 S_{ut} = 0.5(120 \text{ kpsi}) = 60 \text{ kpsi} \qquad (3.89)$$

where S_{ut} is as listed in Table C.5.

The surface factor, c_{surf}, is found from equation (3.76) using the parameters from Table 3.3 for hot-rolled steel as

$$c_{\text{surf}} = 14.5(120)^{-0.719} = 0.46 \qquad (3.90)$$

The equivalent diameter of the rectangular cross section is needed to calculate the size factor. Equation (3.81) results in

$$d_e = 0.808\sqrt{bh} = 0.808\sqrt{(3 \text{ in.})(0.2 \text{ in.})} = 0.63 \text{ in.} \qquad (3.91)$$

Substituting into equation (3.77) results in a size factor, c_{size}, of

$$c_{\text{size}} = \left(\frac{d}{0.3}\right)^{-0.1133} = \left(\frac{0.63 \text{ in.}}{0.3}\right)^{-0.1133} = 0.92 \qquad (3.92)$$

The load factor, c_{load}, is found from equation (3.83) as

$$c_{\text{load}} = 1.0 \qquad (3.93)$$

Assuming a reliability of 99.99%, Table 3.4 lists a reliability factor, c_{reliab}, of

$$c_{\text{reliab}} = 0.702 \qquad (3.94)$$

The stress concentration factor for a ductile material is taken into account by reducing the endurance limit. This is done by using the miscellaneous effects factor, c_{misc}, from equation (3.84) as

$$c_{\text{misc}} = \frac{1}{K_f} = \frac{1}{1.4} = 0.71 \qquad (3.95)$$

The corrected endurance limit is found from equation (3.74) as

$$\begin{aligned} S_e &= c_{\text{surf}} c_{\text{size}} c_{\text{load}} c_{\text{reliab}} c_{\text{misc}} S_{e'} \\ &= (0.46)(0.92)(1.0)(0.702)(0.71)(60 \text{ kpsi}) \\ &= 12.7 \text{ kpsi} \end{aligned} \qquad (3.96)$$

The safety factor for completely reversed loading is calculated from equation (3.86) as

$$SF = \frac{S_e}{\sigma_{\max}} = \frac{12.7 \text{ kpsi}}{8 \text{ kpsi}} = 1.6 \qquad (3.97)$$

Fatigue Failure

(b) If the load were doubled to $F = 16$ lb, the stress would also double to

$$\sigma_{max} = 16 \text{ kpsi} \tag{3.98}$$

which is greater than the endurance limit and finite life is expected. Assuming that it will still be high cycle fatigue (i.e., greater than 1000 cycles), equation (3.58) can be rearranged as

$$\log N = \frac{1}{b_f} \log \frac{S_f}{a_f} \tag{3.99}$$

where a_f is found from equation (3.64) as

$$a_f = \frac{(c_f S_{ut})^2}{S_e} = \frac{[(0.9)(120 \text{ kpsi})]^2}{12.7 \text{ kpsi}} = 918 \text{ kpsi} \tag{3.100}$$

and c_f was determined from equation (3.57). b_f is calculated from equation (3.63) as

$$b_f = -\frac{1}{3} \log \frac{c_f S_{ut}}{S_e} = -\frac{1}{3} \log \frac{(0.9)(120 \text{ kpsi})}{12.7 \text{ kpsi}} = -0.31 \tag{3.101}$$

Substituting these values into equation (3.99) and realizing that the fatigue strength, S_f, is equal to the maximum stress for this case results in

$$\log N = \frac{1}{b_f} \log \frac{S_f}{a_f} = \frac{1}{-0.31} \log \frac{16 \text{ kpsi}}{918 \text{ kpsi}} = 5.67 \tag{3.102}$$

Solving for N yields

$$N = 471{,}000 \text{ cycles} \tag{3.103}$$

Example: Completely Reversed Bending of an Electrostatic Resonator. The electrostatic drive shown in Figure 3.12 can be driven in either direction, depending on which side the voltage difference is applied to. It is possible to switch the voltage from side to side. Resonance will result if this switching is done at the natural frequency of the device. This has been used as a method to determine the material properties of the beams by tuning the switching until resonance is achieved, measuring the frequency and back calculating Young's modulus required to have that natural frequency given the geometry. Calculate the required minimum endurance limit for infinite life. The device has dimensions of $L = 250$ μm, $b = 3.5$ μm, and $h = 3$ μm, and a shuttle displacement of $\delta = 10$ μm. The material is polycrystalline silicon (polysilicon), and assume properties of

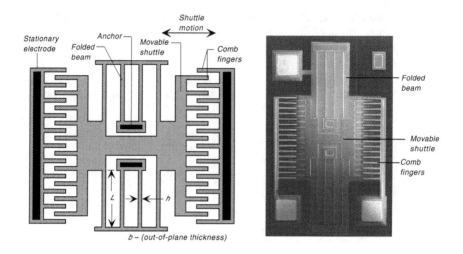

Figure 3.12. Electrostatic drive that can be driven in either direction. A scanning electron microscope photo of such a device is shown on the right.

$E = 170$ GPa and $S_{ut} = 0.93$ GPa. Assume a stress concentration factor of $K_f = 1.1$.

Solution: The first step is to calculate the maximum stress. This calculation was performed in Section 3.2.2, and the maximum nominal stress (the stress calculated without taking the stress concentrations into account) was calculated as

$$\sigma_{o(\max)} = 122 \text{ MPa} \tag{3.104}$$

Because this is a brittle material, the stress concentration will be taken into account by increasing the stress term rather than decreasing the endurance limit. Multiplying the maximum nominal stress by the stress concentration results in a maximum stress of

$$\sigma_{\max} = K_f \sigma_{o(\max)} = 1.1(122 \text{ MPa}) = 134 \text{ MPa} \tag{3.105}$$

Next, equation (3.86) is rearranged to solve for the endurance limit to obtain

$$S_e = \text{SF}\sigma_{\max} = 134 \text{ MPa} \tag{3.106}$$

because SF = 1 at the minimum required endurance limit. Note that the endurance limit is about 14% of the ultimate strength. This is low enough that it is reasonable to assume that it is possible to obtain infinite life with the given geometry and displacements.

Fatigue Failure

3.3.11 Fluctuating Stresses

Fluctuating stresses that are not completely reversed are a more general, and a more common, loading condition in compliant mechanisms. Figure 3.13 shows two loading conditions that are typical given the mechanical constraints and motions to which the compliant members are exposed. These can be expressed in terms of a mean stress, σ_m, and an alternating stress, σ_a, as illustrated in Figure 3.13. These stresses can be written in terms of the maximum stress, σ_{max}, and minimum stress, σ_{min}, as

$$\sigma_m = \frac{\sigma_{max} + \sigma_{min}}{2} \tag{3.107}$$

and

$$\sigma_a = \frac{\sigma_{max} - \sigma_{min}}{2} \tag{3.108}$$

Several approaches are available to predict fatigue failure from not completely reversed fluctuating loads, some of which are described in [103], [104], and [109]. A fairly conservative theory, *modified Goodman*, is common in design and will be used here. In this approach the mean stress, σ_m, is plotted on the *x*-axis and the alternating stress, σ_a, is plotted on the *y*-axis, as shown in Figure 3.14. The line connecting the endurance limit, S_e, plotted on the *y*-axis (or fatigue strength S_f), and the ultimate strength, S_{ut}, plotted on the *x*-axis, is called the *modified Goodman line*. If the stress condition is below this line when plotted on the diagram, it is expected to have infinite life (or an acceptable life based on the fatigue strength).

It is also important to check that the part does not fail under static conditions. To do this, the maximum stress, σ_{max}, is compared to the yield strength, where the maximum stress is

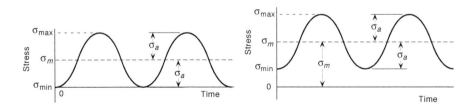

Figure 3.13. Examples of fluctuating stress loading conditions common in compliant mechanisms.

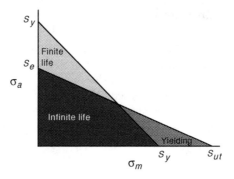

Figure 3.14. Fatigue diagram for fluctuating stresses.

$$\sigma_{max} = \sigma_m + \sigma_a \qquad (3.109)$$

The line connecting the yield strength on the x-axis and the same value on the y-axis on the fatigue diagram represents the yielding failure line. Points that lie outside of this line will yield in the first application of the load.

The fatigue diagram of Figure 3.14 shows the modified Goodman and yielding lines and the regions associated with infinite life (or acceptable life based on a fatigue strength), yielding, and finite life. The safety factors for the modified Goodman and yielding lines are

$$\frac{1}{SF} = \frac{\sigma_a}{S_e} + \frac{\sigma_m}{S_{ut}} \qquad \text{modified Goodman } (S_e) \qquad (3.110)$$

or

$$\frac{1}{SF} = \frac{\sigma_a}{S_f} + \frac{\sigma_m}{S_{ut}} \qquad \text{modified Goodman } (S_f) \qquad (3.111)$$

and

$$SF = \frac{S_y}{\sigma_m + \sigma_a} \qquad \text{yielding} \qquad (3.112)$$

When σ_m is compressive (plotted on the negative x-axis), it is assumed that failure does not occur so long as the alternating stress is less than the endurance limit (or fatigue strength) and the maximum stress is less than the yield strength. This is because the compressive stresses tend to close fatigue cracks.

For cases of multiaxial stress, σ_a and σ_m are the von Mises alternating and mean stresses calculated using equation (3.28) or (3.29), but using the alternating and mean stress components.

Fatigue Failure

Example: Infinite Life of a Cantilever Beam. Consider the cantilever beam that has length $L = 20$ in., width $b = 3$ in., thickness $h = 0.2$ in., a fatigue stress concentration of $K_f = 1.4$, ultimate strength $S_{ut} = 120$ kpsi, and a completely reversed load of $F = 8$ lb. Now suppose that in addition to the completely reversed load, a weight of 2 lb is permanently attached to the free end as shown in Figure 3.15. Calculate the safety factor for infinite life.

Solution: Using the stress relation of equation (3.88), the mean load of 2 lb results in a mean stress of

$$\sigma_m = \frac{6FL}{bh^2} = \frac{6(2\text{ lb})(20\text{ in.})}{(3\text{ in.})(0.2\text{ in.})^2} = 2\text{ kpsi} \quad (3.113)$$

and the alternating stress is calculated with the load of 8 lb, or

$$\sigma_a = \frac{6FL}{bh^2} = \frac{6(8\text{ lb})(20\text{ in.})}{(3\text{ in.})(0.2\text{ in.})^2} = 8\text{ kpsi} \quad (3.114)$$

The endurance limit was calculated previously to be $S_e = 12.7$ kpsi. The modified Goodman relation of equation (3.110) results in

$$\frac{1}{SF} = \frac{\sigma_a}{S_e} + \frac{\sigma_m}{S_{ut}} = \frac{8\text{ kpsi}}{12.7\text{ kpsi}} + \frac{2\text{ kpsi}}{120\text{ kpsi}} = 0.65 \quad (3.115)$$

or

$$SF = 1.5 \quad (3.116)$$

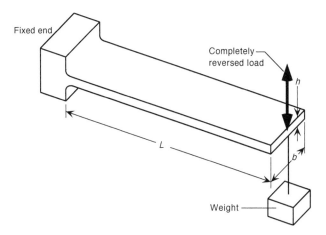

Figure 3.15. Cantilever beam with a static load and a reversed load.

Example: Infinite Life of an Orthoplanar Spring. The principal stresses and safety factor for static failure were analyzed for a compliant orthoplanar spring in Sections 3.1.1 and 3.2.1. The spring is shown again in Figure 3.16 for convenience. The spring is 0.01 in. thick, type 410 heat-treated stainless steel with an ultimate strength of 180 kpsi (1241 MPa). Assume a stress concentration of $K_f = 1.2$ at the point of highest stress. (a) Estimate the safety factor for infinite life when the spring displacement is $\delta = 0.03$ in., then (b) approximate the maximum displacement that the spring could undergo and still have infinite life.

Solution: (a) The equations for steel will be used as a rough approximation for the fatigue of the stainless steel part. The uncorrected endurance limit is approximated from equation (3.70) as

$$S_{e'} = 0.5 S_{ut} = 0.5(180 \text{ kpsi}) = 90 \text{ kpsi} \tag{3.117}$$

This endurance limit must be modified to obtain an approximation of the endurance limit for the actual part. The surface factor, c_{surf}, is approximated from equation (3.76) using the parameters from Table 3.3 (assuming cold-rolled) as

$$c_{\text{surf}} = a S_{ut}^b = 2.67(180)^{-0.265} = 0.67 \tag{3.118}$$

The equivalent diameter, d_e, for the rectangular cross section is needed to calculate the size factor. Equation (3.81) results in

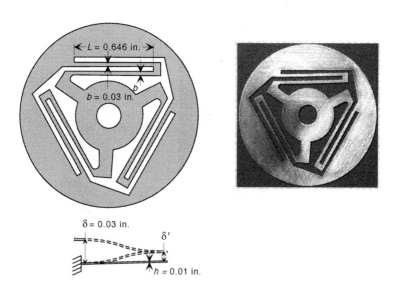

Figure 3.16. Orthoplanar spring experiencing fluctuating loads. A photo of a spring is shown on the right.

Fatigue Failure

$$d_e = 0.808\sqrt{bh} = 0.808\sqrt{(0.03 \text{ in.})(0.01 \text{ in.})} = 0.014 \text{ in.} \quad (3.119)$$

Substituting the results of equation (3.119) into equation (3.77) results in

$$c_{size} = 1 \quad (3.120)$$

because $d_e < 0.11$.

The load factor is found from equation (3.83) as $c_{load} = 1.0$ because the loading is bending. The reliability factor can be found from Table 3.4. Assuming a reliability of 99.9% results in $c_{reliab} = 0.753$.

The stress concentration can be taken into account through use of equation (3.84) as

$$c_{misc} = \frac{1}{K_f} = \frac{1}{1.2} = 0.83 \quad (3.121)$$

Now that the modification factors have been determined, the endurance limit can be approximated using equation (3.74) as

$$S_e = c_{surf} c_{size} c_{load} c_{reliab} c_{misc} S_{e'} \\ = (0.67)(1)(1)(0.753)(0.83)(90 \text{ kpsi}) = 37.7 \text{ kpsi} \quad (3.122)$$

The orthoplanar spring fluctuates between displacements of $\delta = 0$ to $\delta = 0.03$ in. Neglecting the small torsional effects, the maximum stress occurs at the maximum displacement and was calculated in Section 3.2.1 as $\sigma_{max} = 29.7$ kpsi. The minimum stress occurs when it is undeflected such that $\sigma_{min} = 0$. The mean and alternating stresses are found from equations (3.107) and (3.108), respectively, as

$$\sigma_m = \frac{\sigma_{max} + \sigma_{min}}{2} = \frac{29.7 \text{ kpsi} + 0}{2} = 14.85 \text{ kpsi} \quad (3.123)$$

and

$$\sigma_a = \frac{\sigma_{max} - \sigma_{min}}{2} = \frac{29.7 \text{ kpsi} - 0}{2} = 14.85 \text{ kpsi} \quad (3.124)$$

The safety factor for infinite life can be found using the modified Goodman equation in equation (3.110) as

$$\frac{1}{SF} = \frac{\sigma_a}{S_e} + \frac{\sigma_m}{S_{ut}} = \frac{14.85 \text{ kpsi}}{37.7 \text{ kpsi}} + \frac{14.85 \text{ kpsi}}{180 \text{ kpsi}} = 0.48 \quad (3.125)$$

or

$$SF = 2.1 \quad (3.126)$$

(b) Yielding should also be checked when using modified Goodman. The results of this check are found in equation (3.42).

Next, find the maximum deflection for which infinite life would still be expected. This is found by setting SF = 1, realizing from equations (3.123) and (3.124) that $\sigma_m = \sigma_a = \sigma_{max}/2$, and substituting into equation (3.110) to get

$$1 = \frac{\sigma_{max}}{2}\left(\frac{1}{S_e} + \frac{1}{S_{ut}}\right) \quad (3.127)$$

Solving for σ_{max} results in

$$\sigma_{max} = 2\left(\frac{S_e S_{ut}}{S_e + S_{ut}}\right) = 2\frac{(37.7 \text{ kpsi})(180 \text{ kpsi})}{(37.7 \text{ kpsi} + 180 \text{ kpsi})} = 62.3 \text{ kpsi} \quad (3.128)$$

Because the torsional shear stresses are very small compared to the other stresses, it is safe to assume that the bending stress is the maximum stress, σ_{max}, of

$$\sigma_{max} = \frac{3\delta Eh}{2L^2} \quad (3.129)$$

See equation (3.18) and Section 3.1.1 for details of the equation above. Substituting equation (3.129) into equation (3.128) and solving for the deflection results in a maximum spring displacement for infinite life, δ_{max}, of

$$\delta_{max} = \frac{4L^2}{3Eh}\frac{S_e S_{ut}}{S_e + S_{ut}} \quad (3.130)$$

or for this example

$$\delta_{max} = \frac{4(0.646 \text{ in.})^2}{3(27.5 \text{ Mpsi})(0.01 \text{ in.})}\left(\frac{(37.7 \text{ kpsi})(180 \text{ kpsi})}{37.7 \text{ kpsi} + 180 \text{ kpsi}}\right) \quad (3.131)$$

$$= 0.063 \text{ in.}$$

This is less than half the deflection calculated when only yielding failure was considered.

3.3.12 Fatigue of Polymers

The development of the fatigue analysis methods discussed above has centered around metals. However, polymers are becoming increasingly popular in engineering applications. This is particularly true for compliant mechanisms because many compliant mechanisms can be molded of one piece of plastic, which can result in a significant cost reduction compared to fabricating and assembling many parts. Advances in materials have made it possible to design reliable components using engineering polymers.

The ability to predict the behavior of polymers has lagged behind the advances in the materials. One reason for this is that polymers are significantly more difficult to analyze. This difficulty stems from nonlinearities inherent in their behavior, time dependence of properties, and many other factors that are usually not an issue with metals such as steel and aluminum. A few of the factors that influence the fatigue of polymers are:

- Viscoelasticity
- Creep
- Stress relaxation
- Temperature dependence
- Dependence on frequency of loading
- Damping
- Nonlinear elastic material properties
- Orientation of polymers
- Different properties for different loading conditions

Creep, stress relaxation, and viscoelasticity are related and cause the mechanical behavior of the material to vary with time. The material will flow when a polymer is under load because of viscoelastic material properties. This is a nonlinear, time-dependent phenomenon. The viscoelastic properties will vary depending on the material, how much glass reinforcement is included, and the temperature. Creep occurs when a material flows due to a load, causing the deflection to increase with time. Stress relaxation occurs when a displacement is applied and the reaction force decreases over time. Vendors should be asked to supply material property data for the materials used, and testing should be performed for the particular application in mind.

The mechanical properties of polymers are very sensitive to temperature. As a general rule, strength and Young's modulus decrease as temperature increases. This is important to consider in high-temperature environments or in cases where the temperature will fluctuate. An issue related to temperature is the frequency of loading. If a part is loaded in fatigue such that the load is changing at a high frequency, the damping caused by the viscoelasticity will cause the material to heat up and the material properties will change. This dependence on the frequency of the load is different than would be experienced with metals and many other materials.

Another challenge for many polymers is that their material properties are nonlinear, even in the elastic range. This means that even when the temperature is constant, the loading frequency is low, and the period of interest is small enough that time-dependent properties do not have an effect, the stress-strain curve for the material is still nonlinear. One consequence of this is that there is not a constant value that can be used for Young's modulus in the elastic range, and otherwise simple equations (such as small-deflection beam equations) become nonlinear and much more difficult to solve. It is not uncommon to simplify these nonlinearities by choosing a constant value for Young's modulus that is the slope of the best linear fit of the nonlinear stress-strain curve.

The orientation of the polymers in the manufacturing process of a part can have a significant influence on the part's behavior. An extreme example of this is a polypropylene living hinge. A living hinge is a small, flexible segment that when designed with specific geometry, and with the polymers aligned appropriately, can undergo millions of cycles. Living hinges are discussed in more detail in Section 5.8.1.

Another challenge with the analysis and design of polymers is that the mechanical properties such as strength and modulus depend on the type of loading applied. For example, many polymers will have a different value for tensile modulus than for flexural modulus, as illustrated in Table C.2. The properties will also be different at various rates of loading. The stiffness and strength tend to increase as the loading rate increases. The material also tends to be more brittle as the loading rate increases. The exception to this is if in addition to the load being applied quickly it also is being cycled quickly. In such a case the material may heat up and temperature effects will have an influence in decreasing the strength and modulus, as discussed above.

More detail on fatigue of polymers can be found in [117] through [119].

Design of Polymer Components Experiencing Cyclic Loading. The behavior of polymers is much more difficult to analyze than that of metals, but in many applications the advantages of using polymers far exceed this disadvantage. There is not currently a good method for fatigue analysis that is practical for initial design. Despite this, the fact remains that polymer components must be designed to have acceptable life. One approach is to use a simplified analysis method to obtain dimensions for the design and test the device under conditions expected in its operation.

One simplified method of analysis [118] is to get an *S-N* diagram for the material using test conditions as similar to the operating conditions as possible. The data from this diagram may come from the literature provided by suppliers, from technical literature in handbooks and journals, or from in-house testing. If no test data are available and a rough design is desired that will be followed by testing, the *S-N* diagram can be obtained using the methods discussed above for metals. In this case the endurance limit is approximated to be in the range [119]

$$S_{e'} = 0.2 S_{ut} \text{ to } S_{e'} = 0.4 S_{ut} \text{ (polymers)} \tag{3.132}$$

The endurance limit estimate is modified for variations in temperature, surface finish, and other factors, as explained above for metals. The analysis procedure described in previous sections is used to estimate the fatigue life or the factor of

Fatigue Failure

safety for fully reversed loading. Special care must be taken for fluctuating loads where the mean stress is not zero because it will tend to cause stress relaxation or creep [120].

Example: Bistable Switch. The fully compliant bistable switch shown in Figure 3.17 can be used as an electrical switch with contacts placed at the positions shown. A force applied to the handle will cause it to move to a position where the contacts close a circuit. A force in the other direction causes it to snap back to the original position where the circuit is open. Bistable mechanisms are discussed in more detail in Chapter 11. The material is polypropylene with a Young's modulus of $E = 170,000$ lb/in². The small-length flexural pivot has length $L = 0.13$ in., out-of-plane thickness $b = 0.25$ in., and in-plane thickness $h = 0.04$ in. The small-length flexural pivot goes through a maximum angular deflection of $\theta_o = 0.15$ rad during its motion between stable equilibrium positions. For an ultimate strength of 8 kpsi, estimate the minimum endurance limit required for infinite life of the small-length flexural pivot.

Solution: The mechanism has three living hinges and a small-length flexural pivot. The living hinges offer very little resistance to motion compared to the small-length flexural pivot. Polypropylene living hinges have special fatigue characteristics when constructed using the geometry discussed in Section 5.8.1 and have been tested to undergo millions of cycles without failure.

Assuming that the bending moment is the predominate loading in the small-length flexural pivot, the angular deflection (θ_o) can be related to the moment (M_o) as (see Section D.4)

$$\theta_o = \frac{M_o L}{EI} \tag{3.133}$$

Rearranging for M_o and substituting into the bending stress equation, $\sigma = M_o c / I$, results in a maximum stress, σ_{max}, of

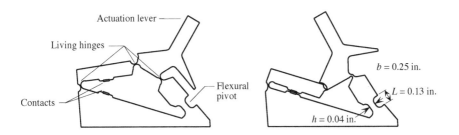

Figure 3.17. Compliant bistable switch shown in its two stable equilibrium positions.

$$\sigma_{max} = \frac{Eh}{2L}\theta_o = \frac{(170 \text{ kpsi})(0.04 \text{ in.})}{2(0.13 \text{ in.})}(0.15 \text{ rad}) = 4 \text{ kpsi} \qquad (3.134)$$

Because the stress fluctuates between zero and the maximum, the modified Goodman equations will be used. The mean and alternating stresses can be found from equations (3.107) and (3.108) as

$$\sigma_m = \frac{4 \text{ kpsi} + 0}{2} = 2 \text{ kpsi} \qquad (3.135)$$

$$\sigma_a = \frac{4 \text{ kpsi} - 0}{2} = 2 \text{ kpsi} \qquad (3.136)$$

The modified Goodman relation of equation (3.110) can be rearranged to solve for the endurance limit with SF = 1 as

$$S_e = \frac{\sigma_a S_{ut}}{S_{ut} - \sigma_m} \qquad (3.137)$$

or

$$S_e = \frac{(2 \text{ kpsi})(8 \text{ kpsi})}{8 \text{ kpsi} - 2 \text{ kpsi}} = 2.7 \text{ kpsi} \qquad (3.138)$$

This is about 33% of the ultimate strength. Although this may be high for an uncorrected endurance limit for many materials, testing has shown that this is not unreasonable for polypropylene.

3.3.13 Testing

Testing is an important part of designing components for acceptable fatigue life. The analysis in the design phase is used to obtain dimensions that are most likely to provide desired results and to minimize costly and time-consuming iterations in prototyping and testing. Types of testing range from standard fatigue specimen tests to testing the actual device under operating conditions.

A sketch of an R. R. Moore rotating-beam fatigue testing device is shown in Figure 3.18. The test specimen has dimensions as shown and is carefully machined and polished. The part is loaded in pure bending by means of the weights, and the specimen is usually rotated at about 1750 rpm. Axial fatigue tests can be performed in tensile test machines that have fatigue loading control algorithms.

The testing of the actual device usually requires more work than for a standard test specimen. Two major challenges accompany the testing of compliant mechanisms. First, a new test must be designed for each new type of device. Second, because of the large number of cycles required for fatigue testing, the test device

Fatigue Failure

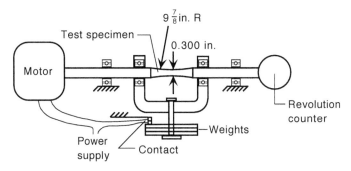

Figure 3.18. R. R. Moore rotating-beam fatigue-testing machine with a test specimen.

may also fail due to fatigue. In many companies these challenges are easily overcome because a machine can be designed for the particular products produced by the company, and once a test machine is designed, the staff can be trained on its use and repair. Research labs and new product development groups may have additional challenges because of the wide diversity of devices tested.

It is possible to use general test devices to test variations of a base product or to test different types of devices. The schematics of two such devices are shown in Figures 3.19 and 3.20. These devices have been used successfully at Brigham Young University to test a number of different compliant mechanisms. Figure 3.19 uses a slider–crank mechanism that has an adjustable crank length to account for different stroke lengths needed for different devices. A computer-controlled VCR is used to record the device for a few seconds after a specified number of cycles. The resolution of the cycles at failure is the same as the number of cycles between recordings. One advantage of this device is that several identical mechanisms can be tested at the same time by placing them side by side with the slider. The test

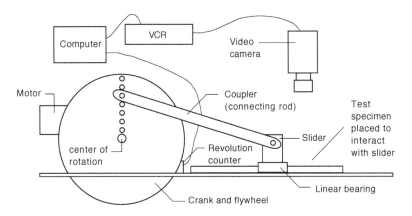

Figure 3.19. General fatigue test device based on a slider–crank mechanism.

Failure Prevention

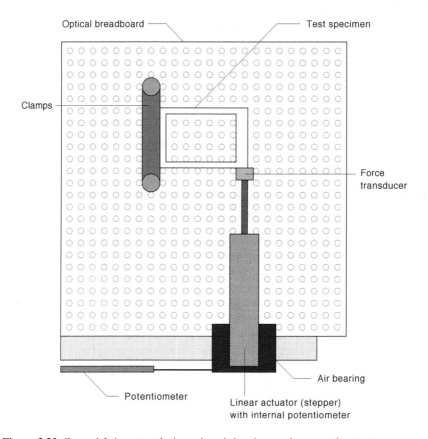

Figure 3.20. General fatigue test device using air bearings and pneumatic actuators.

apparatus in Figure 3.20 allows for force–displacement tests of compliant mechanisms. The air bearing allows the actuator to translate in one direction with little resistance. A force transducer is placed at the end of the actuator. It is possible to test devices for large numbers of force applications.

PROBLEMS

3.1 Calculate the following for the stress element shown in Figure P3.1:
   (a) The principal stresses.
   (b) Safety factor for ductile failure if the material is type 301 annealed stainless steel. Provide answers for both maximum shear stress and distortion energy theories.

Fatigue Failure

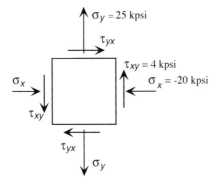

Figure P3.1. Figure for Problem 3.1.

(c) Safety factor for brittle failure if the material is class 40 gray cast iron (ultimate tensile and compressive strengths of 290 MPa and 965 MPa, respectively).

3.2 Calculate the principal stresses for the cases described below:
(a) An element at the top of a beam in bending (positive M).
(b) An element at the top of a beam in pure torsion.
(c) A cube (three-dimensional) element in hydrostatic compression (equal pressure on all sides).

3.3 A round cantilever beam is subjected to an axial force (F), a vertical load (P), and a torque (T), all acting at the free end. The diameter of the bar is (d) and the length is (L).
(a) Determine the location of the critical element.
(b) Draw the loads on the critical stress element.
(c) Determine the principal stresses in terms of F, P, T, d, and L.

3.4 The two cantilevered beams shown in Figure P3.4 are made of steel with b = 5 cm, h = 15 cm, and l = 1 m. The beam on the left has a force applied at the end. The beam on the right has the same geometry but has a displacement load at the end.

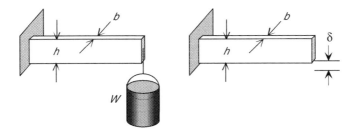

Figure P3.4. Figure for Problem 3.4.

(a) Calculate the maximum bending stress in the beam with the force load. Assume that the load is $W = 800$ N.
(b) Calculate the maximum bending stress for the beam on the right if the end has a displacement of $\delta = 1.2$ cm.
(c) Calculate the stresses in (a) and (b) if the beams are placed on their sides (i.e., if $b = 15$ cm and $h = 5$ cm).
(d) Is the stress higher in (a) or (c) for the force loaded beam? Is the stress higher in (b) or (c) for the displacement loaded beam? Explain.

3.5 If $d = 1.95$ cm, $h = 1$ cm, and $b = 3$ cm in Figure P3.5, the cylindrical beam and the rectangular beam have equal cross-sectional areas. Determine which will have the greater torsional stress if they both experience the same torque.

3.6 Figure 2.10 shows a compliant mechanism ice-cream scoop. If the ice cream is very hard and packed very tightly into the scoop, the rotating blade won't be able to rotate, and the rotating shaft will experience torsional stress if an input force is applied to the flexible handle. Calculate the magnitude of the torsional stress on the outer surface of the rotating shaft if the length of the rotating shaft is 3 in., its diameter 0.125 in., the pinion has a pitch diameter of 0.25 in., and the force is 2 lb.

3.7 Consider the structure shown in Figure P3.7.
(a) Find the location of the maximum stress and draw the critical stress element.
(b) Calculate the principal stresses symbolically.
(c) Develop the equation for the safety factor for static failure if the material is ductile.
(d) Develop the equation for the safety factor for static failure if the material is brittle.
(e) Suppose that the material is 2024 T3 aluminum, $d_1 = 4$ cm, $l_1 = 16$ cm, $l_2 = 10$ cm, $F_1 = 20$ N. Calculate the maximum value of F_2 before yielding.
(f) Repeat (e) using the same dimensions, but for class 20 gray cast iron (ultimate tensile and compressive strengths of 152 MPa and 572 MPa, respectively).

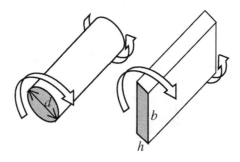

Figure P3.5. Figure for Problem 3.5.

Fatigue Failure

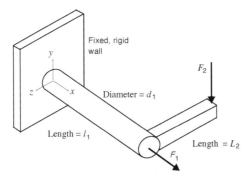

Figure P3.7. Figure for Problem 3.7.

3.8 Consider the parallel mechanism shown in Figure P3.8 with dimensions $l_1 = 10$ in., $l_2 = 12$ in., $h_1 = 0.03$ in., $b_1 = 1.2$ in., $h_2 = 1.0$ in., and $b_2 = 1.2$ in., where b_1 and b_2 are the out-of-plane widths of the flexible and rigid segments, respectively. The material is 1010 HR steel. The shuttle must undergo a horizontal deflection of $\delta = 0.05$ in. (assume small deflections).
   (a) Identify the location of the maximum stress and draw the critical stress element.
   (b) Calculate the principal stresses.
   (c) Calculate the factor of safety for static failure.
   (d) How would your answer in (c) change if h_1 were doubled? Explain.
   (e) How would your answer in (c) change if b_1 were doubled? Explain.

3.9 The 40-kg bucket in Figure P3.9 is suspended from a rope tied around the drum of a cantilevered shaft. The drum has a diameter of 1 m, and the shaft has a diameter of 40 cm and a length of 60 cm.
   (a) Determine the maximum bending and shear stresses and where they act on the shaft.
   (b) If the shaft is made of 1100 O aluminum, determine whether the shaft will fail. If it will not fail, what is its safety factor?

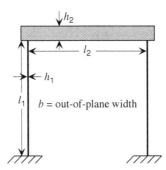

Figure P3.8. Figure for Problems 3.8 and 3.13.

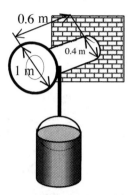

Figure P3.9. Figure for Problem 3.9.

(c) If the shaft is made of gray cast iron, determine whether the shaft will fail. If it will not fail, what is its safety factor? (S_{ut} = 152 Mpa, S_{uc} = 572 MPa).

3.10 A cantilever beam made from 2024 T4 aluminum has a completely reversed vertical end load of $F = 30$ N as shown in Figure P3.10. The beam has a length of $L = 50$ cm, a width of $b = 7.5$ cm, and a thickness of $h = 0.5$ cm. The fixed end is connected such that the fatigue stress concentration is $K_f = 1.2$.
(a) Estimate the factor of safety for a fatigue of life of 5×10^8.
(b) What is the expected life if the load were $F = 60$ N?

3.11 The machine component illustrated in Figure P3.11 has a bending load as it rotates (this causes completely reversed stress). It is made of AISI 1050 cold-drawn steel, and has dimensions $D = 1.05$, $d = 1$ in., $r = 0.05$ in. Calculate the maximum allowable moment that would still allow infinite life.

3.12 A 1020 hot-rolled steel part has a rectangular cross section (2 in. × 1 in.), a stress concentration factor of $K_f = 2.17$, and a completely reversed axially load.

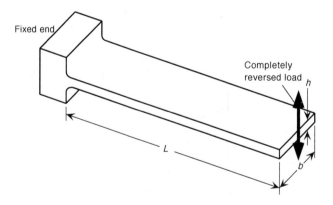

Figure P3.10. Figure for Problem 3.10.

Fatigue Failure

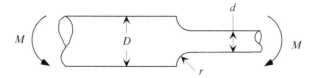

Figure P3.11. Figure for Problem 3.11.

(a) Estimate the load that would result in a life of 200,000 cycles.
(b) How would you have treated the stress concentration factor if the part were a brittle material?

3.13 Consider the parallel mechanism shown in Figure P3.8 with dimensions $l_1 = 10$ in., $l_2 = 12$ in., $h_1 = 0.03$ in., $b_1 = 1.2$ in., $h_2 = 1.0$ in., and $b_2 = 1.2$ in., where b_1 and b_2 are the out-of-plane widths of the flexible and rigid segments, respectively. The material is 1010 HR steel. Use small-deflection beam equations to estimate the fatigue life for the conditions listed below. Assume a reliability of 99.9% and a stress concentration at the ends of $K_f = 1.75$.
(a) Calculate the maximum possible completely reversed horizontal shuttle deflection, δ for infinite life.
(b) Estimate the maximum possible completely reversed horizontal shuttle deflection, δ for a finite life of 100,000 cycles.
(c) Calculate the maximum possible horizontal shuttle deflection, δ, for infinite life, where the deflection is only in one direction.

3.14 The stress in a compliant connector was calculated in Section 2.6. Determine the life of this connector for a reliability of 99.9% if its surface factor is equal to 1 and if it is made of polystyrene.

3.15 Consider the fingernail clippers in Figure P2.6. If the minimum and maximum stresses endured by the flexible segments are 0 and 30 MPa, respectively, determine the life of the clippers for a reliability of 99.9%. The flexible segments have a cross-sectional area of 0.25 in. by 0.05 in. At the pinned end, the flexible segments have a stress concentration with a notch radius of 0.1 in. Assume that the material is 1020 CD steel.

3.16 Because a paper clamp (shown in Figure P3.16) is opened and closed many times during its life, it is said to be loaded in fatigue. While the handles experience a minimum force of zero (in the position shown in Figure P3.16a) and a maximum force of 10 N (when the clamp is pulled open), the clamp experiences a minimum nonzero stress even in the closed position shown in Figure P3.16a, which means that the clamp stresses must be divided into a mean and an alternating stress, σ_m and σ_a, as shown in Figure P3.16b. The minimum stress is equal to 20 MPa, and the maximum stress is equal to 50 MPa. There exists a stress concentration where the base of the clamp meets the handles. The notch radius there is 2 mm. Assume that the clamp is made of 1040 hot-

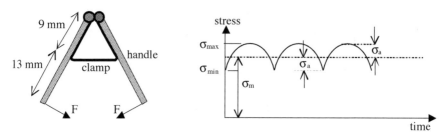

Figure P3.16. Figure for Problem 3.16.

rolled steel, and the handles are made of 2024 T3 aluminum. The equilateral sides of the clamp are 9 mm high (as shown), 15 mm deep, and 0.5 mm thick. The handles have a circular cross-section diameter $D = 1$ mm. For a reliability of 99.9%:
(a) What is the life of the handles?
(b) What is the life of the clamp?
(c) Which will fail sooner (if at all), the clamp or the handles?

3.17 A configuration of the overrunning clutch shown in Figure 1.3a is constructed of 4130 Q&T 400 steel to make a bicycle freewheel. The stresses at the critical stress element fluctuate between zero and a maximum stress (including stress concentration) of 49 kpsi (340 MPa). It is machined using a wire EDM process. Calculate the safety factor for infinite life.

3.18 A compliant mechanism similar to the one shown in Figure P3.18 has its flexible member constructed of polypropylene. The motion is such that the maximum stress in the flexible member is 1500 psi.
(a) Calculate the minimum endurance limit that would allow infinite life for completely reversed stress.
(b) Calculate the minimum endurance limit that would allow infinite life for fluctuating stress between zero and the maximum stress.
(c) Do the answers above seem reasonable for polypropylene? Explain.

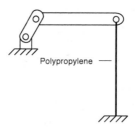

Figure P3.18. Figure for Problem 3.18.

CHAPTER 4

RIGID-LINK MECHANISMS

4.1 INTRODUCTION

Traditional mechanism analysis assumes that the deflections of a mechanism's parts are negligible compared to the overall motion of the mechanism. If the parts are rigid, the mechanism motion is not a function of the shape of the links or the forces applied. This allows motion analysis (kinematics), and the analysis of motion and the forces that produce it (kinetics), to be analyzed independently, thus simplifying the analysis.

A few major concepts from rigid-body mechanisms are reviewed in this chapter to serve as a reference and to build a foundation for later discussions. Many good texts deal with this subject, including [1], [2], and [121] through [124].

4.1.1 Mobility

The minimum number of variables required to describe the configuration of a mechanism completely is called its *degrees of freedom*. An unconstrained planar rigid link has three degrees of freedom because three displacement variables are required to describe its position and orientation. Therefore, the total possible degrees of freedom of n unconstrained links is $3n$. By definition, a mechanism has one fixed link with zero degrees of freedom. The maximum possible degrees of freedom of an n-link mechanism is then $3(n-1)$.

A lower kinematic pair, such as a revolute or slider joint, removes two degrees of freedom from the system. This is easily seen by considering two links, one of which is fixed (grounded) and the other unconstrained, as shown in Figure 4.1a.

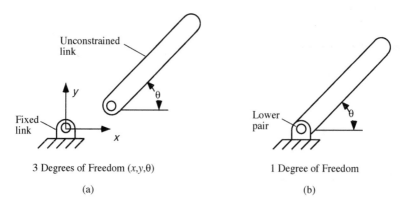

Figure 4.1. (a) Unconstrained and fixed link, and (b) link connected to a fixed link by a lower pair.

The location and orientation of the unconstrained link can be described by three displacement variables (x, y, and θ), and the system has three degrees of freedom. If the unconstrained link is joined to the ground link by means of a pin joint (a lower pair), as shown in Figure 4.1b, the system's mobility can be described with only one variable, the angle of the moving link. The lower pair decreases the system degrees of freedom by two. The same reasoning may be used to illustrate that a higher pair (e.g., a cam and follower pair) removes a single degree of freedom. This discussion may be summarized analytically in the form of Gruebler's equation:

$$F = 3(n-1) - 2J_1 - J_2 \tag{4.1}$$

where F is the degrees of freedom, n the number of links, and J_1 and J_2 are the number of lower and higher pairs, respectively.

As an example, the Vise-Grip mechanism shown in Figure 1.1b has four links ($n = 4$), four lower pairs ($J_1 = 4$), and no higher pairs ($J_2 = 0$). Substituting these values into equation (4.1) yields $F = 1$. The slider–crank mechanism shown in Figure 1.1a has four links, since the slider is a link. The sliding joint is also a lower pair, and the mechanism has one degree of freedom.

4.1.2 Kinematic Chains and Inversions

Links connected together by joints are called a *kinematic chain*. The chain is considered a mechanism if one of the links is considered to be the *fixed link*, which means that it is chosen as the reference link. The fixed link is usually the frame or base link connected to ground. The basic kinematic chain has the same relative motion between links, regardless of which link is fixed. A *kinematic inversion* is obtained when a different link is fixed. This does not change the relative motion between links but can drastically change the absolute motion of the mechanism.

Introduction

4.1.3 Classification of Four-Bar Mechanisms

Grashof's law states that for at least one link of a four-bar mechanism to have full rotation, the following inequality must hold:

$$s + l \leq p + q \tag{4.2}$$

where s is the length of the shortest link, l the length of the longest link, and p and q the lengths of the remaining links. The shortest link of a Grashofian mechanism is allowed full rotation relative to its adjacent links. Table 4.1 lists the different types of mechanisms based on which link is the shortest link. For example, if a side link is the shortest link in a Grashofian mechanism, it is called a *crank–rocker mechanism*; the shorter side link (the crank) is able to revolve, and the other side link (the rocker) rocks between two limit positions. Three types of Grashofian mechanisms are shown in Figure 4.2. The mechanisms shown are inversions of the same basic kinematic chain.

If all the link lengths are equal, at one position all centerlines of links become collinear. This position is called the *change-point position*. On leaving the change-point position, the mechanism may go into one of two possible configurations. If Grashof's law is not satisfied, the mechanism is a triple rocker and no links can revolve completely.

It is also useful to note that for the links to be assembled, the following inequality must be satisfied:

$$s + p + q > l \tag{4.3}$$

4.1.4 Mechanical Advantage

The mechanical advantage of a mechanism is the ratio of the output torque/force to the input torque/force required. The mechanical advantage may be calculated assuming that power is conserved between the input and the output. For the simple

TABLE 4.1. Classifications of four-bar mechanisms

$s + l(<,=,>)p + q$	Shortest link	Type
<	Fixed	Double-crank
<	Side	Crank–rocker
<	Coupler	Double-rocker
=	Any	Change-point
>	Any	Triple-rocker

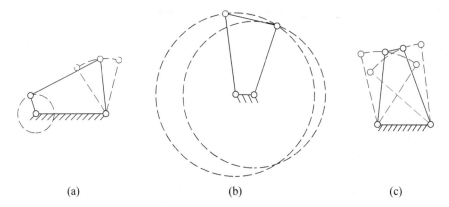

(a) (b) (c)

Figure 4.2. (a) Crank–rocker, (b) double-crank, and (c) double-rocker mechanisms.

lever illustrated in Figure 4.3, the mechanical advantage may be found by assuming the system is in static equilibrium and summing the moments about the fulcrum, that is,

$$F_{in} r_{in} = F_{out} r_{out} \tag{4.4}$$

The mechanical advantage, MA, is the ratio of the output force, F_{out}, to the input force, F_{in}, or

$$MA = \frac{F_{out}}{F_{in}} = \frac{r_{in}}{r_{out}} \tag{4.5}$$

Similar methods can be used to calculate the mechanical advantage for more complicated mechanisms. One method of performing such an analysis, the principle of virtual work, is discussed in more detail in Chapter 6.

The mechanical advantage, MA, of the four-bar mechanism shown in Figure 4.4 is

$$MA = \frac{T_4}{T_2} = \frac{r_4 \sin(\theta_4 - \theta_3)}{r_2 \sin(\theta_2 - \theta_3)} \tag{4.6}$$

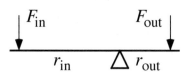

Figure 4.3. Mechanical advantage for a lever.

Position Analysis

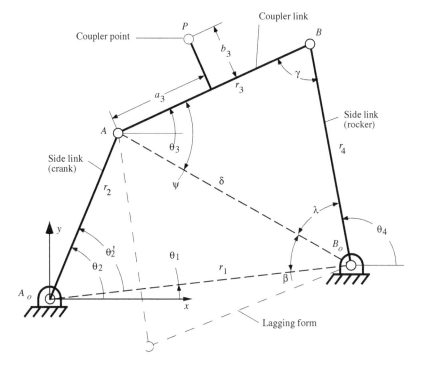

Figure 4.4. Rigid-link four-bar mechanism.

where T_4 and T_2 are the output and input torques, respectively. The numerator is related to the sine of the transmission angle, γ, and a transmission angle near 90° is usually desirable. Note also that if links 2 and 3 are collinear (i.e., $\theta_2 - \theta_3 = 0$), the mechanical advantage tends to infinity. Such a position is called a *toggle position*. This phenomenon is experienced in many devices, such as the Vise Grip shown in Figure 1.1b.

4.2 POSITION ANALYSIS

Many methods are available for the position and displacement analysis of mechanisms. Graphical methods often provide a fast and efficient means of analyzing mechanisms. Analytical methods are currently more common because of the ease in which they are programmed. A number of commercial software packages are also available that are powerful tools in mechanism analysis and design.

Some common analytical methods that are easily programmed are reviewed in this section. Closed-form solutions for four-bar crank–rocker and slider–crank mechanisms are presented as a reference for analyzing these common mechanisms.

Then the complex number method is introduced as a general method capable of analyzing and synthesizing many types of mechanisms.

4.2.1 Four-Bar Mechanism: Closed-Form Equations

Consider a four-bar crank–rocker mechanism with link lengths and angles as shown in Figure 4.4. The crank angle, θ_2, is the input. The crank angle measured from r_1, θ_2', is

$$\theta_2' = \theta_2 - \theta_1 \tag{4.7}$$

The law of cosines may be used to determine the length of δ and the internal angles β, ψ, and λ as

$$\delta = (r_1^2 + r_2^2 - 2r_1 r_2 \cos \theta_2')^{1/2} \tag{4.8}$$

$$\beta = \operatorname{acos} \frac{r_1^2 + \delta^2 - r_2^2}{2 r_1 \delta} \tag{4.9}$$

$$\psi = \operatorname{acos} \frac{r_3^2 + \delta^2 - r_4^2}{2 r_3 \delta} \tag{4.10}$$

$$\lambda = \operatorname{acos} \frac{r_4^2 + \delta^2 - r_3^2}{2 r_4 \delta} \tag{4.11}$$

Two possible values exist for each angle for a given θ_2. One set of angles represents the leading form, and the other the lagging form of the mechanism (Figure 4.4). The link angles are calculated as follows: For $0 \le \theta_2' \le \pi$,

$$\theta_3 = \psi - (\beta - \theta_1) \tag{4.12}$$

$$\theta_4 = \pi - \lambda - (\beta - \theta_1) \tag{4.13}$$

and for $\pi \le \theta_2' \le 2\pi$,

$$\theta_3 = \psi + (\beta + \theta_1) \tag{4.14}$$

$$\theta_4 = \pi - \lambda + (\beta + \theta_1) \tag{4.15}$$

The transmission angle, γ, may be used to help determine the mechanical advantage of a mechanism, and is found as

Position Analysis

$$\gamma = \pm a\cos\frac{r_3^2 + r_4^2 - \delta^2}{2r_3 r_4} \qquad (4.16)$$

The position of the coupler point, P, is

$$x_P = r_2 \cos\theta_2 + a_3 \cos\theta_3 - b_3 \sin\theta_3 \qquad (4.17)$$

$$y_P = r_2 \sin\theta_2 + a_3 \sin\theta_3 + b_3 \cos\theta_3 \qquad (4.18)$$

4.2.2 Slider–Crank Mechanism: Closed-Form Equations

The replacement of one of a four-bar's turning joints by a sliding joint results in a slider–crank mechanism, as shown in Figure 4.5. This is one of the easiest mechanisms to analyze, since the links form a triangle. If the crank angle is the input, the law of sines is used to determine the following:

$$\theta_3 = a\sin\frac{r_4 - r_2 \sin\theta_2}{r_3} \qquad (4.19)$$

$$r_1 = r_2 \cos\theta_2 + r_3 \cos\theta_3 \qquad (4.20)$$

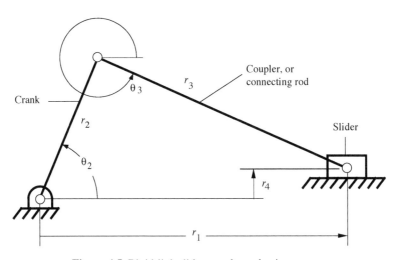

Figure 4.5. Rigid-link slider–crank mechanism.

If the slider is the input and $r_4 = 0$, the law of cosines is used to determine the crank angle:

$$\theta_2 = \mathrm{acos}\frac{r_1^2 + r_2^2 - r_3^2}{2r_1 r_2} \qquad (4.21)$$

4.2.3 Complex Number Method

A number of kinematic analysis and synthesis methods are available. The complex number method is very general, and many other methods based on complex number modeling may be derived from it. The general nature of the method also allows its use for linear and nonlinear synthesis problems and for single- and multiple-loop mechanisms. This method forms the basis of the generalized loop-closure method for the analysis and design of compliant mechanisms discussed in Chapter 8.

Complex Numbers. Complex numbers are often used to represent vectors in the modeling of mechanisms. A complex number $\vec{Z}$ may be written in Cartesian, or rectangular, form as

$$\vec{Z} = a + ib \qquad (4.22)$$

where i is the imaginary number $\sqrt{-1}$, a is called the real part, and b is the imaginary part. This form may be used to represent a vector $\vec{Z}$ in a plane, in which the x-axis is the real axis, and the y-axis the imaginary axis, as shown in Figure 4.6a.

The complex number of equation (4.22) may also be written as

$$\begin{aligned}\vec{Z} &= r(\cos\theta + i\sin\theta) \\ &= re^{i\theta}\end{aligned} \qquad (4.23)$$

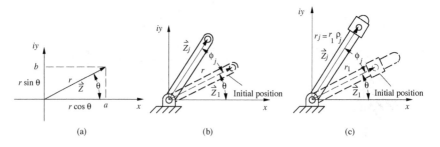

Figure 4.6. (a) Complex plane, (b) link in two positions, and (c) link and slider in two positions.

Position Analysis

The link in Figure 4.6b is shown in its initial and final positions, following a rotation of ϕ_j. The complex numbers that describe these positions may be written as

$$\vec{Z}_1 = re^{i\theta} \tag{4.24}$$

and

$$\vec{Z}_j = re^{i(\theta + \phi_j)} = re^{i\theta}e^{i\phi_j} \tag{4.25}$$

Note that the second position vector may be written in terms of the first as

$$\vec{Z}_j = \vec{Z}_1 e^{i\phi_j} \tag{4.26}$$

Consider the link and slider shown in Figure 4.6c. The two positions shown may be described as

$$\vec{Z}_1 = r_1 e^{i\theta} \tag{4.27}$$

and

$$\vec{Z}_j = r_j e^{i(\theta + \phi_j)} \tag{4.28}$$

Again, the second position vector may be written in terms of the first as

$$\vec{Z}_j = \vec{Z}_1 \rho_j e^{i\phi_j} \tag{4.29}$$

where

$$\rho_j = \frac{r_j}{r_1} \tag{4.30}$$

Loop-Closure Equations. Complex numbers may be used to model a mechanism by creating an equation for each closed loop in the mechanism. Consider the four-bar mechanism with coupler point P shown in Figure 4.7.

The vector loop in the figure results in the following equation:

$$\vec{Z}_2 + \vec{Z}_3 - \vec{Z}_5 - \vec{Z}_4 - \vec{Z}_1 = 0 \tag{4.31}$$

where $\vec{Z}_j = r_j e^{i\theta_j}$. Each loop equation may be expressed in terms of its real and imaginary parts, resulting in two equations. The four-bar loop equation represented by equation (4.31) may be rewritten in scalar form as

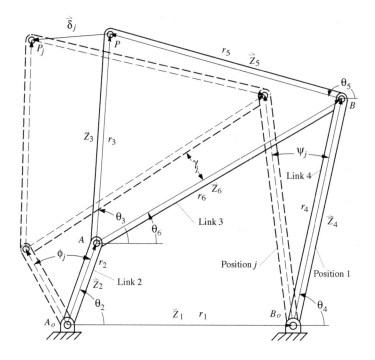

Figure 4.7. Vector loop for a four-bar mechanism with coupler point P.

$$r_2 \cos\theta_2 + r_3 \cos\theta_3 - r_5 \cos\theta_5 - r_4 \cos\theta_4 - r_1 \cos\theta_1 = 0 \quad (4.32)$$

$$r_2 \sin\theta_2 + r_3 \sin\theta_3 - r_5 \sin\theta_5 - r_4 \sin\theta_4 - r_1 \sin\theta_1 = 0 \quad (4.33)$$

These equations are also known as *Freudenstein's equations*.

Figure 4.7 shows a four-bar mechanism in two positions. The *j*th position may be expressed as

$$\vec{Z}_2 e^{i\phi_j} + \vec{Z}_3 e^{i\gamma_j} - \vec{Z}_5 e^{i\gamma_j} - \vec{Z}_4 e^{i\psi_j} - \vec{Z}_1 = 0 \quad (4.34)$$

where ϕ_j, γ_j, and ψ_j are the changes in the angles of links 2, 3, and 4, respectively, from position 1 to position *j*. These equations may be used for function, path, and motion generation, as described in Section 4.5.

Example: Slider–Crank Mechanism. Consider again the slider-crank mechanism shown in Figure 4.5. The vector loop is shown in Figure 4.8 and the associated equations can be written as

Position Analysis

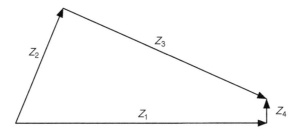

Figure 4.8. Vector loop for the slider–crank mechanism in Figure 4.5.

$$\vec{Z}_2 + \vec{Z}_3 = \vec{Z}_1 + \vec{Z}_4 \qquad (4.35)$$

or

$$r_2 e^{i\theta_2} + r_3 e^{i\theta_3} = r_1 + r_4 i \qquad (4.36)$$

Separating into real and imaginary portions components results in

$$r_2 \cos\theta_2 + r_3 \cos\theta_3 = r_1 \qquad (4.37)$$

and

$$r_2 \sin\theta_2 + r_3 \sin\theta_3 = r_4 \qquad (4.38)$$

These two equations can be solved for the unknowns as

$$\theta_3 = \mathrm{asin}\,\frac{r_4 - r_2 \sin\theta_2}{r_3} \qquad (4.39)$$

$$r_1 = r_3 \cos\theta_3 + r_2 \cos\theta_2 \qquad (4.40)$$

Example: Slider Mechanism Displacement. The unknowns for the slider mechanism shown in Figure 4.9 are θ_4 and r_3. The complex vector loop equation is

$$\vec{Z}_2 + \vec{Z}_3 = \vec{Z}_1 + \vec{Z}_4 \qquad (4.41)$$

or

$$r_2 e^{i\theta_2} + r_3 e^{i(\theta_2 + \pi/2)} = r_1 e^{i\theta_1} + r_4 e^{i\theta_4} \qquad (4.42)$$

The real and imaginary components are

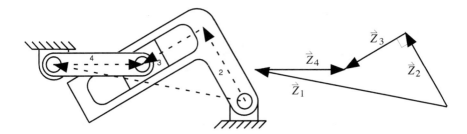

Figure 4.9. Slider mechanism example and vector loop.

$$r_2 \cos \theta_2 + r_3 \cos (\theta_2 + \pi/2) = r_1 \cos \theta_1 + r_4 \cos \theta_4 \qquad (4.43)$$

$$r_2 \sin \theta_2 + r_3 \sin (\theta_2 + \pi/2) = r_1 \sin \theta_1 + r_4 \sin \theta_4 \qquad (4.44)$$

The values of θ_4 and r_3 are found by solving the two nonlinear equations simultaneously using an approach such as the Newton–Raphson method.

Example: Multiloop Mechanism. The loop-closure method applies to the displacement of multiple-loop mechanisms also. Consider the six-bar mechanism shown in Figure 4.10. Any two of the three possible vector loops may be used to generate the equations. Using loops 1 and 2 results in two loop equations, that is,

$$\vec{Z}_2 + \vec{Z}_3 + \vec{Z}_4 - \vec{Z}_1 = 0 \qquad (4.45)$$

and

$$\vec{Z}_4 + \vec{Z}_5 + \vec{Z}_6 = 0 \qquad (4.46)$$

Expressing equations (4.45) and (4.46) in terms of their real and imaginary parts results in four scalar equations for the mechanism:

$$r_2 \cos \theta_2 + r_3 \cos \theta_3 + r_4 \cos \theta_4 - r_1 \cos \theta_1 = 0 \qquad (4.47)$$

$$r_2 \sin \theta_2 + r_3 \sin \theta_3 + r_4 \sin \theta_4 - r_1 \sin \theta_1 = 0 \qquad (4.48)$$

$$r_4 \cos \theta_4 + r_5 \cos \theta_5 + r_6 \cos \theta_6 = 0 \qquad (4.49)$$

$$r_4 \sin \theta_4 + r_5 \sin \theta_5 + r_6 \sin \theta_6 = 0 \qquad (4.50)$$

Velocity and Acceleration

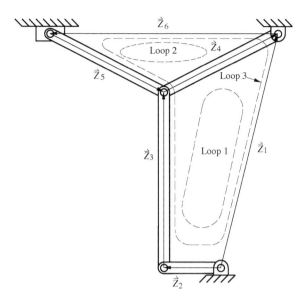

Figure 4.10. Six-bar mechanism and its vector loops.

Suppose that the links have lengths $r_2 = 1$ cm, $r_3 = 3$ cm, and $r_4 = r_5 = 9$ cm. The distance between the two pin joints that connect links 2 and 4 to ground is defined by $\vec{Z}_1 = r_1 e^{i\theta_1}$, where $r_1 = 16.5$ cm and $\theta_1 = 1.33$ rad. The slider moves horizontally, and $\theta_6 = \pi$. Link 2 is the input and has an initial angle of $\theta_2 = \pi$.

The unknowns for the displacement analysis are the angles of links 3, 4, and 5 (θ_3, θ_4, and θ_5) and the position of the slider (r_6). This results in four equations and four unknowns. These equations may be solved simultaneously using a nonlinear equation solving method, such as the Newton–Raphson method. For the example at hand, the equations result in $\theta_3 = 1.58$ rad, $\theta_4 = 0.462$ rad, $\theta_5 = 5.82$ rad, and $r_6 = 16.1$ cm.

4.3 VELOCITY AND ACCELERATION

The velocities of the links of a rigid-link mechanism may be calculated by differentiating the position equations with respect to time and solving for the unknown velocities. If the position analysis is completed and the number of known velocity inputs is equal to the degrees of freedom, the velocity of the system can be determined. Assuming that the input velocity of link 2 ($d\theta_2/dt$) is known for a four-bar mechanism, the angular velocities of the links 3 and 4 are

Rigid-Link Mechanisms

$$\omega_3 = -\omega_2 \frac{r_2 \sin(\theta_4 - \theta_2)}{r_3 \sin(\theta_4 - \theta_3)} \tag{4.51}$$

and

$$\omega_4 = \omega_2 \frac{r_2 \sin(\theta_3 - \theta_2)}{r_4 \sin(\theta_3 - \theta_4)} \tag{4.52}$$

where

$$\omega_i = \frac{d\theta_i}{dt} \tag{4.53}$$

The velocity of point P, $\vec{V}_P$, on the coupler can be written in complex form as

$$\vec{V}_P = \omega_2 i r_2 e^{i\theta_2} + \omega_3 e^{i\theta_3}(ia_3 - b_3) \tag{4.54}$$

where a_3 and b_3 are as defined in Figure 4.4. The velocity of other mechanisms can be found by differentiating the position equations. For example, for the mechanism shown in Figure 4.9, equation (4.42) may be differentiated with respect to time to obtain

$$r_2 i \omega_2 e^{i\theta_2} + r_{B2} i \omega_2 e^{i(\theta_2 + \alpha)} + \frac{dr_{B2}}{dt} e^{i(\theta_2 + \alpha)} = r_{B4} i \omega_4 e^{i\theta_4} \tag{4.55}$$

If link 2 is the input and the input velocity (ω_2) is known, the only unknowns in equation (4.55) are ω_4 and dr_{B2}/dt. The unknowns can be solved for by writing the real and imaginary equations associated with equation (4.55), and solving the resulting linear equations.

The acceleration of the rigid links may be determined by differentiating the velocity equations with respect to time and solving for the unknowns. This assumes that the unknowns in the position and velocity analysis have been solved previously.

For a four-bar mechanism with a known input acceleration of link 2 ($d^2\theta_2/dt^2$), the angular accelerations of links 3 and 4 are

$$\alpha_3 = \frac{-r_2 \alpha_2 \sin(\theta_4 - \theta_2) + r_2 \omega_2^2 \cos(\theta_4 - \theta_2) + r_3 \omega_3^2 \cos(\theta_4 - \theta_3) - r_4 \omega_4^2}{r_3 \sin(\theta_3 - \theta_4)} \tag{4.56}$$

and

$$\alpha_4 = \frac{r_2 \alpha_2 \sin(\theta_3 - \theta_2) - r_2 \omega_2^2 \cos(\theta_3 - \theta_2) + r_4 \omega_4^2 \cos(\theta_3 - \theta_4) - r_3 \omega_3^2}{r_4 \sin(\theta_3 - \theta_4)} \tag{4.57}$$

where

$$\alpha_i = \frac{d^2\theta_i}{dt^2} \tag{4.58}$$

The acceleration of point P on the coupler is

$$\vec{A}_P = \alpha_2 i r_2 e^{i\theta_2} - \omega_2^2 r_2 e^{i\theta_2} + (\alpha_3 e^{i\theta_3} + \omega_3^2 i e^{i\theta_3})(ia_3 - b_3) \tag{4.59}$$

The acceleration of other mechanisms can be found by differentiating the velocity equations. For example, for the mechanism shown in Figure 4.9, equation (4.55) may be differentiated with respect to time to obtain

$$r_2 e^{i\theta_2}(i\alpha_2 - \omega_2^2) + e^{i(\theta_2 + \pi/2)}\left(2\frac{dr_{B2}}{dt}i\omega_2 - r_{B2}\omega_2^2 + \frac{d^2 r_{B2}}{dt^2}\right) = r_{B4} e^{i\theta_4}(i\alpha_4 - \omega_4^2) \tag{4.60}$$

The unknowns in this equation are $d^2 r_{B2}/dt^2$ and α_4. They can be solved for by writing the real and imaginary components of equation (4.60) and solving the resulting linear equations.

4.4 KINEMATIC COEFFICIENTS

Other derivatives, often called *kinematic coefficients*, will be used in later chapters. These are provided below for a four-bar chain and a slider–crank mechanism.

4.4.1 Four-Bar Kinematic Coefficients

Kinematic coefficients are often identified as h_{ij} for a four-bar where

$$h_{ij} = \frac{d\theta_i}{d\theta_j} \tag{4.61}$$

where

$$\frac{d\theta_3}{d\theta_2} = h_{32} = \frac{r_2 \sin(\theta_4 - \theta_2)}{r_3 \sin(\theta_3 - \theta_4)} \tag{4.62}$$

$$\frac{d\theta_4}{d\theta_2} = h_{42} = \frac{r_2 \sin(\theta_3 - \theta_2)}{r_4 \sin(\theta_3 - \theta_4)} \tag{4.63}$$

$$\frac{d\theta_4}{d\theta_3} = h_{43} = \frac{r_3 \sin(\theta_3 - \theta_2)}{r_4 \sin(\theta_4 - \theta_2)} \quad (4.64)$$

and

$$h_{ij} = \frac{1}{h_{ji}} \quad (4.65)$$

4.4.2 Slider–Crank Kinematic Coefficients

Similar equations may be found for slider–crank mechanisms, where the kinematic coefficient, g_{ij}, is

$$g_{ij} = \frac{d\theta_i}{d\theta_j} \quad (4.66)$$

where

$$\begin{aligned} g_{21} &= \frac{d\theta_2}{dr_1} = \frac{-\cos\theta_3}{r_2 \sin(\theta_2 - \theta_3)} \\ g_{31} &= \frac{d\theta_3}{dr_1} = \frac{\cos\theta_2}{r_3 \sin(\theta_2 - \theta_3)} \\ g_{32} &= \frac{d\theta_3}{d\theta_2} = \frac{-r_2 \cos\theta_2}{r_3 \cos\theta_3} \end{aligned} \quad (4.67)$$

and

$$g_{ij} = \frac{1}{g_{ji}} \quad (4.68)$$

4.5 MECHANISM SYNTHESIS

Kinematic analysis is used to determine the motion characteristics of a given mechanism with known geometry. Kinematic synthesis is the means used to design mechanisms for specified motion [1]. A given design problem typically has many different solutions. Mechanisms are proposed and then analyzed to determine if the design objectives are met. Mechanism design therefore often involves iteration between kinematic synthesis and analysis [122].

The major categories of synthesis include type, number, and dimensional synthesis. *Type synthesis* is defined as "the definition of the proper type of mechanism best suited to the problem" [1]. A review of rigid-body type synthesis can be found in [126], and methods of type synthesis for use in compliant mechanisms in [62] through [67]. *Number synthesis*, which may be considered a subset of type synthesis, involves determination of the number of links and degrees of freedom a mechanism should have to perform a specified task. *Dimensional synthesis* is the determination of mechanism geometry to accomplish a specified task. Dimensional synthesis is the focus of the following work.

Three common tasks for kinematic synthesis are function, path, and motion generation. *Function generation* is the correlation of the input and output. In other words, for a given value of the input, an output value is also specified. In *path generation*, a point is required to travel through a specified path. If the position of the point on the path is correlated to the input, it is called *path generation with prescribed timing*. In *motion generation* (also called *rigid-body guidance*), a body is moved through a specified motion. The functions, paths, and motions specified cannot, in general, be generated exactly for the entire mechanism motion. There are two ways to account for this. First, the values of the input and output may be specified and met exactly at a certain number of points called the *precision points*. As an alternative, numerical methods such as optimization or least squares could be used to minimize the overall deviation from the desired output. The precision point method is the method we review.

4.5.1 Function Generation

In function generation, the positions of the output link are prescribed corresponding to given values of the input link position. Only one vector is required to represent link 3, and equations (4.31) and (4.34) may be written as

$$\vec{Z}_2 + \vec{Z}_6 - \vec{Z}_4 - \vec{Z}_1 = 0 \tag{4.69}$$

$$\vec{Z}_2 e^{i\phi_j} + \vec{Z}_6 e^{i\gamma_j} - \vec{Z}_4 e^{i\psi_j} - \vec{Z}_1 = 0 \tag{4.70}$$

where

$$\vec{Z}_6 = \vec{Z}_3 - \vec{Z}_5 \tag{4.71}$$

Vector $\vec{Z}_1$ may be eliminated by subtracting equation (4.69) from equation (4.70), resulting in

$$\vec{Z}_2(e^{i\phi_j} - 1) + \vec{Z}_6(e^{i\gamma_j} - 1) - \vec{Z}_4(e^{i\psi_j} - 1) = 0 \tag{4.72}$$

Values of ϕ_j and ψ_j are prescribed according to the desired function, $\psi_j = f(\phi_j)$. The magnitude and direction of $\vec{Z}_2$, $\vec{Z}_6$, and $\vec{Z}_5$ are undefined, as is the angle γ_j. Therefore, the total number of unknowns is $6 + (n - 1)$, where n is the total number of precision points. A vector-loop equation in the form of equation (4.72) may be written for each precision point from $j = 2$ to $j = n$. Since each vector equation results in two scalar equations, the total number of scalar equations available is $2(n-1)$. For most problems, the total number of unknowns exceeds the number of equations. The result is that values may need to be assigned to a number of variables so that the problem may be solved deterministically. The variables for which values are specified are called *free variables* or *free choices*. The number of free choices is the difference between the number of unknowns and the number of equations. For the four-bar function generator, the number of free choices is given by

$$6 + (n-1) - 2(n-1) = 7 - n \tag{4.73}$$

This approach is summarized in Table 4.2. Theoretically, each free choice has an infinite number of possible values; the number of solutions for n positions of the mechanism for $(7-n)$ free choices is described to be $O(\infty^{7-n})$, or of the order of infinity to the power $(7-n)$ [1].

Although any of the unknown variable values may be chosen as a free choice, the decision may affect the nature of the solution. It is possible for the solution to be linear for two, three, and four precision points. This is done by choosing all values of γ_i as the free choices. For two and three precision points, additional free choices remain after all the values of γ_i are chosen.

As an example, consider a four-bar function generation problem for three precision points. To obtain a linear solution, γ_2 and γ_3 are selected as free choices. The remaining two free choices are the values defining $\vec{Z}_2$. Vectors $\vec{Z}_6$ and $\vec{Z}_4$ may then be found by solving the following set of linear (complex) equations:

$$\begin{bmatrix} e^{i\gamma_2} - 1 & e^{i\psi_2} - 1 \\ e^{i\gamma_3} - 1 & e^{i\psi_3} - 1 \end{bmatrix} \begin{bmatrix} \vec{Z}_6 \\ \vec{Z}_4 \end{bmatrix} = \begin{bmatrix} -\vec{Z}_2(e^{i\phi_2} - 1) \\ -\vec{Z}_2(e^{i\phi_3} - 1) \end{bmatrix} \tag{4.74}$$

TABLE 4.2. Number of free choices for various synthesis problems

Task	No. of unknowns	No. of equations	No. of free choices	No. of solutions	
Function	$5 + n$	$2n - 2$	$7 - n$	$O(\infty^{7-n})$	$n = 2, ..., 7$
Path without prescribed timing	$5 + 3n$	$4n - 4$	$9 - n$	$O(\infty^{9-n})$	$n = 2, ...9$
Path with prescribed timing	$6 + 2n$	$4n - 4$	$10 - 2n$	$O(\infty^{10-2n})$	$n = 2, ..., 5$
Motion	$6 + 2n$	$4n - 4$	$10 - 2n$	$O(\infty^{10-2n})$	$n = 2, ..., 5$

Mechanism Synthesis

For problems with more precision points, or where different variables are chosen for free choices such that at least one angle is an unknown, a system of nonlinear equations is the result. The use of a nonlinear equation solver is more general in that it may be used to solve both linear and nonlinear problems.

4.5.2 Path Generation

If the position of point P shown in Figure 4.7 is required to traverse a prescribed path, the synthesis problem is called *path generation*. If the position of point P is also correlated to the input, it is referred to as *path generation with prescribed timing*. The change in position of point P from the initial position to position j may be described by a vector, $\vec{\delta}_j$, as shown in Figure 4.7.

The path generation equations may be derived from the general loop-closure equations. Subtracting equation (4.31) from equation (4.34) results in

$$\vec{Z}_2(e^{i\phi_j} - 1) + \vec{Z}_3(e^{i\gamma_j} - 1) - \vec{Z}_5(e^{i\gamma_j} - 1) - \vec{Z}_4(e^{i\psi_j} - 1) = 0 \tag{4.75}$$

Rearranging yields

$$\vec{Z}_2(e^{i\phi_j} - 1) + \vec{Z}_3(e^{i\gamma_j} - 1) = \vec{Z}_5(e^{i\gamma_j} - 1) + \vec{Z}_4(e^{i\psi_j} - 1) = \vec{\delta}_j \tag{4.76}$$

This results in the *standard form* equations [1] of dyadic construction, written as

$$\vec{Z}_2(e^{i\phi_j} - 1) + \vec{Z}_3(e^{i\gamma_j} - 1) = \vec{\delta}_j \tag{4.77}$$

$$\vec{Z}_5(e^{i\gamma_j} - 1) + \vec{Z}_4(e^{i\psi_j} - 1) = \vec{\delta}_j \tag{4.78}$$

These equations may also be derived using vector pairs, called *dyads*. For example, equation (4.77) may be found by deriving the equation for the vector loop created by $\vec{\delta}_j$ and the vectors of the dyad of links 2 and 3 in positions 1 and j.

In path generation, $\vec{\delta}_j$ are prescribed. The unknowns are $\vec{Z}_2$, $\vec{Z}_3$, $\vec{Z}_4$, $\vec{Z}_5$, ϕ_j, γ_j, ψ_j, for a total number of unknowns of $8 + 3(n - 1) = 5 + 3n$, where n is the number of precision points. The total number of equations is $4(n - 1)$. The number of free choices is the difference between the number of unknowns and the number of equations, or $9 - n$. The number of solutions is then $O(\infty^{9-n})$. This also implies that the maximum number of precision points possible for this method is 9. This is summarized in Table 4.2.

For path generation with prescribed timing, both $\vec{\delta}_j$ and either ϕ_j or ψ_j are prescribed. The total number of unknowns is $8 + 2(n - 1) = 6 + 2n$, and the number of equations is still $4(n - 1)$. This results in $10 - 2n$ free choices. The maximum number of precision points that can be prescribed for this method is 5, and the order of the number of solutions is $O(\infty^{10-2n})$.

A linear solution may be obtained for two and three precision points by specifying values for all unknown angles as free choices. A nonlinear solution is required if one or more angles are chosen as unknowns. A nonlinear solution is also required for four or more precision points, since the number of unknown angles exceeds the number of free choices.

4.5.3 Motion Generation

Recall that in motion generation, or rigid-body guidance, a body is moved through a specified motion. For example, suppose that a container was to be moved from one conveyer to another without spilling its contents. A four-bar mechanism may be designed such that the container rests on the coupler (link 3), and both the translational and rotational displacements of the coupler are specified. Motion generation may also be viewed as prescribing the motion of a line in a plane [1], whereas path generation prescribes the motion of a point in a plane.

Consider the mechanism of Figure 4.7 for motion generation. The values prescribed are $\vec{\delta}_j$ and γ_j. The unknowns are, therefore, $\vec{Z}_2$, $\vec{Z}_3$, $\vec{Z}_4$, $\vec{Z}_5$, ϕ_j, ψ_j. It is important to note that this is the same as path generation with prescribed timing, except that γ_j is prescribed instead of ϕ_j or ψ_j. Because of this, the numbers of unknowns, equations, and free choices, and the order of the number of solutions, are the same for motion generation as for path generation with prescribed timing. The method of solution is also similar, and a linear solution is possible for two and three precision points if the unknown angle values are specified as free choices.

Example: Synthesis. A mechanism is to be designed for three-precision-point path generation with prescribed timing. The coupler point is to have the displacements of $\vec{\delta}_2 = -5 + 3i$ for $\phi_2 = 10°$ and $\vec{\delta}_3 = -8 + 10i$ for $\phi_3 = 25°$. If the remaining unknown angles are chosen as free choices, the solution is linear. Equations (4.77) and (4.78) may be expanded with $j = 2$ (for change from position 1 to position 2) and $j = 3$ (change from position 1 to position 3) and rearranged as

$$\begin{bmatrix} e^{i\phi_2} - 1 & e^{i\gamma_2} - 1 \\ e^{i\phi_3} - 1 & e^{i\gamma_3} - 1 \end{bmatrix} \begin{bmatrix} \vec{Z}_2 \\ \vec{Z}_3 \end{bmatrix} = \begin{bmatrix} \vec{\delta}_2 \\ \vec{\delta}_3 \end{bmatrix} \quad (4.79)$$

$$\begin{bmatrix} e^{i\psi_2} - 1 & e^{i\gamma_2} - 1 \\ e^{i\psi_3} - 1 & e^{i\gamma_3} - 1 \end{bmatrix} \begin{bmatrix} \vec{Z}_4 \\ \vec{Z}_5 \end{bmatrix} = \begin{bmatrix} \vec{\delta}_2 \\ \vec{\delta}_3 \end{bmatrix} \quad (4.80)$$

With the free choices $\gamma_2 = -8°$, $\gamma_3 = -25°$, $\psi_2 = 10°$, and $\psi_3 = 15°$, equations (4.79) and (4.80) may be solved to obtain

Mechanism Synthesis

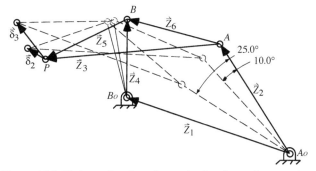

Figure 4.11. Skeleton drawing of a mechanism for path generation.

$$\vec{Z}_1 = -45.75 + 15.26i = 48.23e^{161.6°}$$
$$\vec{Z}_2 = -19.48 + 29.85i = 35.65e^{123.1°}$$
$$\vec{Z}_3 = -48.82 + -4.22i = 49.01e^{-175.1°}$$
$$\vec{Z}_4 = 0.25 + 21.30i = 21.30e^{89.3°} \quad (4.81)$$
$$\vec{Z}_5 = -22.80 - 10.92i = 25.28e^{154.4°}$$
$$\vec{Z}_6 = -26.02 + 6.70i = 26.87e^{165.6°}$$

Figure 4.11 shows a skeleton drawing of this mechanism. If this configuration is not acceptable, different free choices can be used to generation another mechanism that also undergoes the prescribed motion.

PROBLEMS

4.1 Use Gruebler's equation to determine the number of degrees of freedom of a slider–crank mechanism.

4.2 Calculate the number of degrees of freedom for the mechanism in Figure P4.2.

4.3 Use Grashof's criteria to determine the classification of the four-bar mechanisms with link lengths listed in the table. Assume that link 1 is ground, links 2 and 4 are side links, and link 3 is a coupler link.

	Link			
Mechanism	1	2	3	4
A	3	1	2.5	3
B	5	4	2	5
C	2	1	1.5	5

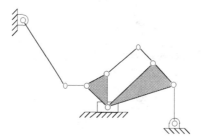

Figure P4.2. Figure for Problem 4.2.

4.4 Develop a program to calculate the position of a four-bar mechanism. The angular deflection of links 3 and 4 (θ_3 and θ_4), coupler point position (x_p and y_p), and the mechanical advantage (MA) should be calculated at every 10 degrees of input rotation. Link 2 (θ_2) may be considered to be the input link. Print out the results for $r_1 = 3$, $r_2 = 1$, $r_3 = 2.5$, $r_4 = 3$, $a_3 = 1$, and $b_3 = 1$.

4.5 Develop a program to calculate the position of a slider–crank mechanism. The angular deflection of link 3 (θ_3), and the position of the slider (r_1) should be calculated at every 10 degrees of input rotation. Print out the results for $r_4 = 0$, $r_2 = 1$, and $r_3 = 2.5$.

4.6 A camping water filter pump mechanism is illustrated in Figure P4.6 with link lengths $r_2 = 4$ cm, $r_3 = 3$ cm, and $r_4 = 4$ cm. At this position $\theta_2 = 60$ degrees measure from the horizontal. (The drawing is not to scale.)
   (a) Draw a vector loop and write the associated complex equations for displacement analysis.
   (b) Solve for the unknown displacements and give correct signs and units.

4.7 Perform the velocity analysis of the mechanism described in Figure P4.6 if the input velocity is $\omega_{in} = 0.5$ rad/s, clockwise.

4.8 The mechanism illustrated in Figure P4.8 is known as a Scotch-yoke mechanism. For this problem, assume that the crank has a length of 4 cm and is at an angle of 35°.

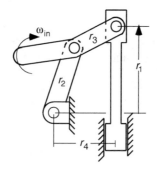

Figure P4.6. Figure for Problems 4.6 and 4.7.

Mechanism Synthesis

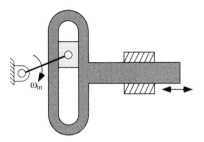

Figure P4.8. Scotch-yoke mechanism for Problem 4.8.

(a) Draw a vector loop and write the associated complex equations for displacement analysis.
(b) Solve for the unknown displacement of the slider.
(c) If the input velocity is $\omega_{in} = 0.5$ rad/s, clockwise, solve for the velocity of the slider.

4.9 The linkage shown in Figure P4.9 has the following dimensions: $r_2 = 2$ cm, $r_3 = 6$ cm, $r_4 = 6$ cm, $x = 7.5$ cm, $y = 5$ cm, and $\theta_2 = \pi/4$.
(a) Calculate the degrees of freedom of the mechanism.
(b) Solve for the unknown displacements of the linkage (i.e., find θ_3, θ_4, θ_5, and the position of the slider, C) by combining the closed-form equations for a four-bar mechanism and a slider–crank. (Note that links 1, 2, 3, and 4 for a four-bar linkage, and that links 1, 4, 5 and slider C form a slider mechanism.)

4.10 The linkage shown in Figure P4.9 has the following dimensions: $r_2 = 2$ cm, $r_3 = 6$ cm, $r_4 = 6$ cm, $x = 7.5$ cm, $y = 5$ cm, and $\theta_2 = \pi/4$.
(a) Find appropriate vector loops to be used to solve for the displacement of the mechanism, and write the corresponding vector-loop equations in complex form.
(b) Solve for the unknown displacements of the linkage (i.e., find θ_3, θ_4, θ_5, and the position of the slider, C) by solving the complex equations above.

4.11 The linkage shown in Figure P4.9 has the following dimensions: $r_2 = 2$ cm, $r_3 = 6$ cm, $r_4 = 6$ cm, $x = 7.5$ cm, $y = 5$ cm, $\theta_2 = \pi/4$, $\omega_2 = 1000$ rpm (revolutions per minute), and $\alpha_2 = 0$ rad/s^2. Note that the complex equations for the displacement of this mechanism were found in Problem 4.10.
(a) Perform a velocity analysis of the linkage (i.e., find ω_3, ω_4, ω_5, and V_C).
(b) Perform an acceleration analysis of the linkage (i.e., find α_3, α_4, α_5, and A_C).

4.12 Consider the mechanism illustrated in Figure P4.12. Assume that the link lengths are known and that the angle of the input link is given.
(a) Draw the vector loops used to solve for the displacement of the mechanism and write the vector loop equations in complex form. State the unknowns for the displacement analysis.

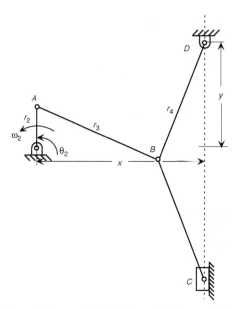

Figure P4.9. Figure for Problems 4.9, 4.10, and 4.11.

(b) Develop the equations for the velocity analysis of the mechanism, assuming that the input velocity is known. State the unknowns for the velocity analysis.

(c) Develop the equations for the acceleration analysis of the mechanism, assuming that the input acceleration is known. State the unknowns for the acceleration analysis.

4.13 Design a mechanism for path generation with the following specified path: $\vec{\delta}_2 = -10 + 5i$, and $\vec{\delta}_3 = -15 - 2i$.

4.14 Design a mechanism for motion generation with the following specified motion: $\gamma_2 = -10°$, and $\gamma_3 = -20°$, $\vec{\delta}_2 = -10 + 5i$, and $\vec{\delta}_3 = -15 - 2i$.

4.15 Design a function generation mechanism with the following specified motion: $\psi_2 = 50°$, $\psi_3 = 80°$, $\phi_2 = 20°$, and $\phi_3 = 30°$.

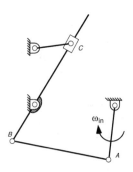

Figure P4.12. Figure for Problem 4.12.

CHAPTER 5

PSEUDO-RIGID-BODY MODEL

The large-deflection analysis methods discussed in Chapter 2 are valuable analysis tools that are particularly useful when a design has been chosen and the geometry, material properties, and load conditions are available to input into the algorithms. There is a need, however, for more efficient methods to arrive at and improve these initial designs. The purpose of the pseudo-rigid-body model is to provide a simple method of analyzing systems that undergo large, nonlinear deflections. The pseudo-rigid-body model concept is used to model the deflection of flexible members using rigid-body components that have equivalent force–deflection characteristics. Rigid-link mechanism theory may then be used to analyze the compliant mechanism. In this way, the pseudo-rigid-body model is a bridge that connects rigid-body mechanism theory and compliant mechanism theory. The method is particularly useful in the design of compliant mechanisms. Pseudo-rigid-body models that accurately describe the behavior of compliant mechanisms are discussed in this chapter. Different types of segments require different models; thus a number of models for individual segments are presented, followed by a discussion of how they may be applied to compliant mechanisms.

For each flexible segment, a pseudo-rigid-body model predicts the deflection path and force–deflection relationships of a flexible segment. The motion is modeled by rigid links attached at pin joints. Springs are added to the model to accurately predict the force–deflection relationships of the compliant segments. The key for each pseudo-rigid-body model is to decide where to place the pin joints and what value to assign the spring constants. This is described in the following sections.

5.1 SMALL-LENGTH FLEXURAL PIVOTS[*]

The first flexible segment to be discussed is the small-length flexural pivot. Consider the cantilever beam shown in Figure 5.1a. The beam has two segments; one is short and flexible, and the other is longer and rigid. If the small segment is significantly shorter and more flexible than the large segment, that is,

$$L \gg l \tag{5.1}$$

$$(EI)_L \gg (EI)_l \tag{5.2}$$

the small segment is called a *small-length flexural pivot*. (For clarity, Figure 5.1 shows l with an exaggerated length. Usually, L is 10 or more times larger than l.) The deflection equations for the flexible segment with a moment at the end were derived in Chapter 2 as

$$\theta_o = \frac{M_o l}{EI} \tag{5.3}$$

$$\frac{\delta_y}{l} = \frac{1 - \cos \theta_o}{\theta_o} \tag{5.4}$$

$$\frac{\delta_x}{l} = 1 - \frac{\sin \theta_o}{\theta_o} \tag{5.5}$$

and may be used to define a simple pseudo-rigid-body model for small-length flexural pivots. Since the flexible section is much shorter than the rigid section, the motion of the system may be modeled as two rigid links joined at a pin joint, called the *characteristic pivot*. The characteristic pivot is located at the center of the flexural pivot, as shown in Figure 5.1b. This is an accurate assumption because the deflection occurs at the flexible segment and it is small compared to the length of the rigid segment. For this same reason, nearly any point along the flexible segment would represent an acceptable position for the characteristic pivot and the center point is used for convenience. The angle of the pseudo-rigid link is the pseudo-rigid-body angle, Θ. For small-length flexural pivots, the pseudo-rigid-body angle is equal to the beam end angle:

$$\Theta = \theta_o \quad \text{(small-length flexural pivots)} \tag{5.6}$$

---

[*]See [3].

Small-Length Flexural Pivots

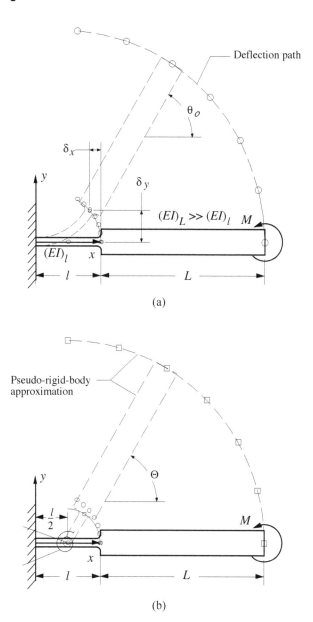

Figure 5.1. (a) Small-length flexural pivot, and (b) its pseudo-rigid-body model. (The length of the small-length flexural pivot is exaggerated in the figure for clarity, but for an accurate model, $L \gg l$.)

The x and y coordinates of the beam's end (a and b, respectively) are approximated as

$$a = \frac{l}{2} + \left(L + \frac{l}{2}\right)\cos\Theta \qquad (5.7)$$

and

$$b = \left(L + \frac{l}{2}\right)\sin\Theta \qquad (5.8)$$

or, expressed in nondimensional form,

$$\frac{a}{l} = \frac{1}{2} + \left(\frac{L}{l} + \frac{1}{2}\right)\cos\Theta \qquad (5.9)$$

and

$$\frac{b}{l} = \left(\frac{L}{l} + \frac{1}{2}\right)\sin\Theta \qquad (5.10)$$

For a given angular deflection, the rigid sections of the calculated and approximated beams will be parallel, causing the distance between the corresponding path points, which were determined by the two methods, to be the same at both ends of the rigid section. This distance, d, decreases as l decreases (see Figure 5.2). For this reason, the accuracy of the pseudo-rigid-body model is improved because the lengths of the flexural pivots are small compared with the lengths of the rigid sections.

The beam's resistance to deflection is modeled using a torsional spring with spring constant K. The torque required to deflect the torsional spring through an angle of Θ is

$$T = K\Theta \qquad (5.11)$$

The spring constant, K, may be found from elementary beam theory. The end angle for a beam with a moment at the end is

$$\theta_o = \frac{Ml}{(EI)_l} \qquad (5.12)$$

Rearranging to solve for M results in

$$M = \frac{(EI)_l}{l}\theta_o \qquad (5.13)$$

Small-Length Flexural Pivots

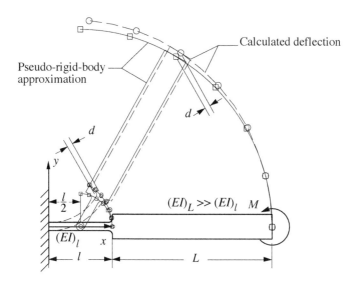

Figure 5.2. Error associated with the small-length flexural pivot approximation.

Comparing equations (5.11) and (5.13), and recalling that $M = T$ and $\theta_o = \Theta$, reveals that the spring constant, K, is

$$K = \frac{(EI)_l}{l} \tag{5.14}$$

This model is more accurate if bending is the dominant loading in the flexural pivot. If transverse and axial loads are larger than the bending moment, greater error will be introduced into the model. An advantage of this simple model is that for pure bending, equations (5.3) through (5.5) are accurate even for large deflections, because no assumptions about small deflection were made in their derivation.

The nature of small-length flexural pivots ensures that the assumption that bending is the predominate loading is accurate in most applications. The reason for this is illustrated in Figure 5.3, which shows the beam loaded with a vertical force at the free end. The reaction moments at the ends of the flexible segment shown are $M_1 = F(L+l)$ and $M_2 = FL$. For a pure moment loading,

$$\frac{M_1}{M_2} = 1 \quad \text{(moment at free end)} \tag{5.15}$$

and for the small-length flexural pivot shown in Figure 5.3, assuming small deflections,

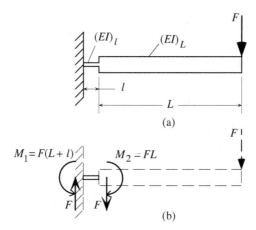

Figure 5.3. Two-segment beam with a vertical end force at the free end.

$$\frac{M_1}{M_2} = 1 + \frac{l}{L} \quad \text{(force at free end)} \tag{5.16}$$

Since L is much greater than l ($L \gg l$), the ratios of equations (5.15) and (5.16) are nearly equal. The torque at the torsional spring in the pseudo-rigid-body model, T, is taken as the moment at the midpoint of the flexible segment. This is discussed in more detail in the following section.

5.1.1 Active and Passive Forces

Figure 5.4 illustrates a beam with a small-length flexural pivot and a force at the free end. Both the magnitude and direction of the force must be known to define the force, F. The angle may be defined using the angle ϕ, as shown in the figure, or it may be expressed in terms of the vertical component (P) and the horizontal component (nP). This is a *nonfollower* force in that it remains at the same angle regardless of the deflection of the beam. In this case

$$F = P\sqrt{n^2 + 1} \tag{5.17}$$

$$\phi = \operatorname{atan}\frac{1}{-n} \tag{5.18}$$

It is also useful to view the force in terms of the components of the force that are normal and tangential to the path of the beam end. The component tangent to the

Small-Length Flexural Pivots

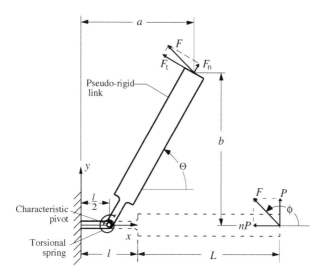

Figure 5.4. Pseudo-rigid-body model of a beam with a small-length flexural pivot and a force at the free end.

path (or normal to the pseudo-rigid link), F_t, causes a moment at the torsional spring of

$$T = F_t \left(L + \frac{l}{2} \right) \quad (5.19)$$

Since this component of the force contributes to the deflection of the pseudo-rigid link, it is called an *active force*.

The force normal to the path, F_n, is called a *passive force* because it does not contribute to the deflection. If the angle of the applied force remains the same throughout the beam deflection, the active and passive components change when the deflection changes. The active force is

$$F_t = F \sin(\phi - \Theta) \quad (5.20)$$

If F is vertical (i.e., $\phi = \pi/2$), then $F_t = F \cos \Theta$.

5.1.2 Stress

The maximum stress, σ_{max}, for a cantilever beam with a force at the free end occurs at the fixed end. The maximum magnitude may occur at the top or bottom of the beam, depending on the direction of the beam bending and the direction of the force. For the beam shown in Figure 5.4, the maximum moment, M_{max}, is

$$M_{max} = Pa + nPb \tag{5.21}$$

The stress at the top and bottom of the beam is

$$\sigma_{top} = \frac{-(Pa + nPb)c}{I} - \frac{nP}{A} \tag{5.22}$$

$$\sigma_{bot} = \frac{(Pa + nPb)c}{I} - \frac{nP}{A} \tag{5.23}$$

For a beam with cross-section width w and height h,

$$\sigma_{top} = \frac{-6(Pa + nPb)}{wh^2} - \frac{nP}{wh} \tag{5.24}$$

$$\sigma_{bot} = \frac{6(Pa + nPb)}{wh^2} - \frac{nP}{wh} \tag{5.25}$$

Example: Small-Length Flexural Pivot. Suppose that the small-length flexural pivot shown in Figure 5.5 is required to have a vertical deflection of $b = 3$ in. Determine the corresponding horizontal deflection, maximum stress and vertical force required.

Solution: The moment of inertia for the flexible segment is

$$I = \frac{bh^3}{12} = \frac{(0.75 \text{ in.})(0.03 \text{ in.})^3}{12} = 1.7 \times 10^6 \text{ in.}^4 \tag{5.26}$$

The torsional spring constant is calculated from equation (5.14):

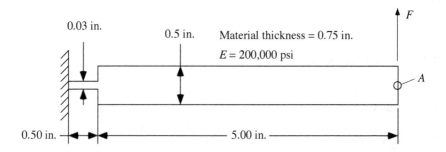

Figure 5.5. Example of a small-length flexural pivot.

Small-Length Flexural Pivots

$$K = \frac{EI}{l} = \frac{(2 \times 10^5 \text{ lb/in.}^2)(1.7 \times 10^{-6} \text{ in.}^4)}{0.5 \text{ in.}} = 0.68 \text{ in.-lb} \quad (5.27)$$

Rearranging equation (5.8) to solve for Θ yields

$$\Theta = \text{asin} \frac{b}{L + l/2} = \text{asin} \frac{3 \text{ in.}}{5 \text{ in.} + (0.5 \text{ in.})/2} = 0.608 \text{ rad} \quad (5.28)$$

The horizontal coordinate of the beam end, a, is found from equation (5.7) as

$$\begin{aligned} a &= \frac{l}{2} + \left(L + \frac{l}{2}\right)\cos\Theta \\ &= \frac{0.5 \text{ in.}}{2} + \left(5.0 \text{ in.} + \frac{0.5 \text{ in.}}{2}\right)\cos(0.608) = 4.56 \text{ in.} \end{aligned} \quad (5.29)$$

The horizontal deflection is

$$\delta_x = L + l - a = 5.0 \text{ in.} + 0.5 \text{ in.} - 4.56 \text{ in.} = 0.94 \text{ in.} \quad (5.30)$$

The force required to cause the deflection may also be calculated. Solving equation (5.20) for F yields

$$F = \frac{F_t}{\sin(\phi - \Theta)} \quad (5.31)$$

Solving equation (5.19) for F_t results in

$$F_t = \frac{T}{L + l/2} \quad (5.32)$$

Substituting equations (5.32) and (5.11) into equation (5.31) yields a value of the force of

$$\begin{aligned} F &= \frac{K\Theta}{(L + l/2)\sin(\phi - \Theta)} \\ &= \frac{(0.68 \text{ in.-lb})(0.608)}{[5 \text{ in.} + (0.5 \text{ in.})/2]\sin(\pi/2 - 0.608)} = 0.096 \text{ lb} \end{aligned} \quad (5.33)$$

The maximum stress, σ_{max}, occurs at the fixed end and can be calculated from equation (5.23) with $P = F$ and $n = 0$ as

$$\sigma_{max} = \frac{Fac}{I} = \frac{(0.096 \text{ lb})(4.56 \text{ in.})(0.015 \text{ in.})}{1.7 \times 10^{-6} \text{ in.}^4} = 3900 \text{ lb/in.}^2 \quad (5.34)$$

Example: Iterative Solution. Consider the same beam as used in the preceding example, but suppose that the force is known to be $F = 0.1$ lb and the deflection is to be calculated.

Solution: Unlike the previous example, this analysis will result in a nonlinear equation that will need to be solved iteratively. The pseudo-rigid-body angle, Θ, must be calculated before the horizontal and vertical deflections can be determined. Combining equations (5.11), (5.31), and (5.32) results in an equation similar to equation (5.33):

$$K\Theta - F\left(L + \frac{l}{2}\right)\sin(\phi - \Theta) = 0 \tag{5.35}$$

Equation (5.35) cannot be solved intrisically for Θ and an iterative method (such as Newton's method) is required to obtain a solution. For this example, this results in

$$\Theta = 0.626 \text{ rad} \tag{5.36}$$

The x and y coordinates of the beam end (a and b) are

$$\begin{aligned} a &= \frac{l}{2} + \left(L + \frac{l}{2}\right)\cos\Theta \\ &= \frac{0.5 \text{ in.}}{2} + \left(5.0 \text{ in.} + \frac{0.5 \text{ in.}}{2}\right)\cos(0.626) = 4.5 \text{ in.} \end{aligned} \tag{5.37}$$

and

$$b = \left(L + \frac{l}{2}\right)\sin\Theta = \left(5.0 \text{ in.} + \frac{0.5 \text{ in.}}{2}\right)\sin(0.626) = 3.08 \text{ in.} \tag{5.38}$$

The maximum stress, σ_{max}, occurs at the fixed end and has a value of

$$\sigma_{max} = \frac{Fac}{I} = \frac{(0.1 \text{ lb})(4.71 \text{ in.})(0.03 \text{ in.}/2)}{1.7 \times 10^{-6} \text{ in.}^4} = 4200 \text{ lb/in.}^2 \tag{5.39}$$

5.1.3 Living Hinges

Extremely short and thin small-length flexural pivots are often called *living hinges*. The pseudo-rigid-body model of a pin joint at the center of the flexible segment is highly accurate for living hinges. In systems with both living hinges and other compliant segments, the rigidity of the living hinges is often so low compared with the other flexible segments in a system that their torsional springs are ignored. However, if a system contains only living hinges, their rigidity should be

Cantilever Beam with a Force at the Free End (Fixed–Pinned)

considered in the analysis. Living hinges are discussed in more detail in Section 5.8.1.

Living hinges are used in many products. One of the most common applications is in closures, such as lids for consumer products. They are also used to simplify the manufacture of containers. A container may be molded as one flat piece of plastic with flexural hinges; the plastic is then folded along the hinge lines to create the container. In this application, the living hinge is deflected only once in the life of the product and, therefore, fatigue is not a concern.

5.2 CANTILEVER BEAM WITH A FORCE AT THE FREE END (FIXED–PINNED)*

Consider the flexible cantilever beam with constant cross section and linear material properties that is shown in Figure 5.6. If the deflections are large, they may be out of the range of linearized beam deflection equations, and elliptic integral solutions or nonlinear finite element analysis may be used to perform the analysis, as discussed in Chapter 2. However, these methods are cumbersome for the early design phases of compliant mechanisms. The pseudo-rigid-body model of the flexible beam provides a simplified but accurate method of analyzing the deflection of flexible beams and provides the designer a means of visualizing the deflection. The loading of a force at the free end of a cantilever beam commonly occurs in fixed–pinned segments of compliant mechanisms.

Large-deflection elliptic-integral equations show that for a flexible cantilever beam with a force at the free end, the free end follows a nearly circular path, with some radius of curvature along the beam's length. This idea will be used to develop parametric approximations for the beam's deflection path.

Figure 5.6 shows a pseudo-rigid-body model of a large-deflection beam for which it is assumed that the nearly circular path can be accurately modeled by two rigid links that are joined at a pivot along the beam. A torsional spring at the pivot represents the beam's resistance to deflection. The location of this pseudo-rigid-body *characteristic pivot* is measured from the beam's end as a fraction of the beam's length, where the fractional distance is γl and γ is the *characteristic radius factor*. The product γl, the *characteristic radius*, is the radius of the circular deflection path traversed by the end of the pseudo-rigid-body link. It is also the length of the pseudo-rigid-body link.

The preceding pseudo-rigid-body approximation will be used to parameterize the deflection path, the angular deflection of the beam's end, and load–deflection relationships in Θ, the *pseudo-rigid-body angle*. The pseudo-rigid-body angle is the angle between the pseudo-rigid-body link and its undeflected position, as shown in Figure 5.6.

---

*See [50].

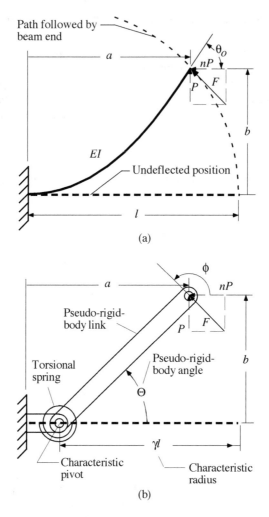

Figure 5.6. (a) Cantilevered segment with forces at the free end, and (b) its pseudo-rigid-body model.

Nomenclature in this chapter is consistent with that used for large-deflection beams in Chapter 2. Specifically, the x and y coordinates of the beam deflection are a and b, respectively; the vertical component of the force is P; the axial force is nP (where a positive value of n represents a force that will cause compression in the undeflected beam); F is the total force, or

$$F = P\sqrt{n^2 + 1} \tag{5.40}$$

and ϕ is the angle of this force:

Cantilever Beam with a Force at the Free End (Fixed–Pinned)

$$\phi = \operatorname{atan} \frac{1}{-n} \qquad (5.41)$$

The negative sign is explicitly defined to be with n to allow the correct quadrant to be determined for the arctangent function.

5.2.1 Parametric Approximation of the Beam's Deflection Path

An acceptable value for the characteristic radius factor, γ, may be found by first determining the maximum acceptable percentage error in deflection. The value of γ that would allow the maximum pseudo-rigid-body angle, Θ, while still satisfying the maximum error constraint is then determined. The problem may be formally stated as follows: *Find the value of the characteristic radius factor, γ, which maximizes the pseudo-rigid-body angle, Θ, where*

$$\Theta = \operatorname{atan} \frac{b}{a - l(1-\gamma)} \qquad (5.42)$$

which is subject to the parametric constraint

$$g(\Theta) = \frac{\text{error}}{\delta_e} \le \left(\frac{\text{error}}{\delta_e}\right)_{\max} \qquad \text{for } 0 < \Theta < \Theta_{\max} \qquad (5.43)$$

where error/δ_e is the relative deflection error, and a and b are the respective horizontal and vertical coordinates of the deflected end, which are calculated using the elliptic integral approach. Figure 5.7 illustrates the error in the deflection approximation. The deflections calculated using the elliptic integral approach, δ_e, and the pseudo-rigid-body approximation, δ_a, are given as

$$\delta_e = \sqrt{(l-a)^2 + b^2} \qquad (5.44)$$

and

$$\delta_a = \sqrt{[\gamma l(1 - \cos\Theta)]^2 + (\gamma l \sin\Theta)^2} \qquad (5.45)$$

The error in the deflection is calculated as

$$\frac{\text{error}}{l} = \sqrt{\left\{\frac{a}{l} - [1 - \gamma(1 - \cos\Theta)]\right\}^2 + \left(\frac{b}{l} - \gamma \sin\Theta\right)^2} \qquad (5.46)$$

and

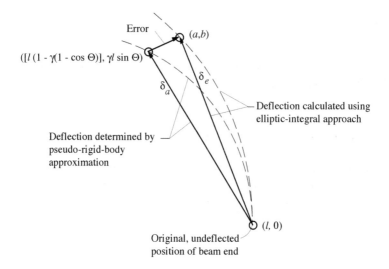

Figure 5.7. Determination of error in deflection approximations.

$$\frac{\text{error}}{\delta_e} = \frac{\sqrt{\{(a/l) - [1 - \gamma(1 - \cos\Theta)]\}^2 + [(b/l) - \gamma\sin\Theta]^2}}{\sqrt{(1 - a/l)^2 + (b/l)^2}} \quad (5.47)$$

The value of the angular deflection of the beam's end, θ_o, at the point at which the error equals or exceeds an acceptable amount, is the maximum angular deflection of the beam's end, or the *parameterization limit*, $\theta_{o\,\text{max}}$. The value of the pseudo-rigid-body angle that corresponds to the parameterization limit is Θ_{max}.

5.2.2 Characteristic Radius Factor

The equations above are used to find the optimal value of the characteristic radius factor, γ. Figure 5.8 shows the results for $n = 0$, for which the optimal γ is 0.8517. This represents an approximation associated with an error smaller than 0.5%, a maximum angular deflection of $\theta_{o\,\text{max}} = 77°$, and a vertical deflection equal to nearly 80% of the beam length.

Figure 5.9 plots γ versus n, a relationship that may be expressed as the equations

$$\gamma = \begin{cases} 0.841655 - 0.0067807n + 0.000438n^2 & (0.5 < n < 10.0) \\ (0.852144 - 0.0182867n) & (-1.8316 < n < 0.5) \\ (0.912364 + 0.0145928n) & (-5 < n < -1.8316) \end{cases} \quad (5.48)$$

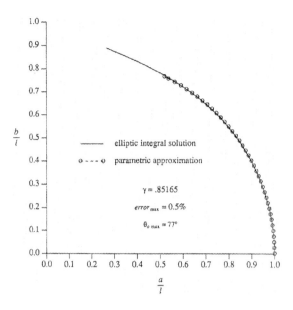

Figure 5.8. Parametric approximation of the beam's deflection path for $n = 0$.

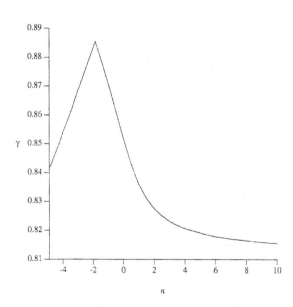

Figure 5.9. Plot of characteristic radius factor, γ, versus n.

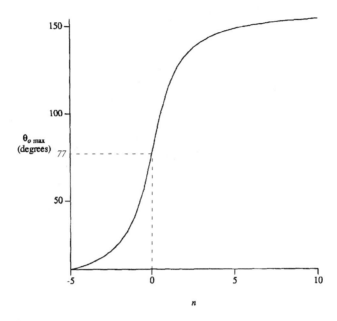

Figure 5.10. Beam-end angular deflection parameterization limit, $\theta_{o\,max}$, versus n.

The approximation is accurate only to a certain limit, which may be listed in terms of the maximum angle of the beam's end for which the approximation is accurate, $\theta_{o\,max}$, or the *beam end angular deflection parameterization limit*. Figure 5.10 plots $\theta_{o\,max}$ versus n. Table 5.1 lists numerical values of γ for various values of n. The maximum pseudo-rigid-body angle Θ_{max} is also listed.

5.2.3 Coordinates of Beam End

The coordinates of the end of a deflected beam may be given by equations parameterized in terms of the pseudo-rigid-body angle, Θ (Figure 5.6). These equations are written as

$$\frac{a}{l} = 1 - \gamma(1 - \cos\Theta) \tag{5.49}$$

$$\frac{b}{l} = \gamma \sin\Theta \tag{5.50}$$

5.2.4 Rule of Thumb for Characteristic Radius Factor

Figure 5.9 shows that the characteristic radius factor, γ, does not vary much over a large range of force angles. The value of γ may be roughly approximated by determining its average value over a specified range of n:

$$\gamma_{ave} = \left(\int_{n_1}^{n_2} \gamma \, dn \right) \div \left(\int_{n_1}^{n_2} dn \right) \tag{5.51}$$

For an equivalent load in the range $135.0° \leq \phi \leq 63.4°$ and $-0.5 \leq n \leq 1.0$, γ_{ave} is determined by using equations (5.48), (5.48), and (5.51) to obtain

$$\gamma_{ave} = \frac{\int_{-0.5}^{0.5} (0.851244 - 0.0182867n) \, dn + \int_{0.5}^{1.0} (0.841655 - 0.0067807n^2) \, dn}{\int_{-0.5}^{1.0} dn} \tag{5.52}$$

$$\approx 0.85$$

TABLE 5.1. Numerical data for γ, c_θ, and K_Θ for various angles of force

n	ϕ	γ	$\Theta_{max}(\gamma)$	c_θ	K_Θ	$\Theta_{max}(K_\Theta)$
0.0	90.0	0.8517	64.3	1.2385	2.67617	58.5
0.5	116.6	0.8430	81.8	1.2430	2.63744	64.1
1.0	135.0	0.8360	94.8	1.2467	2.61259	67.5
1.5	146.3	0.8311	103.8	1.2492	2.59289	65.8
2.0	153.4	0.8276	108.9	1.2511	2.59707	69.0
3.0	161.6	0.8232	115.4	1.2534	2.56737	64.6
4.0	166.0	0.8207	119.1	1.2548	2.56506	66.4
5.0	168.7	0.8192	121.4	1.2557	2.56251	67.5
7.5	172.4	0.8168	124.5	1.2570	2.55984	69.0
10.0	174.3	0.8156	126.1	1.2578	2.56597	69.7
-0.5	63.4	0.8612	47.7	1.2348	2.69320	44.4
-1.0	45.0	0.8707	36.3	1.2323	2.72816	31.5
-1.5	33.7	0.8796	28.7	1.2322	2.78081	23.6
-2.0	26.6	0.8813	23.2	1.2293	2.80162	18.6
-3.0	18.4	0.8669	16.0	1.2119	2.68893	12.9
-4.0	14.0	0.8522	11.9	1.1971	2.58991	9.8
-5.0	11.3	0.8391	9.7	1.1788	2.49874	7.9

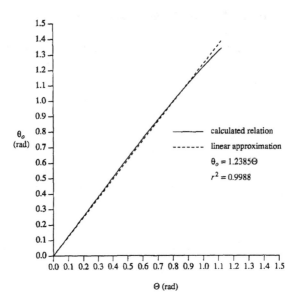

Figure 5.11. Linear approximation of θ_o, versus Θ for $n = 0$.

This approximation is useful for a wide range of force angles and is helpful in making rough calculations.

Burns and Crossley [31] developed a model similar to the pseudo-rigid-body model and used a value of 5/6 for their equivalent of γ.

5.2.5 Angular Deflection Approximation

The nearly linear relationship between θ_o and Θ is approximated by

$$\theta_o = c_\theta \Theta \tag{5.53}$$

where constant c_θ is termed the *parametric angle coefficient*. Figure 5.11 shows a linear curve that fit a large-deflection beam loaded with a single transverse load ($n = 0$). Table 5.1 shows similar information for various values of n and ϕ.

5.2.6 Stiffness Coefficient

The beam's resistance to deflection may be modeled by a nondimensionalized torsional spring constant, K_Θ, called the *stiffness coefficient*. Combined with geometric and material properties, the stiffness coefficient is used to determine the value of the spring constant for a particular beam's pseudo-rigid-body model.

The total force acting on the beam's end, F, is

Cantilever Beam with a Force at the Free End (Fixed–Pinned)

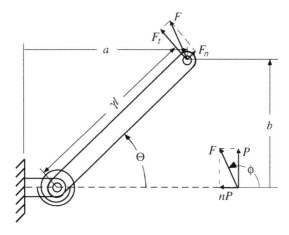

Figure 5.12. Tangential and normal components of the applied force.

$$F = \sqrt{P^2 + (nP)^2} = \eta P \tag{5.54}$$

where

$$\eta = \sqrt{1 + n^2} \tag{5.55}$$

The pseudo-rigid-body model of the cantilever beam suggests that a force perpendicular to the pseudo-rigid-link and tangent to the path of the endpoint, F_t, will generate a torque at the characteristic pivot, and cause the link to deflect. However, a force parallel to the pseudo-rigid link and normal to the path, F_n, is passive and does not contribute to the deflection. This is the same as discussed for small-length flexural pivots in Section 5.1.1. The force components F_t and F_n are illustrated in Figure 5.12.

The transverse (or tangential) component of the load is

$$F_t = F \sin(\phi - \Theta) = \eta P \sin(\phi - \Theta) \tag{5.56}$$

where ϕ is the angle of the applied load, as shown in Figure 5.6 and defined in equation (5.41).

This load may be nondimensionalized as the nondimensionalized transverse load index, $(\alpha^2)_t$, as

$$(\alpha^2)_t = \frac{F_t l^2}{EI} \tag{5.57}$$

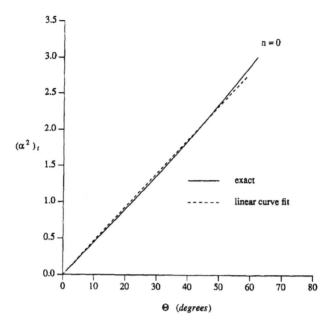

Figure 5.13. Tangential force versus deflection.

Further insight into the pseudo-rigid-body model may be gained by investigating the stiffness characteristics of its equivalent torsional spring [70].

Figure 5.13 plots the nondimensionalized transverse load index, $(\alpha^2)_t$, versus the pseudo-rigid-body angle, Θ, for $n = 0$. The pseudo-rigid-body angle is calculated from

$$\Theta = \operatorname{atan} \frac{b}{a - l(1 - \gamma)} \tag{5.58}$$

A torsional spring constant approximation of these relationships is accurate for most of the pseudo-rigid-body angle. The force–deflection relationship may be written as

$$(\alpha^2)_t = K_\Theta \Theta \tag{5.59}$$

where K_Θ is the *stiffness coefficient*. The torsional spring in Figure 5.6 has a constant stiffness for a given value of n. The relationship in equation (5.59) is very simple; however, it may not be accurate for the entire range of the kinematic model, and its limits must be recognized. The limit is expressed in terms of the maximum pseudo-rigid-body model angle, $\Theta_{\max}$, allowed before exceeding the limits.

The stiffness coefficient, K_Θ, is plotted as a function of n in Figure 5.14. This relationship is described in equation form as

Cantilever Beam with a Force at the Free End (Fixed–Pinned)

$$K_\Theta = 3.024112 + 0.121290n + 0.003169n^2 \quad (-5 < n \leq -2.5) \quad (5.60)$$

$$K_\Theta = 1.967647 - 2.616021n - 3.738166n^2 \\ - 2.649437n^3 - 0.891906n^4 - 0.113063n^5 \quad (5.61) \\ (-2.5 < n \leq -1)$$

$$K_\Theta = 2.654855 - 0.509896 \times 10^{-1}n + 0.126749 \times 10^{-1}n^2 \\ - 0.142039 \times 10^{-2}n^3 + 0.584525 \times 10^{-4}n^4 \quad (5.62) \\ (-1 < n \leq 10)$$

in the range of the pseudo-rigid-body angle, Θ, where

$$\Theta < \Theta_{max} \approx 0.7 \tan^{-1}\frac{1}{-n} = 0.7\phi \quad (-5.0 < n < 10.0) \quad (5.63)$$

The correlation coefficients squared, r^2, for the equations above are 0.99939, 0.97786, and 0.99341, respectively. The curve-fit relations given in equations (5.60) through (5.62) are shown in Figure 5.14. Numerical values for K_Θ are listed in Table 5.1 for various values of n.

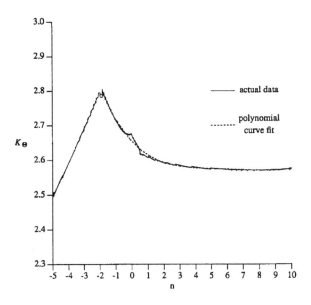

Figure 5.14. Stiffness coefficient versus n.

The value of K_Θ varies only 0.3 between its lowest and highest values for a wide range of loading. This allows an approximation of a constant K_Θ to be made for use in rough calculations. This may be done by calculating the average K_Θ as

$$K_{\Theta_{ave}} = \frac{\int_{n_1}^{n_2} K_\Theta dn}{\int_{n_1}^{n_2} dn} \tag{5.64}$$

For the load-angle range $11.3° < \phi < 174.3°$ or $-5.0 < n < 10.0$, K_Θ is approximated by using equations (5.60) through (5.64), resulting in $K_{\Theta_{ave}} = 2.61$. Considering loads in only the most common range of $63° < \phi < 135°$ or $-0.5 < n < 1.0$ results in

$$K_{\Theta_{ave}} = 2.65 \tag{5.65}$$

Another simple approximation is

$$K_\Theta \approx \pi\gamma \tag{5.66}$$

The π term appears here by chance for a small range and equation (5.66) should only be used as a rough rule of thumb.

5.2.7 Torsional Spring Constant

As mentioned previously, the transverse load, F_t, causes the beam to deflect. For the pseudo-rigid-body model of a flexible beam shown in Figure 5.6, the torque at the pin joint, T, is the torsional spring constant, K, times the angular deflection, Θ:

$$T = K\Theta \tag{5.67}$$

This torque can also be expressed as the transverse force, F_t, multiplied by the moment arm (the length of the pseudo-rigid link):

$$T = F_t \gamma l \tag{5.68}$$

Combining equations (5.67) and (5.68) and solving for F_t yields

$$F_t = \frac{K\Theta}{\gamma l} \tag{5.69}$$

Combining equations (5.57) and (5.59) results in

$$\frac{F_t l^2}{EI} = K_\Theta \Theta \tag{5.70}$$

Substituting equation (5.69) into equation (5.70) yields

$$\frac{K\Theta l}{\gamma EI} = K_\Theta \Theta \tag{5.71}$$

The pseudo-rigid-body angle, Θ, cancels from both sides of the equation, resulting in the following equation for the torsion spring constant:

$$K = \gamma K_\Theta \frac{EI}{l} \tag{5.72}$$

Equation (5.72) shows that the value of the spring constant, K, depends on the pseudo-rigid-body constants (γK_Θ), the geometry (I/l), and the material properties (E). Substituting equation (5.66) into equation (5.72) yields another approximation:

$$K \approx \pi \gamma^2 \frac{EI}{l} \tag{5.73}$$

5.2.8 Stress

The maximum stress for a cantilever beam occurs at the fixed end. The maximum moment, M_{ave}, is

$$M_{ave} = Pa + nPb \tag{5.74}$$

The stress at the top and bottom of the beam is

$$\sigma_{top} = \frac{-(Pa + nPb)c}{I} - \frac{nP}{A} \tag{5.75}$$

$$\sigma_{bot} = \frac{(Pa + nPb)c}{I} - \frac{nP}{A} \tag{5.76}$$

For a beam with cross-section width w and height h,

$$\sigma_{top} = \frac{-6(Pa + nPb)}{wh^2} - \frac{nP}{wh} \tag{5.77}$$

$$\sigma_{bot} = \frac{6(Pa + nPb)}{wh^2} - \frac{nP}{wh} \tag{5.78}$$

Example: Use of the pseudo-rigid-body model for flexible beams that undergo large deflections. A flexible steel beam with modulus $E = 30 \times 10^6$ lb/in.^{2}, length $l = 20$ in., width $w = 1.25$ in., and height $h = 1/(32$ in.$)$ is required to obtain a vertical deflection of $b = 10$ in. (a) Calculate the vertical force required to cause this deflection. (b) Find the horizontal end coordinate, a, the beam end angle, θ_o, and the maximum stress. (c) Solve the problem assuming that the force acts at an angle of $\phi = 135°$.

Solution: (a) Since the beam is loaded with a vertical force, $n = 0$, $\gamma = 0.85$, $K_\Theta = 2.68$, and $c_\theta = 1.24$, the length of the pseudo-rigid link is

$$\gamma l = (0.85)(20 \text{ in.}) = 17 \text{ in.} \quad (5.79)$$

The moment of inertia, I, is

$$I = \frac{wh^3}{12} = \frac{(1.25 \text{ in.})(1/32 \text{ in.})^3}{12} = 3.18 \times 10^{-6} \text{ in.}^4 \quad (5.80)$$

The torsional spring constant is found from equation (5.72) as

$$K = \frac{\gamma K_\Theta EI}{l}$$
$$= \frac{(0.85)(2.68)(30 \times 10^6 \text{lb/in.}^2)(3.18 \times 10^{-6} \text{in.}^4)}{20 \text{ in.}} = 10.9 \text{ in.-lb/rad} \quad (5.81)$$

The pseudo-rigid-body angle, Θ, can be found from equation (5.50) since the vertical deflection is known:

$$\Theta = \operatorname{asin}\frac{b}{\gamma l} = \operatorname{asin}\frac{10 \text{ in.}}{(0.85)(20 \text{ in.})} = 0.629 \text{ rad} = 36° \quad (5.82)$$

The force, P, is found by combining equations (5.56) and (5.69) and rearranging them to form

$$P = \frac{K\Theta}{\eta\gamma l \sin(\pi/2 - \Theta)} = \frac{(10.9 \text{ in.-lb})(0.629 \text{ rad})}{(1)(0.85)(20 \text{ in.})\sin(\pi/2 - 0.629 \text{ rad})} = 0.50 \text{ lb} \quad (5.83)$$

(b) The horizontal coordinate of the end, a, is found from equation (5.49) to be

$$a = l[1 - \gamma(1 - \cos\Theta)]$$
$$= (20 \text{ in.})\{1 - 0.85[1 - \cos(0.629)]\} = 16.7 \text{ in.} \quad (5.84)$$

The magnitude of the stress is the same on top and bottom because $n = 0$. This stress is determined from equation (5.25) to be

Cantilever Beam with a Force at the Free End (Fixed–Pinned)

$$\sigma = \frac{6Pa}{wh^2} = \frac{6(0.50 \text{ lb})(16.7 \text{ in.})}{(1.25 \text{ in.})(1/32 \text{ in.})^2} = 41{,}000 \text{ lb/in.}^2 \qquad (5.85)$$

The beam end angle is

$$\theta_o = c_\theta \Theta = (1.24)(0.629 \text{ rad}) = 0.78 \text{ rad} \qquad (5.86)$$

The accuracy of the results may be assessed by comparing the results to those obtained using other analysis methods. Elliptic-integral equations discussed in Chapter 2 provide results of $F = 0.49$ lb, $a = 16.7$ in., and $\theta_o = 0.79$ rad. Numerical methods, such as nonlinear finite element analysis, are another approach. A commercial nonlinear finite element analysis program capable of nonlinear analysis (ANSYS) with 20 beam elements obtains results of $F = 0.49$ lb, $a = 16.7$ in., and $\theta_o = 0.79$ rad, which are very close to those predicted above. If the predicted force of $F = 0.50$ lb is applied rather than a displacement load and another finite element analysis program is used (ABACUS), the following results are obtained: $b = 10.1$ in. and $a = 16.7$ in. Results for both finite element analysis and the pseudo-rigid-body model are similar, but we will see that the pseudo-rigid-body model has advantages in simplifying compliant mechanism design.

(c) The value of n is found as

$$n = \frac{-1}{\tan \phi} = 1 \qquad (5.87)$$

and

$$\eta = \sqrt{1 + n^2} = \sqrt{2} \qquad (5.88)$$

The characteristic radius factor, γ, is found from equation (5.48):

$$\gamma = 0.841655 - 0.0067807(1) + 0.000438(1)^2 = 0.84 \qquad (5.89)$$

The stiffness coefficient, K_Θ, is found from equation (5.62) to be

$$K_\Theta = 2.62 \qquad (5.90)$$

The torsional spring constant is

$$K = \frac{(0.84)(2.62)(30 \times 10^6 \text{ lb/in.}^2)(3.18 \times 10^{-6} \text{ in.}^4)}{20 \text{ in.}} = 10.5 \text{ in.-lb} \qquad (5.91)$$

The pseudo-rigid-body angle is

$$\Theta = \operatorname{asin}\frac{10 \text{ in.}}{0.84(20 \text{ in.})} = 0.638 \text{ rad} \qquad (5.92)$$

The vertical force, P, is

$$P = \frac{(10.5 \text{ in.-lb})(0.638 \text{ rad})}{\sqrt{2}\sin(3\pi/4 - 0.638)(0.84)(20 \text{ in.})} = 0.285 \text{ lb} \qquad (5.93)$$

and the total force is

$$F = \eta P = \sqrt{2}(0.285 \text{ lb}) = 0.403 \text{ lb} \qquad (5.94)$$

The horizontal coordinate of the end, a, is

$$a = (20 \text{ in.})\{1 - 0.85[1 - \cos(0.638)]\} = 16.7 \text{ in.} \qquad (5.95)$$

Table 5.1 lists of value of c_θ of 1.2467, resulting in an end angle of

$$\theta_o = c_\theta \Theta = 1.2467(0.638) = 0.79 \text{ rad} \qquad (5.96)$$

The stress is

$$\begin{aligned}\sigma_{top} &= \frac{-6[(0.285 \text{ lb})(16.7 \text{ in.}) + (1)(0.285 \text{ lb})(10 \text{ in.})]}{(1.25 \text{ in.})(1/32 \text{ in.})^2} \\ &\quad - \frac{(1)(0.285 \text{ lb})}{(1.25 \text{ in.})(1/32 \text{ in.})} = -37,400 \text{ psi}\end{aligned} \qquad (5.97)$$

The bending term of this stress equation is over 5000 times greater than the axial term. For this and most flexible beams, the bending stress is dominant while the axial stress can be ignored.

The use of constant values for γ and K_Θ in this example would have resulted in only slight changes in the calculated values.

5.2.9 Practical Implementation of Fixed–Pinned Segments

A force at the end of a flexible segment occurs at a pin joint where there are only forces and no moments. The fixed end of a cantilever may be fixed to any rigid segment, including ground or any other link. Examples are given in Figure 5.15. Note that only one of the four segments shown is fixed to ground and the others are fixed to moving rigid segments.

Now consider practical ways of constructing these types of flexible segments. A fixed–pinned beam is shown in Figure 5.16a as a cantilever beam with a force at the free end. The moment diagram for the beam is shown in Figure 5.16b. Note that the

Cantilever Beam with a Force at the Free End (Fixed–Pinned)

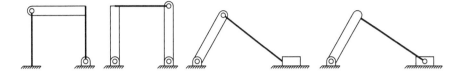

Figure 5.15. Examples of compliant mechanisms with fixed–pinned segments.

moment is zero at the free end and is small near the free end. This will be the case for fixed–pinned segments, such as those in the compliant mechanisms in Figure 5.15. The Bernoulli–Euler theory states that beam curvature is proportional to the applied moment. Therefore, at the pinned end of the fixed–pinned segment, the moment is zero, so the curvature is zero. The moment is small near the pinned end, so the curvature is small also.

Now, suppose that the beam were not a constant thickness but was rigid at the end. Because the moment is small at the end, the change in rigidity will not make a significant difference in the curvature of the beam and therefore will not make a significant difference in the overall beam deflection. This means that fixed–pinned segments of the type shown in Figure 5.17 can be modeled as fixed–pinned segments of constant cross section because the rigid part is located where it has little effect on the overall deflection of the beam. The rigid part of the segment in Figure 5.17a is from adding enough material to allow a hole for the pin of the pin joint, and the segment in Figure 5.17b has a hinge attached at the end such that the other end of the hinge can be attached to another segment.

Another practical implementation of fixed–pinned segments is to use a living hinge to connect a flexible segment to adjoining segments. Living hinges offer very little resistance to bending, and the resulting moment can usually be ignored because they are very small compared to other compliant segments. Examples are shown in Figure 5.18.

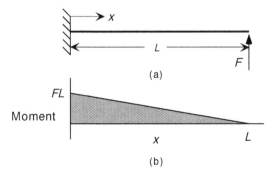

Figure 5.16. (a) Cantilever beam with a vertical force, and (b) its moment diagram.

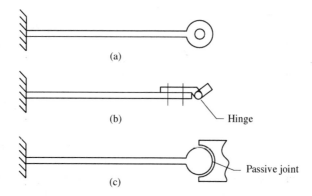

Figure 5.17. Examples of ways to have a pin at the end of a fixed–pinned segment. Because the moment is small at the pinned end of the segment, the addition of a rigid segment at the pin has little effect on the deflection.

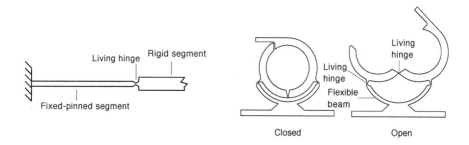

Figure 5.18. Examples of using a living hinge as the pin in a fixed–pinned segment. This is possible because the stiffness of the living hinge is much less than the stiffness of the other flexible segment.

5.3 FIXED–GUIDED FLEXIBLE SEGMENT

Consider the flexible segment with end loads that is shown in Figure 5.19a. One end of the beam is fixed while the other is "guided" in that the angle of that end of the beam does not change. Because the end of this member is to be maintained at a constant angle, a resultant moment must be present at the end (M_o in Figure 5.19a). The resulting deflected shape is antisymmetric at its centerline, where the angular deflection of the beam, θ, is a maximum and the curvature, $d\theta/ds$, is zero. The Bernoulli–Euler assumption states that the moment is directly proportional to the curvature, which implies that the moment is also zero at midlength. If there is no moment at the midpoint, the free-body diagram for one-half of the flexible member is as shown in Figure 5.19b. The half-beam only has a force at the end, so it is

Fixed–Guided Flexible Segment

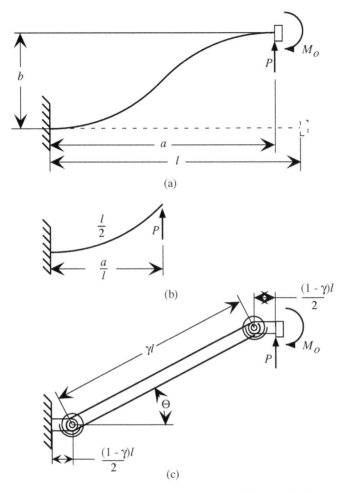

Figure 5.19. (a) Flexible beam with a constant end angle, (b) free-body diagram of one-half of the beam, and (c) pseudo-rigid-body model.

similar to the flexible segment described in Section 5.2. The pseudo-rigid-body model of the half beam is then the same as discussed previously; only half the beam length is used. A pseudo-rigid-body model of the entire segment may be derived by combining the two antisymmetric one-half beams, as shown in Figure 5.19c. (If the end angle were not maintained at a constant value, the inflection point would not occur at the center and two beams of unequal length could be considered.) The value of the characteristic radius factor, γ, was found earlier to be

$$\gamma = 0.8517 \tag{5.98}$$

with a parameterization limit of

$$\Theta_{max} = 64.3° \tag{5.99}$$

For the flexible beam with a constant end angle (Figure 5.19a) and the corresponding pseudo-rigid-body model in Figure 5.19c, the parametric angle coefficient, c_θ, as defined by equation (5.53), is trivial:

$$c_\theta = 0 \tag{5.100}$$

Another equation is needed to predict the reaction moment, M_o (Figure 5.19a), that is required to maintain a constant end angle. Summing moments at either end of the free-body diagram in Figure 5.19b yields

$$Pa - M_o = \frac{Pa}{2} \tag{5.101}$$

or

$$M_o = \frac{Pa}{2} \tag{5.102}$$

Substituting for a from equation (5.49) results in

$$M_o = \frac{Pl}{2}[1 - \gamma(1 - \cos\Theta)] \tag{5.103}$$

Equation (5.59) becomes

$$(\alpha^2)_t = 2K_\Theta\Theta \tag{5.104}$$

for each of the two springs. The torsional spring constant, K, for the springs is

$$K = 2\gamma K_\Theta \frac{EI}{l} \tag{5.105}$$

Since each spring is twice as stiff as for a fixed–free beam, and there are two springs, this segment is four times as stiff as a segment of the same length with fixed–free end conditions. This is consistent with small-deflection beam theory.

The maximum stress occurs at the beam ends where the maximum moment occurs and has a value of

$$\sigma_{max} = \frac{Pac}{2I} \quad \text{at both ends of the beam} \tag{5.106}$$

End-Moment Loading

where c is the distance from the neutral axis to the outer surface of the beam (i.e., half the beam height for rectangular beams, the radius of circular cross-section beams, etc.)

5.4 END-MOMENT LOADING

Consider the flexible beam with an applied moment at the free end illustrated in Figure 5.20. A method similar to that used for the cantilever beam with a force at the free end may be used to find a pseudo-rigid-body model for this type of loading. This results in a characteristic radius factor of

$$\gamma = 0.7346 \tag{5.107}$$

which represents an error of less than 0.5% in the tip deflection up to an end angular deflection of

$$\theta_{o_{max}} = 124.4° \tag{5.108}$$

The parametric angle coefficient is

$$c_\theta = 1.5164 \tag{5.109}$$

The stiffness coefficient is

$$K_\Theta = 2.0643 \tag{5.110}$$

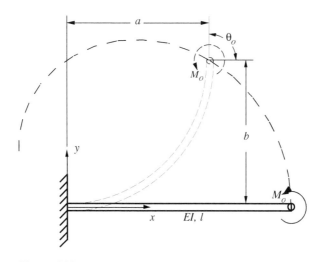

Figure 5.20. Flexible beam with a moment at the free end.

and the spring constant is

$$K = \gamma K_\Theta \frac{EI}{l} \tag{5.111}$$

or, for this special case of an end moment,

$$K = c_\theta \frac{EI}{l} \quad \text{(end moment only)} \tag{5.112}$$

The maximum stress is

$$\sigma_{max} = \frac{M_o c}{I} \tag{5.113}$$

where c is the distance from the neutral axis to the outer surface of the beam (i.e., half the beam height for rectangular beams, the radius of circular cross-section beams, etc.)

5.5 INITIALLY CURVED CANTILEVER BEAM*

The beam shown in Figure 5.21 has an initial radius of curvature of R_i, and therefore a curvature of $1/R_i$. The initial curvature can be related to the length using the nondimensionalized parameter κ_o as

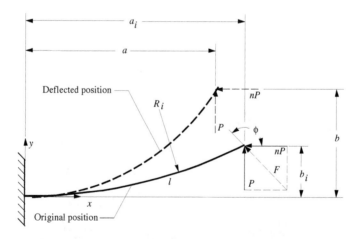

Figure 5.21. Flexible, initially curved beam.

---

*See [52].

Initially Curved Cantilever Beam

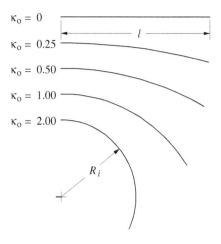

Figure 5.22. Beam shapes for various values of κ_o.

$$\kappa_o = \frac{l}{R_i} \tag{5.114}$$

Figure 5.22 illustrates beam shapes for various values of κ_o. The beams are assumed to be thin; thus the neutral axis is assumed to be at the centroidal axis, even though the beam is curved.

The pseudo-rigid-body model of an initially curved, end-force-loaded cantilever beam is shown in Figure 5.23. The characteristic radius factor, γ, is associated with the beam as if it were initially straight. The characteristic radius factor, γl, is measured along the beam as if it were initially straight (Figure 5.23). To account for the curvature, the length of the rigid-body link is ρl, where ρ is a function of γ and the curvature. Since the segment is initially curved, the pseudo-rigid-body angle, Θ, has a nonzero initial value, Θ_i, such that

$$\Theta_i = \operatorname{atan} \frac{b_i}{a_i - l(1-\gamma)} \tag{5.115}$$

where a_i and b_i are as shown in Figure 5.23. For an initially straight segment, $\rho = \gamma$, $a_i = l$, $b_i = 0$, $\kappa_o = 0$, and $\Theta_i = 0$.

To ensure an accurate representation of the initial, undeflected position of an initially curved segment, the length of the pseudo-rigid-body link (characteristic radius), ρl, is found using

$$\rho = \left\{ \left[\frac{a_i}{l} - (1-\gamma) \right]^2 + \left(\frac{b_i}{l} \right)^2 \right\}^{1/2} \tag{5.116}$$

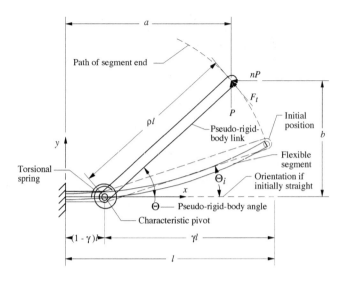

Figure 5.23. Pseudo-rigid-body model of an initially curved, end-force-loaded cantilever beam.

where a_i and b_i are the initial x and y coordinates of the segment end, respectively, and are calculated as

$$\frac{a_i}{l} = \frac{1}{\kappa_o} \sin \kappa_o \qquad (5.117)$$

and

$$\frac{b_i}{l} = \frac{1}{\kappa_o} (1 - \cos \kappa_o) \qquad (5.118)$$

The deflection path of the beam end as approximated by the pseudo-rigid-body model is

$$\frac{a}{l} = 1 - \gamma + \rho \cos \Theta \qquad (5.119)$$

$$\frac{b}{l} = \rho \sin \Theta \qquad (5.120)$$

5.5.1 Stiffness Coefficient for Initially Curved Beams

The component of the applied end force that is tangential to the pseudo-rigid-body link path, F_t, is

$$F_t = F \sin(\phi - \Theta) \tag{5.121}$$

where

$$\phi = \operatorname{atan} \frac{1}{-n} \tag{5.122}$$

The nondimensionalized tangential load factor, α_t^2, is defined as

$$\alpha_t^2 = \frac{F_t l^2}{EI} \tag{5.123}$$

and

$$\alpha_t^2 = K_\Theta (\Theta - \Theta_i) \tag{5.124}$$

where Θ is in radians. The torque at the characteristic pivot is

$$T = \rho l F_t \tag{5.125}$$

This torque may also be expressed as the torsional spring constant times the angular deflection, or

$$T = K(\Theta - \Theta_i) \tag{5.126}$$

Combining equations (5.19) through (5.126) and solving for the torsional spring constant, K, yields

$$K = \rho K_\Theta \frac{EI}{l} \tag{5.127}$$

It is convenient to use a constant value of γ for various values of n. Recall from Sections 5.2.4 and 5.2.6 that values of $\gamma = 0.85$ and $K_\Theta = 2.65$ were recommended for an initially straight beam ($\kappa_o = 0$). Recommended values for γ, ρ, and K_Θ for various values of κ_o are listed in Table 5.2.

TABLE 5.2. Values for γ, ρ, and K_Θ for various values of κ_o

κ_o	γ	ρ	K_Θ
0.00	0.85	0.850	2.65
0.10	0.84	0.840	2.64
0.25	0.83	0.829	2.56
0.50	0.81	0.807	2.52
1.00	0.81	0.797	2.60
1.50	0.80	0.775	2.80
2.00	0.79	0.749	2.99

5.5.2 Stress for Initially Curved Beams

The maximum stress occurs at the fixed end and has a value of

$$\sigma_{max} = \pm \frac{P(a+nb)c}{I} - \frac{nP}{A} \quad \text{at the fixed end} \quad (5.128)$$

where c is the distance from the neutral axis to the outer surface of the beam (i.e., half the beam height for rectangular beams, the radius of circular cross-section beams, etc.)

5.6 PINNED–PINNED SEGMENT*

Consider a pinned–pinned flexible segment, as shown in Figure 5.24a. With the flexibility of the link, the segment may have different motion characteristics, depending on the location and magnitude of the applied loads. This section considers only the segments that are loaded at the rigid-body joints.

A free-body diagram of the pinned–pinned segment is shown in Figure 5.24a. A simple static analysis that sums the forces in the horizontal and vertical directions and moments about point A yields

$$\sum P_x = P_{x_A} + P_{x_B} = 0 \quad (5.129)$$

$$P_{x_A} = -P_{x_B} \quad (5.130)$$

---

*See [35] and [36].

Pinned–Pinned Segment

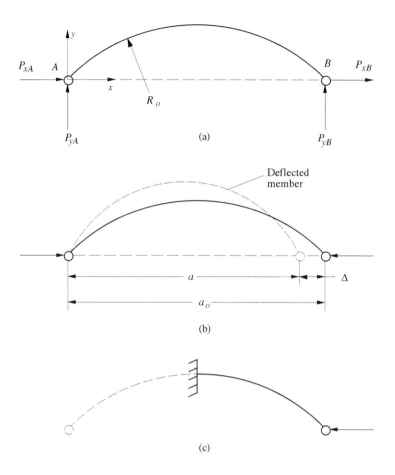

Figure 5.24. (a) Functionally binary, pinned–pinned segment, (b) deflected member, and (c) equivalent half-model.

$$\sum M_A = P_{y_B} x_B = 0 \tag{5.131}$$

$$P_{y_B} = 0 \tag{5.132}$$

$$\sum P_y = P_{y_A} + P_{y_B} = 0 \tag{5.133}$$

$$P_{y_A} = 0 \tag{5.134}$$

These equations indicate that the only forces acting on the simple pinned–pinned segment are the horizontal forces, or the forces acting collinear along the line

172   Pseudo-Rigid-Body Model

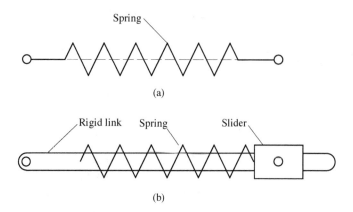

Figure 5.25. (a) Pinned–pinned segment's simple pseudo-rigid-body model, and (b) rigid link and prismatic joint model.

between the two rigid-body joints. The deflections of the segment's endpoints are also along this line. This is true for any flexible beam loaded at pin joints, regardless of its shape.

These results may be used to determine an appropriate pseudo-rigid-body model for the pinned–pinned segment. Since the segment deflection occurs along the line between the two rigid-body joints, an accurate kinematic model may be as simple as the linear spring shown in Figure 5.25a or a prismatic pair (slider) and spring as shown in Figure 5.25b. The spring's stiffness would take into account the stiffness of the material. The stiffness value depends on the geometry and material properties of the segment.

5.6.1 Initially Curved Pinned–Pinned Segments

The discussion above applies to any flexible segment loaded at the pin joints. Since initially curved segments are quite commonly used as pinned–pinned segments, it is worthwhile to present a pseudo-rigid-body model specifically for this type of segment.

Due to the symmetry of the initially curved pinned–pinned beam in Figure 5.24a, a half-model, as shown in Figure 5.24c, is equally applicable to either side of the segment. This beam is similar to the initially curved beam of Section 5.5, only here we know that the force is always horizontal ($n = \infty$). The pseudo-rigid-body model for the half beam is the same as that for the pseudo-rigid-body model for the initially curved beam. The entire segment may be represented in terms of two identical models connected together, as shown in Figure 5.26. The two sides are coupled in that the two angles must always be equal, and the two torsional springs must also be of equal value. The pseudo-rigid-body link lengths are also identical for the two sides.

Pinned–Pinned Segment

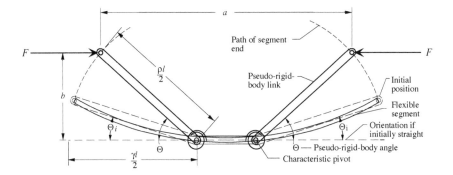

Figure 5.26. Pseudo-rigid-body model for an initially curved, pinned–pinned beam.

The analysis used is similar to that for the initially curved beams discussed in Section 5.5, but a factor of 2 is often required to account for the half model. The length of the pseudo-rigid link is $\rho l/2$, and the characteristic pivot is measured a distance of $\gamma l/2$ (see Figure 5.26). The initial position of the beam end is given by

$$\frac{a_i}{l} = \frac{2}{\kappa_o} \sin \frac{\kappa_o}{2} \tag{5.135}$$

$$\frac{b_i}{l} = \frac{1}{\kappa_o}\left(1 - \cos\frac{\kappa_o}{2}\right) \tag{5.136}$$

The initial angle, Θ_i, and ρ are

$$\Theta_i = \operatorname{atan}\frac{2b_i}{a_i - l(1-\gamma)} \quad \text{and} \quad \rho = \left\{\left[\frac{a_i}{l} - (1-\gamma)\right]^2 + \left(\frac{2b_i}{l}\right)^2\right\}^{1/2} \tag{5.137}$$

The value of γ for various values of κ_o may be found from the equations

$$\gamma = 0.8005 - 0.0173\kappa_0 \quad\quad 0.500 \leq \kappa_0 \leq 0.595 \tag{5.138}$$

$$\gamma = 0.8063 - 0.0265\kappa_0 \quad\quad 0.595 \leq \kappa_0 \leq 1.500 \tag{5.139}$$

A few values are also listed in Table 5.3.

The x and y coordinates of the end are

$$\frac{a}{l} = 1 - \gamma + \rho \cos \Theta \tag{5.140}$$

$$\frac{b}{l} = \frac{\rho}{2} \sin \Theta \tag{5.141}$$

The stiffness coefficient, K_Θ, may be approximated from the equation

$$K_\Theta = 2.568 - 0.028\kappa_0 + 0.137\kappa_0^2 \quad (0.5 \leq \kappa_0 \leq 1.5) \quad (5.142)$$

and values of K_Θ for various values of κ_o are listed in Table 5.3. The recommended maximum value of $\Delta\Theta$ is also listed in the table. Each spring constant has a value of

$$K = 2\rho K_\Theta \frac{EI}{l} \quad (5.143)$$

The torque at the characteristic pivot is

$$T = Fb \quad (5.144)$$

and it is also

$$T = K(\Theta - \Theta_i) \quad (5.145)$$

Equating equations (5.144) and (5.145) and solving for the compression force, F, results in

$$F = \frac{K(\Theta - \Theta_i)}{b} \quad (5.146)$$

The maximum stress occurs midlength of the segment and has a magnitude of

$$\sigma_{max} = \pm\frac{Fbc}{I} - \frac{F}{A} \quad \text{at midlength of segment} \quad (5.147)$$

where c is the distance from the neutral axis to the outer surface of the beam (i.e., half the beam height for rectangular beams, the radius of circular cross-section beams, etc.)

TABLE 5.3. Pseudo-rigid-body link characteristics for initially curved pinned–pinned segment

κ_o	γ	ρ	$(\Delta\Theta)_{max,\gamma}$	K_Θ	$(\Delta\Theta)_{max,K_\Theta}$
0.50	0.793	0.791	1.677	2.59	0.99
0.75	0.787	0.783	1.456	2.62	0.86
1.00	0.783	0.775	1.327	2.68	0.79
1.25	0.779	0.768	1.203	2.75	0.71
1.50	0.775	0.760	1.070	2.83	0.63

5.7 SEGMENT WITH FORCE AND MOMENT (FIXED–FIXED)

Another common flexible segment is one that has a constant cross section, external loads are applied only at the ends of the segment, and both ends are fixed to rigid segments. Because the ends are fixed to other segments, both moment and force loads can be applied at the ends. This is different than for pinned segments, where moments are not applied at the pinned end.

There are three cases of loading for segments with combined end force and moment. These are described so that the physical conditions can be better understood. Pseudo-rigid-body models for these conditions are described, followed by the description of a simplified pseudo-rigid-body model. Because this loading condition is more complex than those discussed previously, the resulting pseudo-rigid-body model is less accurate than the models for other segments, but it is still very useful in the initial design of compliant mechanisms.

5.7.1 Loading Cases

There are three loading cases for these segments: case I—force and moment in the same direction; case II—force and moment loading in opposite directions but no inflection point; and case II—force and moment loading in opposite direction with an inflection point. Each of these is described in more detail below.

Case I: Force and Moment in the Same Direction. Case I may be represented by a cantilever beam with a force and moment applied at the end such that they cause deflection in the same direction, as shown in Figure 5.27. The applied moment creates a curvature in the beam that is continuous throughout the entire beam. The

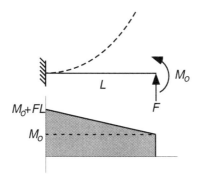

Figure 5.27. Cantilever beam with the case I loading conditions and the corresponding bending moment diagram.

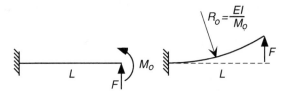

Figure 5.28. Applied moment and its equivalent beam with an initial curvature.

continuous curvature is the same as a beam that is initially curved, where the curvature, κ_o, is defined as

$$\kappa_o = \frac{M_o l}{EI} \tag{5.148}$$

and the equivalent radius of curvature, R_o, is

$$R_o = \frac{EI}{M_o} \tag{5.149}$$

This means that the segment has the same behavior as an initially curved beam with radius R_o and a force applied at its end, as shown in Figure 5.28.

Thus a segment with this type of loading condition can be modeled using the pseudo-rigid-body model for initially curved segments. This means that the model and equations developed in Section 5.5 can be used to model this type of segment, but the curvature due to the applied moment is substituted for the initial curvature. The effect of the force is then applied to the curved beam to determine the final deflection.

Case II: Force and Moment Loading in Opposite Directions, No Inflection Point.
Case II is represented by a cantilever beam with the applied end force and moment acting in opposite directions, as shown in Figure 5.29. However, the magnitude of the applied loads is such that no inflection point is produced in the deflected beam.

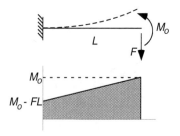

Figure 5.29. Cantilever beam with case II loading conditions and the corresponding bending moment diagram.

Segment with Force and Moment (Fixed–Fixed)

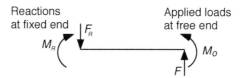

Figure 5.30. Cantilever beam with applied end force and moment in the same direction.

As the reaction forces are examined (Figure 5.30), it becomes apparent that the force and moment at the reactions act in opposite direction of the applied loads at the end. This suggests that the model for a force and moment in opposite directions is given by the same model as case I, with one minor exception: The ground is switched to the opposite side of the beam (Figure 5.31). In mechanism analysis and design, switching the ground link is called an *inversion*. Although not exactly the same, this term will be used to describe the case where a different beam end is fixed. This inversion can make the system easier to model but does not change the relative motion between the ends. From the viewpoint of the inversion, the segment is the same as a case I loading and the same equations from case I may also be applied to this case.

Case III: Force and Moment Loading in Opposite Directions with Inflection Point. Case III loading occurs when a cantilever beam is loaded with an applied end force and end moment acting in opposite directions such that they result in an inflection point in the deflected beam, as shown in Figure 5.32. The fixed–guided segment described in Section 5.3 is a special case of case III loading where the inflection point is at the midpoint of the beam.

The Bernoulli–Euler equation states that the bending moment is proportional to the curvature. At an inflection point there is zero curvature and thus, zero moment. At the inflection point there is only a force acting on the beam. Because of this, the beam can be modeled as two cantilever beams connected at the inflection point, as shown in Figure 5.33.

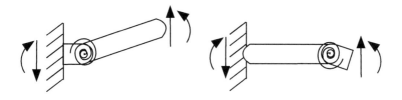

Figure 5.31. "Inversion" of the case I loading condition to represent the case II loading condition.

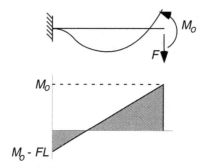

Figure 5.32. Cantilever beam with the case III loading conditions and the corresponding bending moment diagram.

Pseudo-Rigid-Body Models for Fixed–Fixed Segments. Pseudo-rigid-body models have been developed for the various cases of fixed–fixed segments. Some of these models are quite accurate but they are often too complicated to be practical in compliant mechanism design. These models are described briefly here for completeness, while the next section will present a less accurate but more practical method for modeling fixed–fixed segments.

For case I loading the moment causes a constant curvature throughout the length of the beam. This is similar to a constant curvature from an initially curved beam; therefore the pseudo-rigid-body model for an initially curved beam may be used for a fixed–fixed beam with case I loading. When the force and moment loads are known, this is a simple and powerful method. However, in most compliant mechanism analyses, the displacement is known and the loads are unknown. This makes it difficult to know what curvature to use in the pseudo-rigid-body model. As

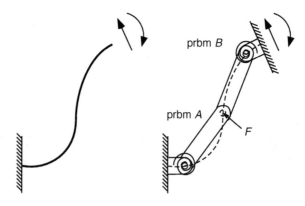

Figure 5.33. Beam with case III loading conditions and viewing it as two beams joined at the inflection point.

Segment with Force and Moment (Fixed–Fixed)

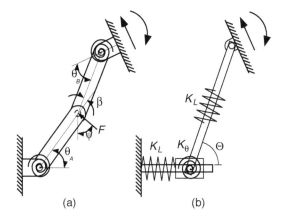

Figure 5.34. Pseudo-rigid-body models for case III loading. Part (a) is described in more detail in [127] and part (b) in [128].

discussed earlier, case II loading can also be modeled using the pseudo-rigid-body model for initially curved beams, but an inversion is performed to begin the analysis. Again, this is very useful when the loads are known, but iteration is required when the displacement is known and the loads are unknown.

Case III flexible segments can be modeled by combining two fixed–pinned pseudo-rigid-body models that are connected at the point of inflection. Unlike fixed–pinned segments, both ends are rigidly attached to other segments and the end angle is important in determining the overall displacement. As the displacement of a mechanism changes, so does the inflection point. Therefore, the two pseudo-rigid-body models must be changed continuously to reflect the actual flexible segment. Because the inflection point is moving and its location is usually unknown at the beginning of the analysis, an iterative method is required to solve for the unknowns. Figure 5.34a illustrates this pseudo-rigid-body model. A pseudo-rigid-body model for fixed–fixed segments was also developed by Saxena and Kramer [128], as shown in Figure 5.34b.

Simplified Pseudo-Rigid-Body Model for Fixed–Fixed Segments. Recall that a fixed-guided segment is a special fixed–fixed segment that maintains a constant angle of the beam ends. This segment is accurately modeled using a pseudo-rigid-body model that has two pin joints located such that they are each located the same distance from their respective end, and the pseudo-rigid link has a length of $r = \gamma l$. Each characteristic pivot has a torsional spring with a torsional spring constant of $K = 2\gamma K_\Theta EI/l$. Although not as accurate as for fixed–guided segments, this model can also be used to model fixed–fixed segments. The only difference in the model is that the ends are not constrained to stay at the same angle and are allowed to move as needed. This model is easy to implement in compliant mechanism

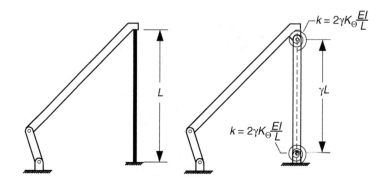

Figure 5.35. Simplified pseudo-rigid-body model for fixed–fixed segments shown as part of a compliant mechanism.

design and is particularly useful in the initial design phase. Once a compliant mechanism design has been developed, it can be further refined and tested.

Figure 5.35 shows the pseudo-rigid-body model of a fixed–fixed segment in relation to a compliant mechanism.

5.8 OTHER METHODS OF PIN JOINT SIMULATION

In rigid-body mechanisms, relative rotation between parts is achieved using pin joints. Although the motion of mechanisms containing pin joints can be assumed to be planar for analysis purposes, the fabrication of the joints requires varying geometry out of the plane. A simple example is shown in Figure 5.36. One advantage of many compliant mechanisms is that the fabrication is simplified by having planar geometry, or in other words, they have geometry that is constant for any cross section in the third dimension.

There are a number of ways to simulate the motion of a pin joint in compliant mechanisms, many of which have been discussed. Small-length flexural pivots can be used, with *living hinges* being a special case that very closely approximates pin joint motion. The pseudo-rigid-body model for fixed–pinned and fixed–guided segments show other ways of using longer flexible segments to achieve such motion. However, these have limitations in some applications, where other methods are useful. For example, if large compressive loads are to be carried through the joint, flexible segments often buckle or yield under the loads. *Passive joints* are one possible method of having pin joint motion with planar geometry and are able to withstand large compressive loads. Another challenge is a pin joint that connects two links at the centers rather than at the ends. Scissors and pliers are simple examples that illustrate where such joints occur. A *Q-joint* can be used for many such cases. Finally, *cross-axis flexural pivots* may also be used to simulate pin joint motion and

Other Methods of Pin Joint Simulation

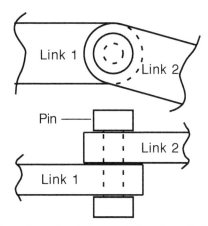

Figure 5.36. Three-dimensional characteristics of a typical pin joint.

are especially common in high-precision instrumentation. Like Q-joints, they also may be used to connect links at their centers as well as at their ends.

5.8.1 Living Hinges

A pin joint allows relative rotation about one axis but does not allow relative rotation in any other axis or relative translation in any direction between connected links. A door hinge is a common example of a pin joint. Small-length flexural pivots have behavior similar to that of pin joints, but they use the deflection of flexible members to obtain motion rather than pure rotation of parts about a pin. The "hinge" of a hardback book cover is an example of a small-length flexural pivot. The rigidity of the flexible portion is much smaller than the more rigid part due to a change in both material and geometry.

Many types of small-length flexural pivots exist, a *living hinge* being a special small-length flexural pivot. They are very small in length, offer little resistance to deflection, and approximate very closely the behavior of a pin joint. Often called *integral hinges*, living hinges are commonly found in consumer products. They offer so little resistance to bending that they are often modeled with the pseudo-rigid-body model as a pin joint without a torsional spring.

Polypropylene is the most commonly used material for living hinges. Other materials may be used but will usually result in a shorter life. In some applications, life is not a major concern since the hinge may only be expected to flex once. For example, many containers are constructed of a single piece of material and then folded at living hinges to make the container. In such cases, the designer has many acceptable options in material and geometry choices. In most compliant mechanism designs, however, living hinges are expected to endure many cycles without failure. The discussion that follows assumes that a long life is required. The recommenda-

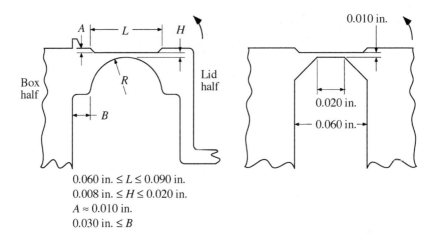

Figure 5.37. Two common types of living hinge geometries used in container lids.

tions are summarized from the experience of several plastics suppliers and other sources, including [25] and [129]. Living hinges made using these methods have been tested to undergo millions of cycles without failure.

Two common geometries for living hinges used for container lids are shown in Figure 5.37. Recommended dimensions are also shown. Hinges that require less than 180 degrees of flex may be designed as shown in Figure 5.38. The ratio $R1/R2$ will influence the degree of travel in flex before binding [129].

Hinges may be made by injection molding, extrusion, hot-stamping, and blow molding. When injection molded, the molten plastic should be caused to flow perpendicular to the hinge. This causes a good fill and also helps align the material in a favorable direction. Extruded hinges will have a much shorter life because the material flow is parallel to the hinge axis.

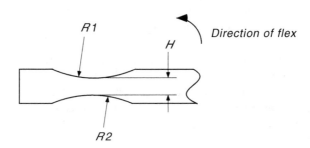

Figure 5.38. Hinge that requires less than 180° of flex.

Other Methods of Pin Joint Simulation

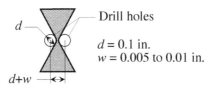

Figure 5.39. Example of a machined living-hinge prototype.

The hinge should be flexed immediately after molding while the heat from the mold is still present. It should be flexed once slowly, then rapidly several times. Flexing will stretch the hinge area considerably (a 0.010-in. thickness may thin down to less than 0.005 in.). The elongation orients the material and increases the tensile strength dramatically. A thin white line will appear on the hinge after flexing. This is normal and does not mean that the hinge has been weakened.

Some molding considerations are as follows [129]: cylinder temperature—450 to 550° F; injection speed—fast; mold temperatures—120 to 150° F; gate opening—if possible, make up to 50% larger than for non hinged parts. If using a single gate, locate it to ensure smooth flow to the hinge area, make the flow perpendicular to the hinge axis, place the gate slightly to the rear of the centerlines of the largest cavity, and center it if the flow to the hinge is greater than 8 in. For multiple gates, ensure that gates on the same side of the hinge are no farther apart than twice the distance from gate to hinge; if the flow on the opposite side of the hinge is greater than 8 in., the part should be gated in both sides; locate so that a weld line does not form at the hinge. The hinge should be an insert machined from hardened steel to resist the stresses of the flowing resin.

It is often not economically feasible to mold living hinges when producing compliant mechanism prototypes, but prototype living hinges can be machined and tested [61]. Although the prototypes will not have as ideal of characteristics as their injection-molded counterparts, it is possible for properly designed and fabricated prototypes to undergo over a million cycles without failure. One successful prototyping method is to drill two holes with a center distance equal to the sum of the diameter and the desired thickness of the living hinge, as shown in Figure 5.39. A good thickness is approximately $w = 0.005$ to 0.010 in. with a diameter of $d = 0.10$ in. The circular holes are blended into their adjacent features when those features are machined. The part should be temporarily adhered to a sacrificial backing (such as plywood) during machining to reduce vibration and part movement.

5.8.2 Passive Joints

A *passive joint* allows relative rotation between rigid segments without using a traditional pin joint. A simple example of a passive joint is shown in Figure 5.40. The two mating parts have circular shapes, as shown, and the center of rotation is about the center of the circles. The *passive cam* rotates in the *socket*. The passive cam gets

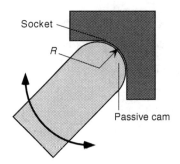

Figure 5.40. Example of a passive joint.

its name from the fact that it is a circular cam with no rise. The joint must be in compression to function because tension loads will pull a passive cam away from the socket. Another limitation is that the angle of rotation is limited and full relative rotation is not allowed.

Consider the crimping mechanism illustrated in Figure 5.41 as an example of a passive joint in a compliant mechanism. Its pseudo-rigid-body model is a four-bar mechanism. A force analysis reveals that the passive joint is always in compression. Another candidate for a passive joint is the compliant segment labeled "A," which is also always in compression. But the rigid segment should not have two passive joints or the rigid segment connecting them would not be connected to the rest of the mechanism. Since only one of the two locations can be a passive joint, the upper one is chosen because it has the largest relative rotation. This choice causes a lower maximum stress in the mechanism for a given displacement.

A second example of a passive joint is the compliant overrunning ratchet and pawl clutch shown in Figure 5.42a. The ratchet overruns (rotates freely) in one direction but transfers the load to the outer hub when it rotates the other direction. To reduce part count, it is desirable to get the spring action to keep the pawls in

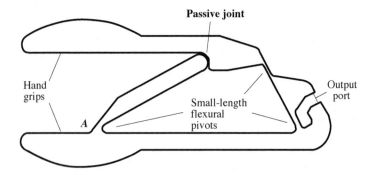

Figure 5.41. Passive joint in a compliant crimping mechanism.

Other Methods of Pin Joint Simulation

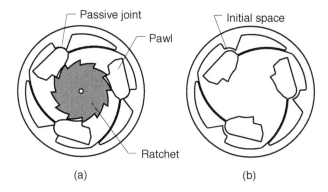

Figure 5.42. (a) Passive joint in a compliant overrunning ratchet and pawl clutch, and (b) one position in which it may be manufactured.

place by using an integrated flexible segment. However, the pawls must carry a high compressive load. The design shown combines a compliant member for spring action and a passive joint for the high compressive load. The pawl may be constructed with a gap between the passive cam and socket to allow for easier manufacture, as shown in Figure 5.42b.

5.8.3 Q-Joints

Sometimes two rigid segments are joined somewhere besides the ends, such as with the scissors and pantograph illustrated in Figure 5.43. It is possible to approximate these types of joints using Q-joints. Q-joints get their name from their quadrilateral shape, in that the pseudo-rigid-body model of the joint forms a quadrilateral. They are fabricated by connecting four rigid segments with four small-length flexural pivots (usually living hinges) such that its pseudo-rigid-body model is a four-bar mechanism. Two basic types of Q-joints are parallelogram Q-joints and deltoid Q-joints.

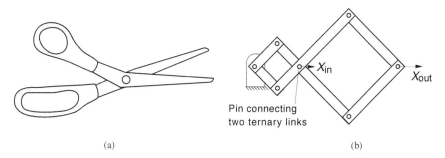

Figure 5.43. Examples of rigid segments joined somewhere besides the ends, including (a) scissors, and (b) a pantograph.

Pseudo-Rigid-Body Model

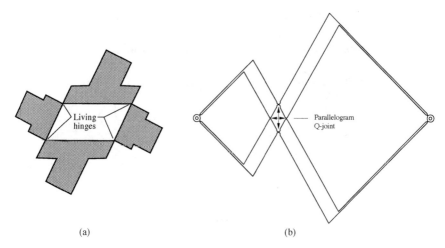

Figure 5.44. (a) Parallelogram Q-joint, and (b) example of its use with a compliant pantograph mechanism.

Parallelogram Joint. When the quadrilateral shape of the Q-joint is a parallelogram (the lengths of the opposite rigid segments are equal), it is a parallelogram-type Q-joint. An example parallelogram Q-joint is shown in Figure 5.44a, and the use of one in a compliant pantograph is illustrated in Figure 5.44b.

A pseudo-rigid-body model of the parallelogram joint in Figure 5.44a is shown in Figure 5.45. To approximate a pin joint in two rigid links accurately, rigid segments 1 and 3 should behave like a single rigid link, and rigid segments 2 and 4 would be the other rigid link.

Note that there are similarities and differences between the motion of this mechanism and the rigid-link pin-joint counterpart. If a rigid link rotates, all parts on the

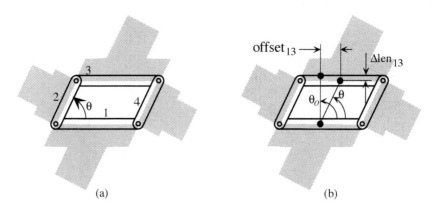

Figure 5.45. Pseudo-rigid-body model of a parallelogram Q-joint.

Other Methods of Pin Joint Simulation

link have the same rotation. Because the pseudo-rigid-body model is a parallelogram linkage, segments 1 and 3 have the same rotation, and segments 2 and 4 have the same rotation. So in this way the parallelogram joint and the rigid-body pin-joint counterpart have similar motion. However, if a rigid link translates, the entire rigid link translates the same amount, but segments 1 and 3 translate relative to each other, as do segments 2 and 4. This translation is illustrated in Figure 5.45. The centerlines of segments 1 and 3 are offset by the amount

$$\text{offset}_{13} = r_2(\cos\theta_o - \cos\theta) \qquad (5.150)$$

where θ_o is the initial angle of link 2 and θ is the final position of link 2. The centers of segments 2 and 4 are offset by

$$\text{offset}_{24} = r_3(\cos\theta_o - \cos\theta) \qquad (5.151)$$

The effective length of the equivalent link represented by segments 1 and 3 changes by an amount

$$\Delta\text{len}_{13} = r_2(\sin\theta_o - \sin\theta) \qquad (5.152)$$

and for the link associated with segments 2 and 4

$$\Delta\text{len}_{24} = r_3(\sin\theta_o - \sin\theta) \qquad (5.153)$$

Limitations of the parallelogram Q-joint include limited rotation and load-bearing capacity.

Deltoid Joint. A deltoid-type Q-joint is constructed when each rigid segment in the quadrilateral is made adjacent to a segment of equal length. An example of a deltoid Q-joint is shown in Figure 5.46a and its use with compliant forceps in Figure 5.46b.

Figure 5.46. (a) Deltoid Q-joint, and (b) example of its use with compliant forceps.

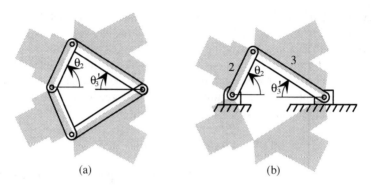

Figure 5.47. Pseudo-rigid-body model of a deltoid Q-joint: (a) four-bar mechanism, and (b) slider–crank mechanism.

The pseudo-rigid-body model of a deltoid joint is shown in Figure 5.47a. The symmetry of the joint also allows it to be modeled as two slider–crank mechanisms, as shown in Figure 5.47b.

Figure 5.48 shows Compliers as an example of a deltoid Q-joint. Note that one vertex of the quadrilateral is a passive joint.

As before, to approximate a pin joint in two rigid links accurately, rigid segments 1 and 3 should behave like a single rigid link, and rigid segments 2 and 4 as the other rigid link. The change in angle between two opposite rigid segments may be determined from the equations for a rigid-link slider–crank mechanism, or from the law of sines as

$$r_2 \sin \theta_2 = r_3 \sin \theta_3' \qquad (5.154)$$

The translational offsets may be calculated from either the four-bar or slider–crank equations presented in Chapter 4.

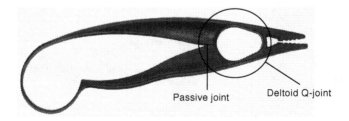

Figure 5.48. Compliars as an example of a deltoid Q-joint, including a passive joint.

Other Methods of Pin Joint Simulation

One advantage of the deltoid Q-joint is that the segment lengths may be modified to provide a desirable mechanical advantage. In this way a motion similar to a pin joint is obtained, but a more desirable mechanical advantage is achieved.

Other Q-Joints. Other Q-joints are possible that have a quadrilateral shape that is not a special case of parallelogram or deltoid joints. These may be used to obtain particular types of motion or force characteristics. The four-bar equations in Chapter 4 may be used to analyze the motion of these types of joints.

5.8.4 Cross-Axis Flexural Pivots

Ideal pin joints allow rotation but not translation. Because small-length flexural pivots are flexible in bending, they allow rotation, but they are stiff in tension and do not allow much axial translation. These characteristics make them very attractive for simulating pin joint motion in compliant mechanisms. However, the stresses in a small-length flexural pivot are often quite high for larger deflections. The stress in a small-length flexural pivot is inversely proportional to length. Therefore, a longer pivot has lower stresses and, consequently, a larger maximum deflection before failure. However, if the pivot is too long, its length is no longer significantly less than the length of the accompanying rigid segments, reducing the accuracy of the pin joint approximation. Therefore, it is desirable to increase the length of the flexible segments without increasing overall pivot length. The *cross-axis flexural pivot*, as shown in Figure 5.49a, helps to decrease stress without increasing pivot length. By

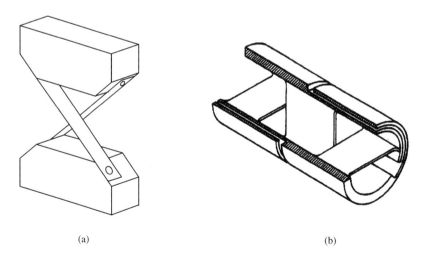

Figure 5.49. (a) Cross-axis flexural pivot, and (b) Bendix Corporation flexural pivot.

having the two flexible beams at an angle, their length is increased without increasing the total effective length of the pivot.

The cross-axis flexural pivot is not limited to two flexible beams, and more may be added for stability. The Bendix Corporation Free-Flex pivot shown in Figure 5.49b has three beams, where two of the beams combined have the same stiffness as the third beam. The flexible members may be made of many types of materials, and steel is commonly used.

Another advantage of cross-axis flexural pivots is that, like Q-joints, they may be used to connect links at points other than their centers. However, this requires special geometry of the cross-axis flexural pivot, such as that shown in Figure 5.49, and the geometry is not planar.

One disadvantage of cross-axis flexural pivots is that they are more difficult to fabricate and are more expensive than other types of compliant segments or passive joints. They also have limited motion and do not allow full rotation. An example of a parallel-guiding mechanism with cross-axis flexural pivots is illustrated in Figure 5.50.

5.8.5 Torsional Hinges

Torsional hinges use the angular displacement of elastic members in torsion to obtain their rotation. Figure 5.51 illustrates two torsional hinges with torsion members that are shafts with circular cross sections. Figure 5.52 illustrates a torsional hinge with rectangular-cross-section torsion members.

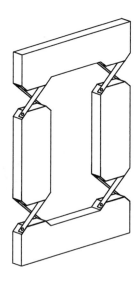

Figure 5.50. Parallel-guiding mechanism constructed using flexural pivots.

Other Methods of Pin Joint Simulation

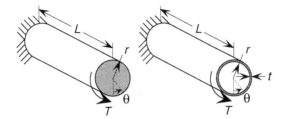

Figure 5.51. Flexible shafts with circular cross sections that undergo an angular deflection with applied torsion.

The angular displacement, θ, for a circular shaft under torsion, as shown in Figure 5.51, is

$$\theta = \frac{TL}{JG} \tag{5.155}$$

where T is the applied torque, L the shaft length, and J the polar moment of inertia (see Appendix B). The modulus of rigidity, G, can be defined in terms of the modulus of elasticity (Young's modulus), E, and Poisson's ratio, ν, as

$$G = \frac{E}{2(1 + \nu)} \tag{5.156}$$

where E and ν are listed in Appendix C for various materials.

The maximum shear stress due to torsion, τ_{max}, of a circular cross section is

$$\tau_{max} = \frac{Tr}{J} \tag{5.157}$$

where r is the circular shaft radius.

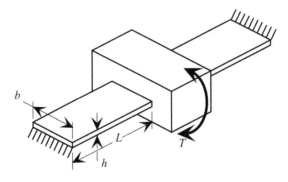

Figure 5.52. Torsional hinge with rectangular cross-section torsion members.

The angular deflection for a rectangular cross section is

$$\theta = \frac{TL}{KG} \qquad (5.158)$$

where K is related to the cross-sectional geometry as [89]

$$K = bh^3\left[\frac{1}{3} - 0.21\frac{h}{b}\left(1 - \frac{h^4}{12b^4}\right)\right] \qquad \text{where } b \geq h \qquad (5.159)$$

The maximum shear stress, τ_{max}, for the rectangular cross section in torsion is [89]

$$\tau_{max} = \frac{3T}{bh^2}\left[1 + 0.6095\frac{h}{b} + 0.8865\left(\frac{h}{b}\right)^2 - 1.8023\left(\frac{h}{b}\right)^3 + 0.9100\left(\frac{h}{b}\right)^4\right] \qquad (5.160)$$

$$\text{where } b \geq h$$

Other types of cross sections have other equations for K and τ_{max} [89]. The lowest possible value of K for a solid section is equal to J for a circular cross section.

Example: Micromirror Device. Micromirror arrays are used in computer projectors and other applications. Each mirror may represent a pixel, and its orientation can determine if the pixel is off or on. Some commercial micromirror arrays (such as Texas Instruments' digital micromirror device) connect the micromirror to torsional hinges, similar to that shown in Figure 5.52. The torsional hinges have a rectangular cross section because of fabrication constraints at the micro level. Suppose that the cross-sectional area is 0.2 by 1.2 µm, and that each hinge is 4 µm long and is made of sputtered aluminum. Calculate the torque required to deflect the mirror 10° and the maximum stress associated with that deflection.

Solution: The torque can be found by rearranging equation (5.158) and solving for the torque, T, as

$$T = \frac{KG}{L}\theta \qquad (5.161)$$

The length was given as $L = 4\mu m$. The angular deflection, θ, must be converted to radians for use in equation (5.161); thus

$$\theta = (10°)\frac{\pi \text{ rad}}{180°} = 0.175 \text{ rad} \qquad (5.162)$$

The material is aluminum; thus $E = 71.7$ GPa and $\nu = 0.34$. These properties can be substituted into equation (5.156) to obtain a modulus of rigidity of

$G = 26.8$ GPa. Because $b \geq h$, $b = 1.2$ μm and $h = 0.2$ μm. Substituting these values into equation (5.159) results in $K = 0.0029$ μm^4. The torque is calculated by substituting the values above into equation (5.161) as

$$T = \frac{(0.0029 \text{ μm}^4)(26.8 \text{ GPa})}{4 \text{ μm}}(0.175 \text{ rad}) = 3.4 \times 10^{-12} \text{ N-m} \qquad (5.163)$$

or, expressed in cgs units, $T = 3.4 \times 10^{-5}$ dyn-cm.

The maximum shear stress can be calculated from equation (5.160) as $\tau_{max} = 230$ MPa.

5.8.6 Split-Tube Flexures*

Goldfarb and Speich [130] presented a "split-tube" flexure which is based on the open-section hollow shaft shown in Figure 5.53. The torsional stiffness of an open section is significantly less than for a closed section, but the bending and compressive mechanics are very similar. This allows the split-tube flexure to be compliant in the desired axis of rotation but stiff about the other axes. Because the slit in the tube dramatically reduces the torsional stiffness but has little effect on the bending and other stiffnesses, they are nearly decoupled.

Assuming that the wall thickness, t, is much less than the tube radius R (i.e., $t \ll R$), the torsional stiffness, k_o, of the split-tube section can be approximated as

$$k_o = \frac{2\pi GRt^3}{3L} \qquad \text{torsion} \qquad (5.164)$$

where G is the modulus of rigidity [see equation (5.156)], and L is the tube length.

The applied torsional moment, M, and the angular deflection, Θ, are related through the torsional spring constant as

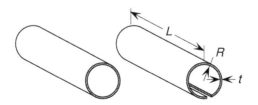

Figure 5.53. Closed- and open-section hollow shafts.

---

*For more detail, see [130].

$$M = k_o \Theta \tag{5.165}$$

The bending stiffness, k_b, is

$$k_b = \frac{\pi E R^3 t}{L} \quad \text{bending} \tag{5.166}$$

where E is Young's modulus. Note that in the limit of $t \ll R$ the torsional stiffness approaches zero while the bending stiffness approaches infinity.

The axis of rotation is not at the center of the hinge but is along the split-tube on a line that joins the links, as shown in Figure 5.54. Ideally, the segments connecting to the tube should be in line contact; thus they should be attached such that depth of the contact is minimized.

The maximum shear stress, τ_{max}, in the split-tube flexure is

$$\tau_{max} = \frac{3M}{2\pi R t^2} \tag{5.167}$$

The stress can also be written in terms of the angular deflection, Θ, as

$$\tau_{max} = \frac{Gt}{L} \Theta \tag{5.168}$$

Advantages of the split-tube flexure are that it can undergo larger angular deflections than many other types of flexures and it may maintain a better center of rotation than other flexural options.

5.9 MODELING OF MECHANISMS

Pseudo-rigid-body models for individual flexible segments offer a simple way to determine the deflections of large-deflection members. The availability of such a method suggests that it be used to model more complex systems that include a flex-

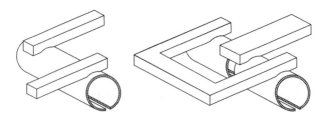

Figure 5.54. Examples of split-tube flexures.

Modeling of Mechanisms

ible segment or several such segments. This pseudo-rigid-body model concept simplifies the analysis and synthesis of compliant mechanisms. One advantage lies in the ability to provide a pseudo-rigid-body model of a compliant mechanism, which allows the designer to apply the knowledge available in the field of rigid-body mechanisms to the design of a compliant mechanism.

The greatest benefit of the pseudo-rigid-body model concept is realized in compliant mechanism design. In the early design stages, the pseudo-rigid-body model is an efficient method of evaluating many different trial designs in order to meet specific design objectives. With this approach, systems may be designed that are capable of more complex tasks than would otherwise be possible. If a designer relies solely on either prototyping or full numerical analysis, an initial design must be obtained before it can be analyzed or built. The pseudo-rigid-body model, on the other hand, may be used to obtain a preliminary design which may then be optimized. Once a design that meets the specified design objectives is obtained, it may be further refined by using a method such as nonlinear finite element analysis, and it may then be prototyped and tested. A sample design method that uses the pseudo-rigid-body model is shown in Figure 5.55. Chapter 8 discusses the design of compliant mechanisms in more detail.

5.9.1 Examples

Using the pseudo-rigid-body model of flexible segments to model compliant mechanisms containing such segments will be illustrated by several simple examples.

Compliant Slider–Crank Mechanism. Consider the compliant slider–crank mechanism shown in a deflected position in Figure 5.56a. It has a flexible segment length of $l = 5.0$ units, a crank length of $r_2 = 3.0$ units, and an offset of $e = -0.5$. It is assumed that the slider always remains in contact with the ground, the frictional resistance to slider motion is negligible, and the flexible segment is straight when undeflected.

Because the slider is free to move in a horizontal direction, there is no horizontal reaction force, and the force on the end of the flexible coupler is therefore vertical ($n = 0$). The values of γ and $\theta_{o\,max}$ are 0.852 and 76.5°, respectively, and a pseudo-rigid-body model is developed as shown in Figure 5.56b. This model may now be analyzed using rigid-body mechanism equations:

$$r_3 = \gamma l \qquad (5.169)$$

$$r_6 = l - r_3 \qquad (5.170)$$

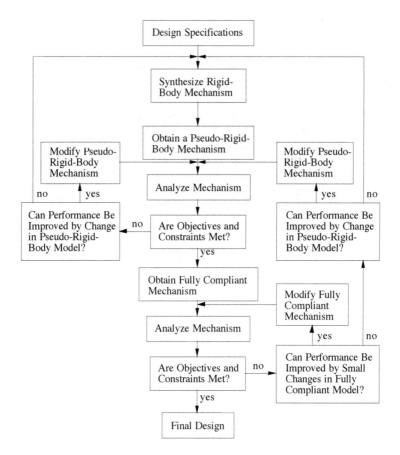

Figure 5.55. Flowchart of a compliant mechanism design process.

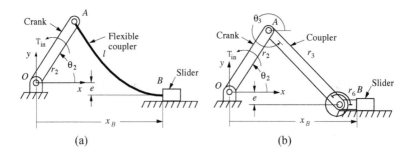

Figure 5.56. (a) Compliant slider–crank mechanism in a deflected position, and (b) its pseudo-rigid-body model.

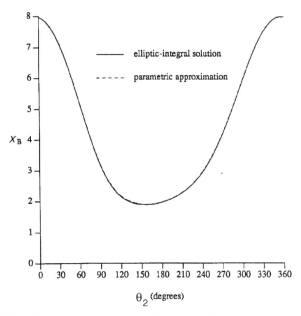

Figure 5.57. Slider displacement, x_B, versus θ_2 for compliant slider–crank mechanism.

$$\theta_3 = \operatorname{asin} \frac{e - r_2 \sin \theta_2}{r_3} \qquad (5.171)$$

$$x_B = r_2 \cos \theta_2 + r_3 \cos \theta_3 + r_6 \qquad (5.172)$$

Note that the value of the offset, e, is negative because it is in the negative y direction.

Figure 5.57 shows the slider coordinate, x_B, versus θ_2, calculated using equations (5.169) through (5.172) and compared with the solution found from the use of the closed-form elliptic-integral equations.

Parallel Mechanism. Some compliant mechanism applications require that the end angle of a flexible member remain constant as it goes through its motion. Consider the compliant mechanism shown in Figure 5.58a. The flexible segments are made of spring steel with a modulus of elasticity of $E = 30 \times 10^6$ lb/in.^{2}, a length of $L = 20$ in., a width of $w = 1.25$ in., and a thickness of $t = 1/32$ in. The rigid segment also has a length of $L = 20$ in. The pseudo-rigid-body model can be used

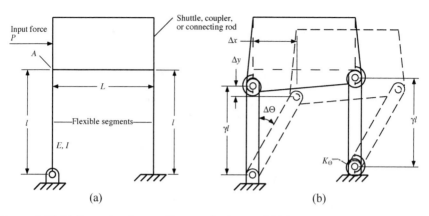

Figure 5.58. (a) Example of a compliant mechanism, and (b) its pseudo-rigid-body model.

to determine the deflection path of the rigid coupler and the horizontal force, P, that is required to obtain this motion.

The pseudo-rigid-body model for the mechanism is shown in Figure 5.58b. The system is such that for each segment, the changing reaction forces due to ground and the coupler yield a different value of n at each mechanism position. This variation of n causes changes in the location of the characteristic pivot and the stiffness coefficient, which may be accounted for in one of two ways. The simplest method is to use the averages of the characteristic radius factor, γ, and the stiffness coefficient, K_Θ, as constant values. Since the values of γ and K_Θ experience relatively small variations, their values are taken as $\gamma = 0.85$ and $K_\Theta = 2.61$. This method is used in the example at hand to illustrate its accuracy and usefulness for even this simplified model.

The second, more accurate method requires updating the changing values of γ and K_Θ at every increment of motion, using equation (5.48) and equations (5.60) through (5.62). The changing values for γ result in varying link lengths in the pseudo-rigid-body model, and it is no longer a parallelogram mechanism. Kinematic equations for a general four-bar linkage are required for the analysis of this type of problem.

A constant value for γ results in a pseudo-rigid-body model that is a parallelogram mechanism, which means that the rigid coupler remains horizontal. The displacement and required force are calculated by imposing an initial displacement on a link, calculating the resulting mechanism motion, and determining the reaction forces. The values of n for each segment are then updated, and the mechanism is incremented to the next displacement.

The force–deflection relationship for the mechanism may be determined in one of several ways. It is possible to make a free-body diagram of each link of the pseudo-rigid-body model and solve for the unknown forces. An alternative method

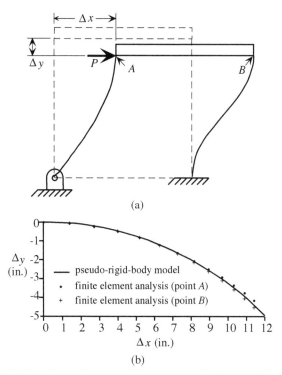

Figure 5.59. (a) Displaced compliant parallel motion mechanism, and (b) deflection path of the rigid-body segment.

is the use of the principle of virtual work. Both of these methods are described in Chapter 6.

The path of the rigid segment in Figure 5.58a is determined by equations (5.49) and (5.50) and is plotted in Figure 5.59. The results are compared to those obtained by a finite element code that is capable of large-deflection analysis. Twenty beam elements were used to model each flexural segment. The required horizontal force, P, versus horizontal deflection, Δ_x, is plotted in Figure 5.60. These results are also compared to those obtained from the finite element solution, and the results compare favorably, even for deflections of Θ greater than the Θ_{max} specified for K_Θ.

The simplified model of a parallel-motion mechanism with constant values of γ and K_Θ resulted in an approximation close to the much more involved finite element solution. The accuracy of the approximation, however, could be improved even more by allowing the values of γ and K_Θ to change throughout the mechanism motion, and combining these values with the kinematic equations of a four-bar mechanism that has arbitrary link lengths. These simplified models are useful in visualizing the motion of large-deflection systems and predicting their

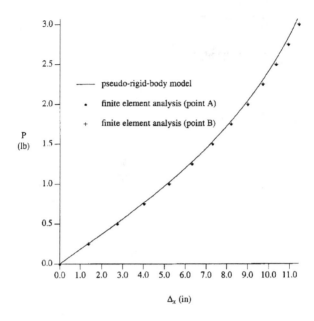

Figure 5.60. Load versus deflection plot of compliant parallel motion mechanisms.

behavior. They are also valuable in the initial design phase, allowing many different designs to be efficiently investigated and optimized.

Fully Compliant Bistable Mechanism. A fully compliant grasping device is shown in its closed (Figure 5.61a) and open (Figure 5.61b) positions. These two positions are stable equilibrium positions, and the mechanism will tend to move to one of these two positions in the absence of external forces. The general theory of bistable mechanisms is discussed in Chapter 11.

The mechanism obtains its motion from the deflection of three living hinges (special cases of small-length flexural pivot) and two flexible beams. However, the symmetry of the mechanism causes the center flexural pivot to travel in a straight-line motion. This motion may be modeled with a slider, and the living hinges may be modeled as pin joints. The living hinges' resistance to motion is small compared with that of flexible beams, so the torsional springs associated with their pseudo-rigid-body models may be ignored. Since the living hinge is modeled as a pin joint, the flexible beam may be modeled as a cantilever beam that is fixed at one end and subject to a force on the other end. The pseudo-rigid-body model of a cantilever beam with a force at the free end is used for this situation. The model of the mechanism is a slider mechanism, as illustrated in Figure 5.61, and rigid-body slider mechanism equations may be used to model the mechanism's motion.

Modeling of Mechanisms

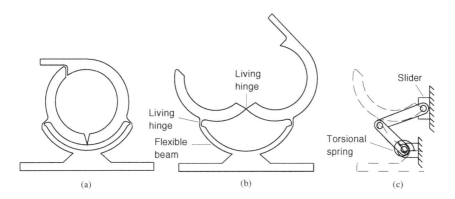

Figure 5.61. Bistable compliant mechanism in its (a) closed position, (b) open position, and (c) pseudo-rigid-body model.

Compliant Gripper Mechanism. The compliant gripper mechanism shown in Figure 5.62a is used to grip silicon chips in various solutions (as shown in Figure 1.6). Most of the gripper is constructed of polytetrafluoroethylene (such as Teflon) so that it is inert to all the harsh chemicals to which it is exposed. The cantilever spring, however, is an insert made of polypropylene, to provide adequate return force. The material has an out-of-plane thickness of b. The dimensions of the

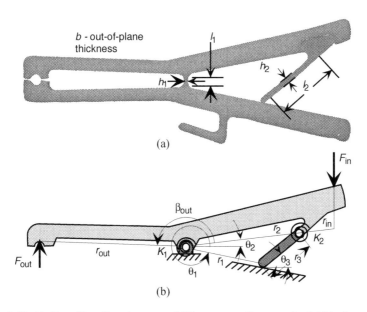

Figure 5.62. (a) Compliant die grippers, and (b) corresponding pseudo-rigid-body model.

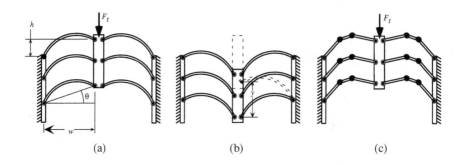

Figure 5.63. (a) Linear displacement mechanism in first stable position. (b) Same mechanism after being snapped down to second stable position. (c) Pseudo-rigid-body model of the mechanism.

small-length flexural pivot are h_1 and l_1. The geometry of the polypropylene insert is h_2 and l_2.

The characteristic pivot for the small length flexural pivot is located at the center point of l_1 and has a spring constant of $K_1 = E_1 I_1 / l_1 = E_1 b h_1^3 / (12 l_1)$. The length of the pseudo-rigid-link associated with the polypropylene insert is $r_3 = \gamma l_2$ in. The torsional spring constant is $K_3 = \gamma K_\Theta E_2 I_2 / l_2 = \gamma K_\Theta E_2 b h_2^3 / (12 l_2)$. The pseudo-rigid-body link length shown as r_2 in Figure 5.62b is the distance between these two characteristic pivots.

With this geometry known, the value of θ_3 can be calculated for a value of θ_2 as

$$\theta_3 = \operatorname{asin} \frac{r_2 \sin \theta_2 - r_1 \sin \theta_1}{r_3} \tag{5.173}$$

Now the position of any point on the mechanism can be found. The force–deflection relationship for the mechanism is developed in Chapter 6.

Microbistable Mechanism [16]. Attaching several functionally binary pinned–pinned segments at an angle to a center slider, with the segments symmetric about the centerline of the slider, results in a bistable mechanism with linear motion as illustrated in Figure 5.63. As the slider moves down, the FBPP segments are compressed, resulting in a tendency for the slider to snap back to its first stable position. The mechanism passes through an unstable equilibrium point when the segment pin joints line up perpendicular to the slider, after which the force exerted by the segments causes the slider to snap down to its second stable position.

Because the pin joint attached to the slider is constrained to move only in the vertical direction, the distance through which the pinned–pinned segments are compressed is known throughout the motion of the slider. As such, a can be determined directly from the geometry as

$$a = \sqrt{w^2 + (h-y)^2} \tag{5.174}$$

where w and h define the initial orientation of the pinned–pinned segment and y is any linear displacement of the slider (Figure 5.63). The maximum displacement between the two stable equilibrium positions is $y = 2h$.

Once a has been determined, it can be used to solve for Θ and b by using equations (5.140) and (5.141), respectively. The compression force, F, on each of the pinned–pinned segments can then be calculated using equation (5.146). The total force, F_t, required to displace the slider a distance y can then be found by summing the vertical component of the force applied by each of the pinned–pinned segments, and is given by

$$F_t = NF_s \sin\theta \tag{5.175}$$

where F is the compressive force defined by equation (5.146), N is the total number of pinned–pinned segments in the mechanism, and θ is the angle of the pinned–pinned segment. For any slider displacement y, the angle θ is given by

$$\theta = \mathrm{asin}\frac{h-y}{\sqrt{w^2+h^2}} \tag{5.176}$$

For equilibrium, the pinned–pinned segments must be symmetric about the centerline of the slider so that the component of the force normal to the slider caused by the left-side segments cancels with the force components normal to the slider from the right-side segments.

Surface micromachining was used to construct the microbistable mechanism shown in the scanning electron photographs shown in Figure 5.64. The mechanism was fabricated with an electrical contact switch on the slider to illustrate a potential application for the device. When the mechanism is switched to its second stable position, it closes the electrical contact between the two contact pads. The pinned–pinned segment radius of curvature, R_o, is 144 µm, with a total length, L, of 85.5 µm. The curvature κ_o can be calculated using equation (5.114) as

$$\kappa_o = 0.5934 \tag{5.177}$$

The fundamental radius factor γ can be approximated from Table 5.2 as

$$\gamma = 0.81 \tag{5.178}$$

The initial coordinates of the beam end, a_i and b_i, can be found by equations (5.135) and (5.136) to be

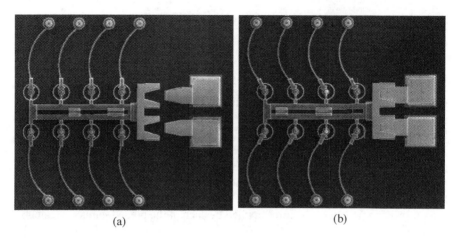

Figure 5.64. Scanning electron microscope photographs of the linear displacement mechanism (a) in the open position, and (b) in the closed position.

$$a_i = 80.5 \ \mu m \qquad (5.179)$$

$$b_i = 24.6 \ \mu m \qquad (5.180)$$

With the initial coordinates known, the characteristic radius factor ρ can be calculated using equation (5.116) to be

$$\rho = 0.7875 \qquad (5.181)$$

The initial pseudo-rigid-body angle, Θ_i, is

$$\Theta_i = 0.3745 \ \text{rad} \qquad (5.182)$$

To calculate the torsional spring constant, the nondimensional parameter K_Θ can be calculated using equation (5.142) as

$$K_\Theta = 2.6 \qquad (5.183)$$

The out-of-plane thickness is 2 µm and the width is 3 µm. Young's modulus of polysilicon can be taken to be $E = 170$ GPa. The spring constant can then be found to be

$$K = 4320 \text{dyn-}\mu m \qquad (5.184)$$

by using equation (5.143). The initial geometry of the pinned–pinned segments as shown in Figure 5.63 is given by

$$w = 210 \text{ μm} \tag{5.185}$$

and

$$h = 55 \text{ μm} \tag{5.186}$$

Because of the limitations of the fabrication process, the flexible segment does not extend the complete length from pin joint to pin joint. However, because the deflection of the beam is a function of the moment, and the moment is very small near the pin-joints, the error introduced is negligible.

With K known, and the geometry of the mechanism defined, its force-deflection properties can now be fully defined with the pseudo-rigid body model by plotting the total force F_t, as defined by equation (5.175), through the entire range of motion, as shown in Figure 5.65. Notice that at a displacement of 55 μm the force transitions from positive to negative. This corresponds to the unstable equilibrium point where the pinned–pinned segments are at their maximum compression and are normal to the slider. As would be expected, the force on the slider is negative after this point as the slider snaps to its second stable equilibrium point.

5.10 USE OF COMMERCIAL MECHANISM ANALYSIS SOFTWARE

A major advantage of the pseudo-rigid-body model is that it makes it possible to analyze compliant mechanisms using the vast amount of knowledge available to

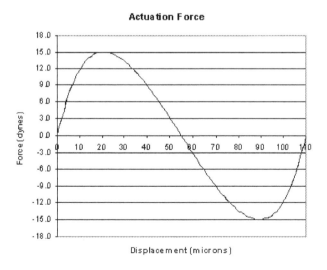

Figure 5.65. Plot of the actuation force over the entire range of motion.

analyze and design rigid-body mechanisms. One important aspect of this is that software used to analyze rigid-body mechanisms may also be used to analyze the pseudo-rigid-body models of compliant mechanisms.

Many commercial software packages are available to analyze rigid-body mechanisms. There are too many to discuss individually and the purpose of this section is to provide exposure to the area so that this advantage of the pseudo-rigid-body model is realized and taken advantage of in compliant mechanism design.

The basic requirements of software for analyzing compliant mechanism pseudo-rigid-body models include the capability of modeling the relative motion of rigid links connected by pin joints, sliding joints, and other kinematic pairs that are in the mechanism (such as cams). It must also have the capability to model torsional springs and linear springs for which the user can define the position, the spring constant, and the undeflected position of the spring.

Some commercial software packages include the ability to analyze flexible members in mechanisms. The user must be very careful to check the analysis method used and determine its accuracy for the given mechanism. The most common problem arises when the model accounts for small, linear deflections when actually they are large, nonlinear deflections.

A typical process to analyze a compliant mechanism using commercial rigid-body mechanism analysis software is as follows:

1. Use the methods described in this chapter to create a pseudo-rigid-body model of a mechanism.
2. Determine the link lengths from the pseudo-rigid-body model and enter them as lengths of rigid links.
3. Determine the spring constants from the pseudo-rigid-body model and place springs at corresponding joints on the computer model.
4. Analyze the resulting mechanism.
5. Carefully monitor the rotation of torsional springs to ensure that the deflection does not exceed acceptable accuracy limits of the pseudo-rigid-body model.
6. After the deflections and forces are known, calculate the stress on the flexible beams using the equations provided previously.

Example: Bistable Switch. Consider the fully compliant bistable switch shown in Figure 5.66a and its pseudo-rigid-body model in Figure 5.66b. Electrical contacts are placed at the positions shown in the figure. A force applied to the handle will cause it to move to the second stable equilibrium position, where the contacts close a circuit. A force in the other direction causes it to snap back to the original position where the circuit is open. Bistable mechanisms are discussed in more detail in Chapter 11.

The mechanism has three living hinges and a small-length flexural pivot. The pseudo-rigid-body model is a four-bar mechanism with pin joints at the living

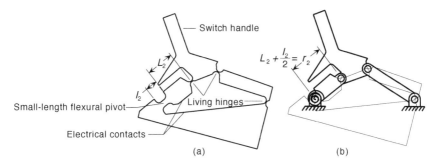

Figure 5.66. (a) Fully compliant bistable mechanism, and (b) its pseudo-rigid-body model (patent pending).

hinges and at the center of the small-length flexural pivot. The link lengths are $r_1 = 1.41$ in., $r_2 = L_2 + l_2/2 = 0.47$ in., $r_3 = 0.4$ in., and $r_4 = 0.78$ in.

The living hinges offer very little resistance to motion compared to the small-length flexural pivot and can be modeled as pin joints without torsional springs. The stiffness of the small-length flexural pivot is modeled by a torsional spring with a spring constant of $K = EI_2/l_2$. The material is polypropylene with a Young's modulus of $E = 200,000$ lb/in². The length of the small-length flexural pivot is $l_2 = 0.13$ in. The out-of-plane thickness is $b = 0.25$ in. and the in-plane thickness is $h_2 = 0.08$ in., resulting in a moment of inertia of $I_2 = bh_2^3/12 = 1.07 \times 10^{-5}$ in⁴. The resulting value of the torsional spring constant is $K = 16.4$ in.-lb/rad.

The links, pin joints, and torsional spring for the pseudo-rigid-body model can be entered into mechanism analysis software to calculate the resulting motion and forces. A snapshot of the pseudo-rigid-body model using commercial mechanism analysis software (Working Model) is shown in Figure 5.67.

Example: Linear Motion Mechanism. A compliant mechanism that was designed to be part of a robot end effector that moves silicon wafers in a clean-room environment is shown in Figure 5.68a. The special geometry of this device allows linear motion in one direction with no other translation or rotation. The flexible beams have fixed–guided end conditions. Recall that a fixed–guided beam can be modeled as a rigid link of length γl with a pin joint at each end, as shown in the pseudo-rigid-body model of Figure 5.68b.

The flexible segments each have a length of $l = 22$ mm, an out-of-plane thickness of $b = 7$ mm, and an in-plane thickness of $h = 0.8$ mm. There is a torsional spring at each pin joint, and each has a torsional spring constant of $K = 2\gamma K_\Theta EI/l$. The material is Delrin (an acetal resin) with a Young's modulus $E = 2.9$ GPa. The moment of inertia is $I = bh^3/12 = 2.99 \times 10^{-13}$ m⁴. Using typical values of $\gamma = 0.85$ and $K_\Theta = 2.65$ results in torsional spring constants of $K = 0.18$ N-m/rad.

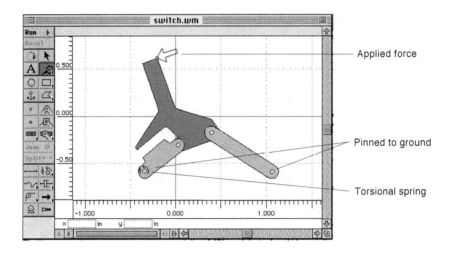

Figure 5.67. Snapshot of a Working Model analysis of the pseudo-rigid-body model for the mechanism shown in Figure 5.66.

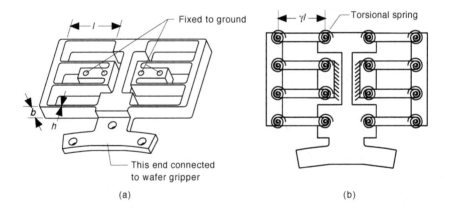

Figure 5.68. (a) Linear motion robot end effector used to transport silicon wafers in a cleanroom environment (courtesy of FSI International), and (b) pseudo-rigid-body model.

Use of Commercial Mechanism Analysis Software

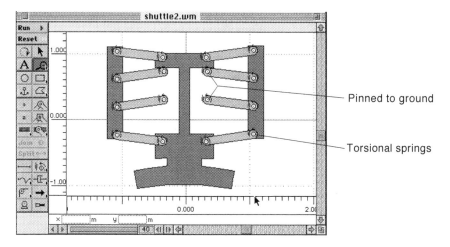

Figure 5.69. Linear motion mechanism of Figure 5.68 shown in a mechanism analysis program.

Figure 5.69 shows the pseudo-rigid-body model of the mechanism in a mechanism analysis program.

PROBLEMS

5.1 Determine the deflection of point A in Figure P5.1 as follows:
   (a) Use the small-length flexural pivot approximation.
   (b) Use the theoretical large-deflection equations.
   (c) Compare the results above. Explain why you may want to use the pseudo-rigid-body model for a case like this, even when the theoretical equations are relatively simple.
5.2 Use the pseudo-rigid-body model to estimate the deflection (a, b, θ_0) for a steel beam with a $1/32$ in. $\times$ 1.25 in. cross section and a length of 24 in.

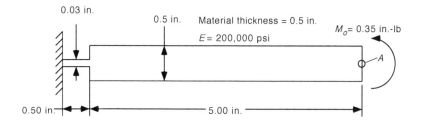

Figure P5.1. Small-length flexural pivot (Problem 5.1).

(a) Assume straight beam with a vertical force of 0.2 lb and a horizontal force (causing compression) of 0.4 lb.
  (i) Use $\gamma = 0.85$ and $K_\Theta = 2.65$.
  (ii) Use equations (5.48) to (5.48) and (5.60) to (5.62) to find more accurate values of γ and K_Θ for this loading.
(b) Find the vertical load that will cause a vertical deflection of 12 in.
(c) Compare the results above to those obtained in Problem 2.9c.

5.3 A cantilevered beam has a moment load, M_o, at the free end. If the beam has a maximum stress of 50,000 psi, calculate the value of the moment (M_o). It is a steel beam with a $1/32$ in. $\times$ 1.25 in. cross section and a length of 24 in.
(a) Use small-deflection theory to calculate the deflection ($y_{max} = M_o L^2 / 2EI$).
(b) Use the pseudo-rigid-body model for a beam with an end moment to approximate the values of the end angle (Θ_o), and the horizontal (a) and vertical (b) coordinates of the endpoint.

5.4 A micro flexible beam is made of polysilicon and has a modulus of elasticity of $E = 1.9 \times 10^{12}$ dyn/cm² (1 dyn = 1 g-cm/s²). It has dimensions $L = 24$ µm, $b = 1$ µm, and $h = 0.2$ µm (1 µm = 1 x 10⁻⁶ m). The beam is loaded with a vertical force at the free end (see Figure P5.4).
(a) Sketch the pseudo-rigid-body model.
(b) What is the length of the pseudo-rigid link? What is the value of the torsional spring constants?
(c) Calculate the force (in dynes) required to obtain a vertical deflection of 8 µm.
(d) What is the horizontal coordinate of the beam end, a, when the vertical deflection is 8 µm?

5.5 A flexible beam is made of polypropylene with $E = 200{,}000$ lb/in.², $L = 20$ in., $w = 0.5$ in., and $h = 0.1$ in. The beam is loaded with a vertical force at the free end (see Figure P5.4).
(a) Sketch the pseudo-rigid-body model.
(b) Label the link lengths and the torsional spring constant.
(c) Calculate the force required to obtain a vertical deflection of 10 in.

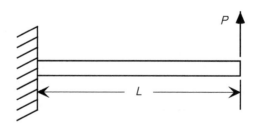

Figure P5.4. Flexible cantilever beam (Problems 5.4 and 5.5).

(d) What is the horizontal coordinate of the beam end, a, when the vertical deflection is 10 in?
(e) Estimate the maximum stress in the beam.

5.6 A flexible beam has one end fixed and the other free, has an initial curvature of $R_i = 10$ in., a length $L = 5$ in., a Young's modulus $E = 30 \times 10^6$ lb/in.^{2}, a width of $w = 1.25$ in., and a thickness of $t = 1/32$ in.
(a) Sketch the pseudo-rigid-body model of the beam.
(b) Label the link lengths and the torsional spring constant.
(c) Calculate the force required to obtain a vertical deflection of 2 in.
(d) What is the horizontal coordinate of the beam end, a, when the vertical deflection is 2 in?

5.7 An initially curved flexible segment is pinned at both ends. It is made of polypropylene, has an initial radius of 55.6 mm, a length of 63 mm, a width of 1.22 mm, and an out-of-plane thickness of 3.23 mm.
(a) Sketch the pseudo-rigid-body model.
(b) Label the link lengths and the torsional spring constant.
(c) Calculate the force required to obtain a deflections of 2.6 mm and 9 mm.
(d) What is the vertical deflection for the two deflections listed above?
(e) Calculate the stress when the deflection is 9 mm.

5.8 A compliant device is required that has the same performance as the pseudo-rigid-body model illustrated in Figure P5.8. Determine dimensions and material that would provide this behavior for the following types of flexible segments:
(a) Small-length flexural pivot
(b) Fixed–pinned segment
(c) Initially curved fixed–pinned beam

5.9 Sketch the pseudo-rigid-body model for the following mechanisms:
(a) The crimping mechanism shown in Figure 1.2a.
(b) The parallel guiding mechanism shown in Figure 1.2b.
(c) The longbow shown in Figure 1.11.
(d) The fingernail clippers shown in Figure P2.6.
(e) The micro positioner shown in Figure 1.5.
(f) The compliant ice-cream scoop shown in Figure 2.10.
(g) The telephone connector shown in Figure P2.4.
(h) A compliant mechanism of your choice.

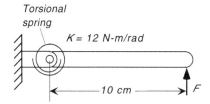

Figure P5.8. Pseudo-rigid-body model associated with Problem 5.8.

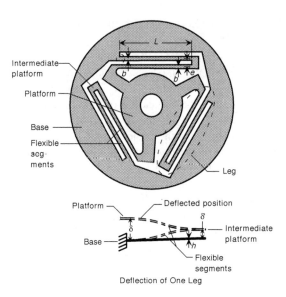

Figure P5.10. Orthoplanar spring (Problem 5.10).

5.10 Consider the orthoplanar spring illustrated in Figure P5.10. It is made from 0.01-in. (0.25-mm)-thick type 410 heat-treated stainless steel. The flexible segments each have a width of $b = 0.03$ in. (0.762 mm) and a length of $L = 0.646$ in. (16.4 mm), and the spacing between their centers is $e = 0.070$ in. (1.778 mm).
  (a) Sketch the pseudo-rigid-body model for one of the three legs of the mechanism.
  (b) Calculate the lengths of the pseudo-rigid links.
  (c) Calculate the spring constants, K, of the torsional springs associated with the pseudo-rigid-body model.
  (d) Write an equation for the force required to obtain a given platform deflection, δ, in terms of the variables given.
  (e) Compare the deflection results calculated using the pseudo-rigid-body model equation and the equation for small deflections in Section 3.1.1 [equation (3.16)].

5.11 Recall the mechanism shown in Figure 5.56.
  (a) Find the equation for input torque as a function of the crank angle for this mechanism. Calculate and plot the input torque for every 10° of crank angle.
  (b) List the appropriate crank angle for two stable equilibrium positions and one unstable equilibrium position.

5.12 Figure P5.12 shows a compliant, parallel-guiding micromechanism with dimensions given in microns (1 micron = 1 μm = 1×10^{-6} m). The material is polysilicon with a modulus of elasticity $E = 1.9 \times 10^{12}$ dyn/cm².
  (a) Sketch the pseudo-rigid-body model for the mechanism.

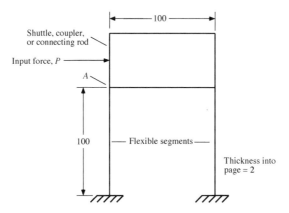

Figure P5.12. Compliant parallel-guiding mechanism (Problem 5.12).

    (b) Calculate the lengths of the pseudo-rigid links.
    (c) Write the equations for the spring constants (K, not K_Θ) of the torsional springs. Write the equations symbolically in terms of known quantities, and calculate their numerical values (in units of dyn-cm/rad).

5.13 Consider the compliant mechanism shown in Figure P5.13.
    (a) Sketch the pseudo-rigid-body model of this mechanism.
    (b) How would you determine the number of degrees of freedom of this mechanism? Calculate the degrees of freedom.

5.14 Develop the equation for the input torque, T_{in}, of the crank–flapper mechanism shown in Figure P5.14 in terms of θ_2. The equation should be stated in

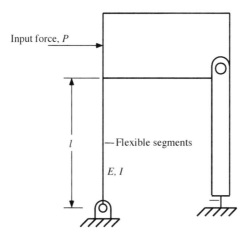

Figure P5.13. Compliant mechanism (Problem 5.13).

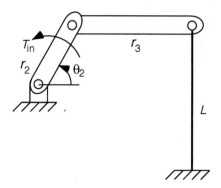

Figure P5.14. Crank–flapper mechanism (Problem 5.14).

terms of both the given variables (r_2, r_3, E, b, h, L, θ_2) and other known variables for which you must specify the numerical value (e.g., γ).

5.15 Consider the flapper mechanism shown in its undeflected position (Figure P5.15).
  (a) Sketch the pseudo-rigid-body model of the mechanism. Symbolically state the length of the pseudo-rigid link and the value of the torsional spring constant.
  (b) Suppose that $r_2 = 2$, $r_3 = 3$, $L = 4$, and the distance between the two connections to ground is 5. How would you determine whether or not link 2 could go through a complete revolution? Will it for this case?

5.16 Consider the mechanism shown in Figure P5.16. Segment 4 is initially curved and is connected to a slider. The mechanism is constructed of steel with $E = 30{,}000{,}000$ lb/in². Each flexible segment has a width (out of the page) of 0.5 in. The small length flexural pivot has a thickness of 0.01 in. and a length of 1 in. The rigid segment connected to it (L_2) has a length of 20 in., and link 3 has a length of 24 in. Segment 4 has a thickness of 0.03 in., a length of 24 in., and an initial radius of curvature of 24 in. Segment 5 has a length of 30 in., and a thickness of 0.03 in.

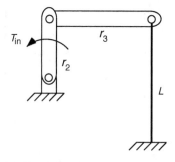

Figure P5.15. Crank–flapper mechanism (Problem 5.15)

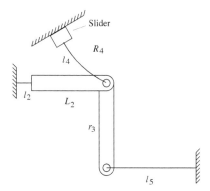

Figure P5.16. Compliant mechanism (Problem 5.16).

    (a) Sketch the pseudo-rigid-body model.
    (b) State the value of the torsional spring constant for the segment associated with l_2. (Show proper units.)
    (c) Is the pseudo-rigid-body model of the mechanism valid for any location of the input force? Explain.

5.17 Consider the mechanism illustrated in Figure P5.17.
    (a) Sketch the pseudo-rigid-body model of this mechanism.
    (b) Write the torsional spring constant symbolically for each torsional spring.

5.18 A compliant, micromechanism is shown in Figure P5.18 with dimensions in microns. The material is polysilicon with a modulus of elasticity of $E = 1.9 \times 10^{12}$ dyn/cm².
    (a) Sketch the pseudo-rigid-body model for the mechanism.
    (b) Calculate the lengths of the pseudo-rigid links.

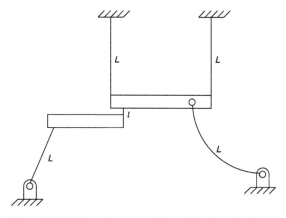

Figure P5.17. Compliant mechanism (Problem 5.17).

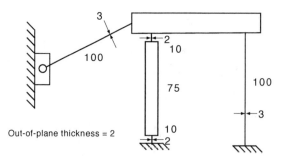

Figure P5.18. Figure for Problem 5.18.

   (c) Label the springs in the pseudo-rigid-body model. Write the equations for the torsional spring constants. Write the equations symbolically and calculate their numerical values. Show proper units.
5.19 The compliant mechanism illustrated in Figure P5.19 has an initially curved flexible segment, as shown. It has an initial curvature of $R_i = 10$ in., a length $L = 5$ in., a Young's modulus $E = 30 \times 10^6$ lb/in.^{2}, a width of $w = 1.25$ in., and a thickness of $t = 1/32$ in. The crank has a length of 0.8 in., and the flexible segment is undeflected when the crank is vertical.
   (a) Sketch the pseudo-rigid-body model of the mechanism.
   (b) Calculate and label the link lengths and the torsional spring constant.
5.20 The bistable mechanism illustrated in Figure P5.20 has two identical initially curved flexible segments that have living hinges at both ends. The device is constrained such that the center shuttle goes through a vertical motion. Because of symmetry, only one of the two pinned–pinned segments need to be analyzed. Each segment is made of polypropylene, has an initial radius of 55.6 mm, a length of 63 mm, a width of 1.22 mm, and an out-of-plane thickness of 3.23 mm. The stiffness of the living hinges is much lower than that of the other elements.
   (a) Sketch the pseudo-rigid-body model.
   (b) Calculate and label the link lengths and the torsional spring constant.

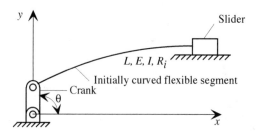

Figure P5.19. Figure for Problem 5.19.

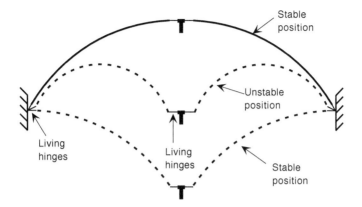

Figure P5.20. Figure for Problem 5.20.

(c) Determine the maximum deflection that the segment must go through before it snaps to its second stable equilibrium position.
(d) Plot the required actuation force as a function of the vertical deflection of the center shuttle.

CHAPTER 6

FORCE–DEFLECTION RELATIONSHIPS

The pseudo-rigid-body-model concept allows compliant mechanisms to be modeled as rigid-body mechanisms. Although easier to analyze than its compliant counterpart, the pseudo-rigid-body model may still be more complex than rigid-body skeleton diagrams. Unlike rigid-body kinematics, the motion of a compliant mechanism depends on the location and magnitude of applied forces. The pseudo-rigid-body model therefore contains not only the rigid links and kinematic pairs necessary to model motion, but also includes appropriate discrete springs to model the compliance of flexible members. This model may become quite unwieldy, making it difficult to determine a mechanism's force–deflection relationships.

Two approaches for determining force–displacement relationships from pseudo-rigid-body models are discussed in this chapter. The first method uses conventional Newtonian methods to obtain free-body diagrams to determine the equations for static equilibrium of each link. The reaction forces may then be found by solving the resulting system of equations. The advantage of this method is that the force system for the entire mechanism is established.

An alternative method of determining force–deflection relationships is the principle of virtual work. This approach lends itself very well to pseudo-rigid-body models. In this method, the system is viewed in its entirety, and it is not necessary to include all the reaction forces. The energy stored in the springs is also easily taken into account. Since most readers will not be familiar with the principle of virtual work, it is described in more detail here than is the more conventional free-body diagram approach.

The decision of which of these methods to use depends on the specific application and what information is needed from the analysis. If the reaction forces at each characteristic pivot is needed, the free-body diagram approach is best. However, if the output displacement or force is needed relative to the input force or displacement, the principle of virtual work usually proves to be a much easier analysis.

6.1 FREE-BODY DIAGRAM APPROACH

One advantage of the pseudo-rigid-body model is that it allows compliant mechanisms to be modeled as rigid-body mechanisms with rigid links and springs. The resulting model is easily analyzed using common mechanism design methods. The pseudo-rigid-body mechanism is drawn with the applied forces included. A free-body diagram is constructed for each link, and the equations of static equilibrium are constructed for each link. For rigid links in a plane, the link is in equilibrium if

$$\Sigma F_x = 0 \tag{6.1}$$

$$\Sigma F_y = 0 \tag{6.2}$$

$$\Sigma M = 0 \tag{6.3}$$

The three equations above are written for each link of the mechanism except the ground link. This results in a system of linear equations that can be solved for the unknowns. This approach is identical to that used for any static equilibrium problem. However, there are some hints that can help obtain the equations more easily. First, define all the forces in the positive direction. If a result is negative, it means that it is in the negative direction. Second, use a consistent nomenclature to write the forces. The approach used in this section will be that a force labeled F_{ij} represents the force on link j where it connects to link i. Third, sum the moments about the point from which the link angle is measured. This allows the angles from the kinematic analysis to be used in the force analysis.

Example: Compliant Slider Mechanism. Perhaps the best way to describe the method is by example. Consider the compliant slider mechanism and its pseudo-rigid-body model shown in Figure 6.1. The input is a moment on link 2, T_{in}, and a torsional spring causes a torque at the connection of links 3 and 4, T_3. The free-body diagrams for links 2 and 3 are shown in Figure 6.2. Since the slider does not resist motion in the x direction, $F_{43x} = 0$. The equations of static equilibrium for link 2 are

$$F_{12x} + F_{32x} = 0 \tag{6.4}$$

$$F_{12y} + F_{32y} = 0 \tag{6.5}$$

$$T_{in} + F_{32y} r_2 \cos\theta_2 - F_{32x} r_2 \sin\theta_2 = 0 \tag{6.6}$$

Free-Body Diagram Approach

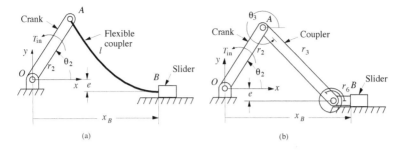

Figure 6.1. (a) Compliant slider mechanism, and (b) its pseudo-rigid-body model.

For link 3

$$F_{23x} = 0 \tag{6.7}$$

$$F_{23y} + F_{43y} = 0 \tag{6.8}$$

$$T_{43} + F_{43y} r_3 \cos \theta_3 = 0 \tag{6.9}$$

where

$$T_{43} = -K_3(\theta_3 - \theta_{3o}) \tag{6.10}$$

We also know that at a pin joint the forces on the connected links have equal magnitude but are in opposite directions, that is,

$$F_{32x} = -F_{23x} \tag{6.11}$$

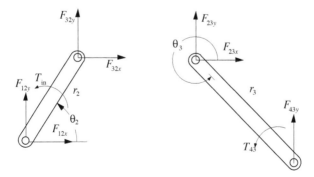

Figure 6.2. Free-body diagrams of links 2 and 3 for the compliant slider mechanism.

222   Force–Deflection Relationships

$$F_{32y} = -F_{23y} \tag{6.12}$$

Substituting equation (6.7) into (6.11) and that result into equation (6.4) results in

$$F_{23x} = F_{32x} = F_{12x} = 0 \tag{6.13}$$

Equation (6.9) may be rearranged to solve for F_{43y} as

$$F_{43y} = \frac{-T_{43}}{r_3 \cos \theta_3} \tag{6.14}$$

Using this result to help solve equation (6.8), then (6.12) and (6.5), results in

$$F_{23y} = F_{12y} = -F_{32y} = -F_{43y} = \frac{T_{43}}{r_3 \cos \theta_3} \tag{6.15}$$

Substituting equations (6.13) and (6.15) into equation (6.6) and solving for T_{in} yields

$$T_{in} = T_{43} \frac{r_2 \cos \theta_2}{r_3 \cos \theta_3} = -\gamma K_\Theta \frac{EI}{l}(\theta_3 - \theta_{3o}) \frac{r_2 \cos \theta_2}{r_3 \cos \theta_3} \tag{6.16}$$

Equation (6.16) may be used to calculate the input torque required to maintain the mechanism at various positions. If the input variable is θ_2, the coupler angle θ_3 may be calculated using the kinematic equations described in Section 4.2.2. The angle θ_{3o} is the value of θ_3 when the torsional spring is undeflected.

Example: Compliant Bicycle Brakes. The compliant bicycle brakes shown in Figure 6.3a have a pseudo-rigid-body model (Figure 6.3b) that is a parallelogram linkage. The purpose of such a configuration is that the brake pads translate but do not rotate. The force–displacement relationships need to be understood so that the flexible beam can be designed in a manner that provides adequate spring force but does not require too much input force. The free-body diagrams for each link are illustrated in Figure 6.4. As mentioned earlier, the brake pad and link 3 do not rotate. The x-axis is chosen to be along link 3 such that $\theta_3 = 0$. For simplicity for the example, it is assumed that F_{in} is in the x direction. The equations of static equilibrium [equations (6.1) to (6.3)] for link 2 are

$$F_{12x} + F_{32x} = 0 \tag{6.17}$$

$$F_{12y} + F_{32y} = 0 \tag{6.18}$$

$$T_{12} + T_{32} + F_{32y} r_2 \cos \theta_2 - F_{32x} r_2 \sin \theta_2 = 0 \tag{6.19}$$

Free-Body Diagram Approach

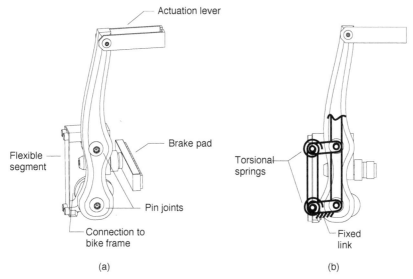

Figure 6.3. (a) Compliant bicycle brakes with brake pads that translate but do not rotate, and (b) the mechanism's pseudo-rigid-body model.

The equations for link 3 are

$$F_{23x} + F_{43x} = 0 \tag{6.20}$$

$$F_{23y} + F_{43y} = 0 \tag{6.21}$$

$$T_{23} + r_3 F_{43y} = 0 \tag{6.22}$$

and the equations associated with link 4 are

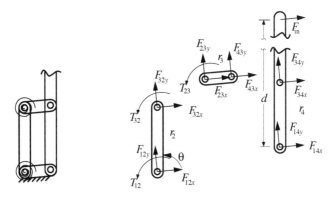

Figure 6.4. Free-body diagrams for the links of the compliant bicycle brakes.

$$F_{34x} + F_{14x} + F_{in} = 0 \qquad (6.23)$$

$$F_{34y} + F_{14y} = 0 \qquad (6.24)$$

$$F_{34y} r_4 \cos\theta_4 - F_{34x} r_4 \sin\theta_4 - F_{in} d \sin\theta_4 = 0 \qquad (6.25)$$

where d is the distance from the ground pin of link 4 to the location of the input force.

The loads on two links connected at a pin joint have forces with equal magnitudes but act in opposite directions, or

$$F_{32x} = -F_{23x} \qquad (6.26)$$

$$F_{32y} = -F_{23y} \qquad (6.27)$$

$$F_{43x} = -F_{34x} \qquad (6.28)$$

$$F_{43y} = -F_{34y} \qquad (6.29)$$

Combining the previous equations with the equilibrium equations results in

$$F_{43x} = -F_{34x} = F_{32x} = -F_{23x} = -F_{12x} \qquad (6.30)$$

and

$$F_{43y} = -F_{34y} = F_{32y} = -F_{23y} = -F_{12y} \qquad (6.31)$$

Because the pseudo-rigid-body model is a parallelogram mechanism, links 2 and 4 have the same rotation angle, θ, or

$$\theta_2 = \theta_4 = \theta \qquad (6.32)$$

The torques at the pin joints due to the springs are

$$T_{12} = T_{32} = -T_{23} = -K(\theta - \theta_0) \qquad (6.33)$$

where θ is the angle of links 2 and 4, and θ_o is the initial angular position of links 2 and 4 when the torsional springs are undeflected. The torsional spring constant, K, for the fixed–guided flexible segment is found from equation (5.105) as

$$K = 2\gamma K_\Theta \frac{EI}{l} \qquad (6.34)$$

Equation (6.22) may be rearranged to obtain

$$F_{43y} = \frac{-T_{23}}{r_3} \tag{6.35}$$

and equation (6.19) may be rearranged to solve for F_{32x} as

$$F_{32x} = \frac{T_{12} + T_{32} + F_{32y}r_2 \cos\theta}{r_2 \sin\theta} \tag{6.36}$$

Combining equations (6.31) and (6.33) to (6.36) results in

$$F_{32x} = \frac{T_{12}[2 + (r_2/r_3)\cos\theta]}{r_2 \sin\theta} \tag{6.37}$$

Combining equations (6.30) and (6.37) provides several of the reaction forces in the x direction. But the input force F_{in} is still needed. Rearranging equation (6.25) results in

$$F_{in} = \frac{F_{34y}r_4 \cos\theta_4 - F_{34x}r_4 \sin\theta}{d \sin\theta} \tag{6.38}$$

Substituting equations (6.30), (6.31), (6.33), (6.35), and (6.37) into equation (6.38) and simplifying results in

$$F_{in} = \frac{-2K(\theta - \theta_o)}{d \sin\theta} \tag{6.39}$$

The principle of virtual work is an alternative method for determining the force–deflection relationships of compliant mechanisms, as explained in the following sections.

6.2 GENERALIZED COORDINATES

Several preliminary concepts are helpful before the principle of virtual work is presented. The first of these is the choice of coordinates and is explained using the mechanism of Figure 6.5 as an example.

The base coordinate system for the work that follows will be in the planar Cartesian coordinates, x and y. The location of point A may be expressed in terms of these coordinates as (x_A, y_A). Lagrangian coordinates are the other coordinates used to describe the system, and may be redundant. A set of Lagrangian coordinates will be expressed as $(\psi_1, \psi_2, ..., \psi_i)$. For the mechanism of Figure 6.5, convenient Lagrangian coordinates are $\psi_1 = \theta$, $\psi_2 = s_A$, and $\psi_3 = s_B$.

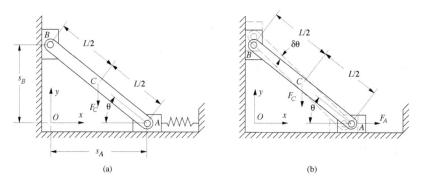

Figure 6.5. (a) Double-slider mechanism and set of Lagrangian coordinates, and (b) virtual displacement.

If the location of all points in a system is expressible in terms of a set of Lagrangian coordinates, the set is complete. The minimum number of Lagrangian coordinates required for a complete set is equal to the number of degrees of freedom of the system. The elements of the complete set are called primary, or generalized, coordinates, and are designated by the symbol q_i. The remaining elements of the set are secondary coordinates. For example, because the mechanism of Figure 6.5 has one degree of freedom, it will have one generalized coordinate. If $q_1 = \theta$ is chosen as the generalized coordinate, then s_A and s_B are the secondary coordinates. The choice here was completely arbitrary; any of the coordinates could have been chosen as the generalized coordinate. As will be demonstrated later, this flexibility in choosing coordinates proves convenient.

6.3 WORK AND ENERGY

The work done on an object, W, by a force, $\vec{F}$, is the dot product of the force and displacement vectors, or

$$dW = \vec{F} \cdot d\vec{z} \tag{6.40}$$

where $d\vec{z}$ is the displacement of the object. The total work done in displacing the object from point 1 to point 2 along its path is

$$W = \int_{z_1}^{z_2} \vec{F} \cdot d\vec{z} \tag{6.41}$$

The work done by a moment, $\vec{M}$, is

$$dW = \vec{M} \cdot d\vec{\theta} \tag{6.42}$$

where $d\vec{\theta}$ is the angular displacement of the object. The work is positive if the moment is in the same direction as the displacement, and negative if it is in the opposite direction.

A force is conservative if the work done by the force is independent of path—that is, dependent on the coordinates of the displacement endpoints only. In this case, the work done is the difference in the potential energy, V, of the system at the two endpoints:

$$W = V_1 - V_2 \qquad (6.43)$$

The work done on a spring fits this category. The strain energy of a spring may be determined from

$$V = \int_{s_o}^{s} f_k(s)\,ds \qquad (6.44)$$

where $f_k(s)$ is the spring force as a function of s and s_o is the value of s for which the spring force is zero. For the special case of a linear spring, $s = x - x_o$, where x and x_o are the deflected and undeflected coordinates of the spring, and $s_o = 0$. The spring force is $f_k(s) = k(x - x_o) = ks$, where k is the spring constant. The strain energy is found from equation (6.44) to be

$$V = \frac{k}{2}(x - x_o)^2 \qquad (6.45)$$

Torsional springs are also common in pseudo-rigid-body models, and their strain energy may be calculated in general form as

$$V = \int_{\psi_o}^{\psi} m_k(\psi)\,d\psi \qquad (6.46)$$

where $m_k(\psi)$ is the spring torque as a function of ψ and ψ_o is the value of ψ for which the spring torque is zero. A linear torsional spring in which $\psi = \theta - \theta_o$, $\psi_o = 0$, and $m_k(\psi) = k_\theta(\theta - \theta_o) = k_\theta \psi$, where k_θ is the torsional spring constant, results in

$$V = \frac{k_\theta}{2}(\theta - \theta_o)^2 \qquad (6.47)$$

The strain energy of the flexible members in a compliant mechanism is typically much greater than the work potential due to gravity. When this is not the case, the work potential due to gravity may be used as $V = mgy_g$, where mg is the weight and y_g is the height of the center of gravity of the mechanism from a fixed datum.

6.4 VIRTUAL DISPLACEMENTS AND VIRTUAL WORK

A virtual displacement is a fictitious or imaginary displacement designated as $\delta \vec{z}$ and is expressed as a function of the generalized coordinates. The value of virtual displacements lies in their use to calculate virtual work. The virtual work, δW, due to a force, $\vec{F}$, and a virtual displacement, $\delta \vec{z}$, is

$$\delta W = \vec{F} \cdot \delta \vec{z} \qquad (6.48)$$

Similarly, the virtual work due to a moment, $\vec{M}$, and a virtual angular displacement, $\delta \vec{\theta}$, is

$$\delta W = \vec{M} \cdot \delta \vec{\theta} \qquad (6.49)$$

A convenient form of virtual work for conservative forces is found from the derivative of the potential energy, V, with respect to the generalized coordinate, q, to be

$$\delta W = -\frac{dV}{dq} \delta q \qquad (6.50)$$

For the general nonlinear springs of equations (6.44) and (6.46),

$$\delta W = -f_k(s) \frac{ds}{dq} \delta q \qquad (6.51)$$

$$\delta W = -m_k(\psi) \frac{d\psi}{dq} \delta q \qquad (6.52)$$

This form is useful because integration and differentiation of the nonlinear spring functions are avoided.

The total virtual work of a system may be written as

$$\delta W = \sum_i \vec{F}_i \cdot \delta \vec{z}_i + \sum_j \vec{M}_j \cdot \delta \vec{\theta}_j - \sum_k \frac{dV_k}{dq_k} \delta q_k \qquad (6.53)$$

Consider again the mechanism of Figure 6.5. The generalized coordinate is chosen to be $q = \theta$. A virtual displacement of $\delta \theta$ is assumed, as shown in Figure 6.5b. The forces performing work to produce this displacement are the vertical applied force at point C, $\vec{F}_C$, and the spring force, $\vec{F}_A$. The vertical reaction force at slider A and the horizontal reaction force at slider B are not included because they do not perform work for the displacement specified. The virtual work of force $\vec{F}_C$ is found to be

Virtual Displacements and Virtual Work

$$\delta W_C = \vec{F}_C \cdot \delta \vec{z}_{C/o} \tag{6.54}$$

where

$$\vec{F}_C = -F_C \hat{j} \tag{6.55}$$

and F_c is the magnitude of the force vector and $\hat{j}$ is the unit vector in the y direction ($\hat{i}$ will be used as the unit vector in the x direction).

The virtual displacement, $\delta \vec{z}_{C/o}$, must be described in terms of the generalized coordinate, θ. This is done by finding the position vector, $\vec{z}_{C/o}$, and using the chain rule of differentiation to find $\delta \vec{z}_{C/o}$:

$$\vec{z}_{C/o} = \frac{L}{2} \cos \theta \hat{i} + \frac{L}{2} \sin \theta \hat{j} \tag{6.56}$$

and

$$\delta \vec{z}_{C/o} = \frac{d \vec{z}_{C/o}}{d \theta} \delta \theta = -\frac{L}{2} \sin \theta \delta \theta \hat{i} + \frac{L}{2} \cos \theta \delta \theta \hat{j} \tag{6.57}$$

The virtual work, δW_C, is

$$\delta W_C = -F_C \hat{j} \cdot \delta \vec{z}_{C/o} = -\frac{F_C L}{2} \cos \theta \delta \theta \tag{6.58}$$

The virtual work of the spring force, $\vec{F}_A$, may be calculated as

$$\vec{F}_A = -k(x_A - x_{Ao}) \hat{i} = -kL(\cos \theta - \cos \theta_o) \hat{i} \tag{6.59}$$

where k is the spring constant, and x_{Ao} and θ_o are the values of x_A and θ when the spring is unstretched. The virtual displacement, $\delta \vec{z}_{A/o}$, is found from

$$\vec{z}_{A/o} = L \cos \theta \hat{i} \tag{6.60}$$

to be

$$\delta \vec{z}_{A/o} = -L \sin \theta \delta \theta \hat{i} \tag{6.61}$$

The virtual work that is caused by $\vec{F}_A$ is

$$\delta W_A = \vec{F}_A \cdot \delta \vec{z}_{A/o} = kL^2 (\cos \theta - \cos \theta_o) \sin \theta \delta \theta \tag{6.62}$$

An alternative method to calculate δW_A is to use the spring potential energy and equation (6.50). The potential energy, V, is

$$V = \frac{k}{2}L^2(\cos\theta - \cos\theta_o)^2 \tag{6.63}$$

and

$$\delta W_A = -\frac{dV}{d\theta}\delta\theta = kL^2(\cos\theta - \cos\theta_o)\sin\theta\,\delta\theta \tag{6.64}$$

which is consistent with the virtual work calculated in equation (6.62).

The total virtual work for the system is

$$\delta W = \delta W_C + \delta W_A = \left[-\frac{F_C L}{2}\cos\theta + kL^2(\cos\theta - \cos\theta_o)\sin\theta\right]\delta\theta \tag{6.65}$$

6.5 PRINCIPLE OF VIRTUAL WORK

The principle of virtual work may be expressed as follows [121]: "The net virtual work of all active forces is zero if and only if an ideal mechanical system is in equilibrium." An ideal mechanical system is one in which the constraints do no work. The mechanisms considered here will be assumed to be ideal.

Applying the principle of virtual work to the previous example and equation (6.65) yields

$$\delta W = \left[-\frac{F_C L}{2}\cos\theta + kL^2(\cos\theta - \cos\theta_o)\sin\theta\right]\delta\theta = 0 \tag{6.66}$$

Simplification results in

$$\frac{F_C}{2}\cos\theta = kL(\cos\theta - \cos\theta_o)\sin\theta \tag{6.67}$$

This equation relates the spring properties, applied force values, and mechanism geometry at the equilibrium position. For instance, the vertical force required to obtain an equilibrium position at an angle θ is found by solving the equation (6.67) for F_C as

$$F_C = 2kL(\cos\theta - \cos\theta_o)\tan\theta \tag{6.68}$$

The principle of virtual work offers considerable flexibility. In the example above, any other Lagrangian coordinate could have been chosen as the generalized coordinate, and another virtual displacement chosen. For instance, s_A and δs_A

Application of the Principle of Virtual Work

could have been chosen as the generalized coordinate and virtual displacement, respectively, and the same answers would have resulted. Another advantage of the method is that only the forces doing work are required in the analysis, whereas Newtonian methods require that all system forces be included.

6.6 APPLICATION OF THE PRINCIPLE OF VIRTUAL WORK

The force–displacement characteristics of compliant mechanisms can be found by applying the principle of virtual work. The method may be described in a series of steps. These steps are described below and provided in the context of the pseudo-rigid-body model of a compliant mechanism, as illustrated in Figure 6.6. The mechanism has a horizontal input force, F_{in}, an output moment, M_{out}, and a torsional spring to represent a small-length flexural pivot with a torsional spring constant of k_4.

Step 1. Choose the generalized coordinate, q.

For this example, it makes sense to choose θ_2 as the generalized coordinate because it is the known input.

Step 2. Express applied forces in vector form.

For the example of Figure 6.6, the input force is horizontal, so

$$\vec{F} = F_{in}\hat{i} \tag{6.69}$$

Step 3. Write a vector from the origin to placement of each force in step 2.

For our example, the origin is chosen as the point where link 2 is pinned to ground. There is only one force, so one displacement vector is needed. This is

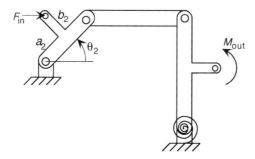

Figure 6.6. Example mechanism for application of the principle of virtual work.

$$\vec{Z} = (a_2 \cos \theta_2 - b_2 \sin \theta_2)\hat{i} + (a_2 \sin \theta_2 + b_2 \cos \theta_2)\hat{j} \qquad (6.70)$$

where a and b are the dimension on the link that locate the force, as shown in Figure 6.6.

Step 4. Determine the virtual displacement by differentiating the position vectors (found in step 3) with respect to the generalized coordinate. $[\delta\vec{Z} = (d\vec{Z}/dq)\delta(q)$, where q is the generalized coordinate.]

If θ_2 is used as the generalized coordinate, then

$$\delta\vec{Z} = \frac{d\vec{Z}}{d\theta_2}\delta\theta_2 = (-a_2 \sin \theta_2 - b_2 \cos \theta_2)\delta\theta_2\hat{i} + (a_2 \cos \theta_2 - b_2 \sin \theta_2)\delta\theta_2\hat{j} \quad (6.71)$$

Step 5. Calculate the virtual work (δW) due to forces by taking the dot product of the force vector (see step 2) and the virtual displacement (see step 4).

For the example, this results in

$$\delta W = \vec{F} \cdot \delta\vec{Z} \qquad (6.72)$$

$$\delta W = -F_{in}(a_2 \sin \theta_2 + b_2 \cos \theta_2)\delta\theta_2 \qquad (6.73)$$

This is the virtual work due to the applied force.

Step 6. Express the applied moments in vector form.

The output moment, M_{out}, can be expressed as

$$M_{out}\hat{k} \qquad (6.74)$$

The moment caused by the torsional spring can either be taken into account in this step, or in the step involving potential energy. Taking into account here results in

$$T_4 = -k_4(\theta_4 - \theta_{40})\hat{k} \qquad (6.75)$$

Step 7. Determine the angles the moments (step 6) act through (the angular displacement), and write in vector form.

For M_{out}

$$\vec{\Theta} = (\theta_4 - \theta_{40})\hat{k} \qquad (6.76)$$

and for T_4

Application of the Principle of Virtual Work

$$\vec{\Theta} = (\theta_4 - \theta_{40})\hat{k} \quad (6.77)$$

In this example $\vec{\Theta}$ happens to be the same for both moments, but this will not always be the case.

Step 8. Find the virtual angular displacements ($\delta\vec{\Theta}$) by differentiating the angular displacements (see step 7) with respect to the generalized coordinate.

For both M_{out} and T_4 this would be

$$\delta\vec{\Theta} = \frac{d}{d\theta_2}\vec{\Theta}\delta\theta_2 = \frac{d\theta_4}{d\theta_2}\delta\theta_2\hat{k} \quad (6.78)$$

where $d\theta_4/d\theta_2$ is the kinematic coefficient described in Chapter 4 as

$$\frac{d\theta_4}{d\theta_2} = \frac{r_2\sin(\theta_3 - \theta_2)}{r_4\sin(\theta_3 - \theta_4)} \quad (6.79)$$

Step 9. Calculate the virtual work (δW) due to moments by taking the dot product of the moment vector (see step 6) and the virtual angular displacement (see step 8).

For M_{out}

$$\delta W = M_{out}\frac{d\theta_4}{d\theta_2}\delta\theta_2 \quad (6.80)$$

and for T_4

$$\delta W = -k_4(\theta_4 - \theta_{40})\frac{d\theta_4}{d\theta_2}\delta\theta_2 \quad (6.81)$$

Step 10. Find sources of potential energy that have not been accounted for in previous steps.

The energy associated with the torsional spring has already been taken care of by including it in the moment terms in previous steps. An alternative approach would be to take into account the potential energy of the system. If this approach was taken, the potential energy for the spring would be

$$V = \frac{1}{2}k_4(\theta_4 - \theta_{40})^2 \quad (6.82)$$

Step 11. Find the virtual work from potential energy (see step 10) by differentiating the potential energy with respect to the generalized coordinate and multiplying by $-\delta q$.

Using the potential energy in step 10 results in

$$\delta W = -k_4(\theta_4 - \theta_{40})\frac{d\theta_4}{d\theta_2}\delta\theta_2 \qquad (6.83)$$

but recall that the virtual work from the spring was already calculated earlier with the moments. Note the equivalence in the two methods—equations (6.81) and (6.83) are identical.

Step 12. Calculate the total virtual work by summing the virtual work terms in steps 5, 9, and 11.

For the example, this results in

$$\delta W = -F_{in}(a_2 \sin\theta_2 + b_2 \cos\theta_2)\delta\theta_2 + M_{out}\frac{d\theta_4}{d\theta_2}(\delta\theta_2) - k_4(\theta_4 - \theta_{40})\frac{d\theta_4}{d\theta_2}\delta\theta_2 \qquad (6.84)$$

Note that the virtual work due to the spring was calculated two ways (steps 9 and 11) to demonstrate the methods, but the virtual work term should only be added once.

Step 13. Apply the principle of virtual work: If in equilibrium, the virtual work is equal to zero ($\delta W = 0$).

If the virtual work in equation (6.84) is set equal to zero, both sides may be divided by $\delta\theta_2$ to obtain

$$-F_{in}(a_2 \sin\theta_2 + b_2 \cos\theta_2) + M_{out}\frac{d\theta_4}{d\theta_2} - k_4(\theta_4 - \theta_{40})\frac{d\theta_4}{d\theta_2} = 0 \qquad (6.85)$$

Step 14. Solve the equation from step 13 for the unknown.

If the input force is known and the output force M_{out} is unknown, then M_{out} may be calculated for a given position. Rearranging equation (6.85) and substituting equation (6.79) for $d\theta_4/d\theta_2$ results in

$$M_{out} = \frac{F_{in}(a_2 \sin\theta_2 + b_2 \cos\theta_2)r_4 \sin(\theta_3 - \theta_4)}{r_2 \sin\theta_3 - \theta_2} + k_4(\theta_4 - \theta_{40}) \qquad (6.86)$$

Application of the Principle of Virtual Work   235

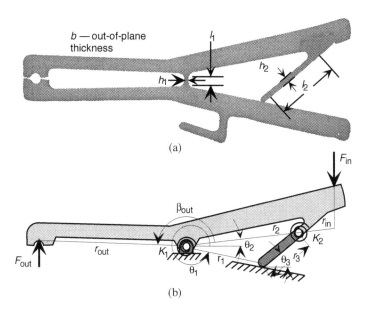

Figure 6.7. (a) Compliant grippers used to transfer silicon chips between various chemicals during processing, and (b) their pseudo-rigid-body model.

Example: Compliant Grippers. Consider the compliant grippers and the corresponding pseudo-rigid-body model shown in Figure 6.7. The displacement equations were developed in Section 5.9.1. The force–deflection relationship can be found using the principle of virtual work by following the steps above.

Step 1: The equations are easily expressed in terms of θ_2 and it is chosen as the generalized coordinate.

Step 2: The applied forces are expressed in vector form as $F_{in}\hat{j}$ and $F_{out}\hat{j}$. Note that F_{in} will have a negative value because it is in the negative $\hat{j}$ direction. The friction force at the sliding connection can be assumed to be small because there is a very low coefficient of friction between Teflon and polypropylene.

Step 3: The vectors from the origin to the placement of the two forces in step 2 are

$$\vec{z}_{in} = r_{in} \cos \theta_2 \hat{i} + r_{in} \sin \theta_2 \hat{j} \tag{6.87}$$

and

$$\vec{z}_{out} = r_{in} \cos(\theta_2 + \beta)\delta\theta_2 \hat{i} + r_{out} \cos(\theta_2 + \beta)\delta\theta \hat{j} \tag{6.88}$$

Step 4: The virtual displacement is found by differentiating the position vectors found in step 3 with respect to the generalized coordinate, θ_2, resulting in

$$\delta \vec{z}_{in} = -r_{in} \sin \theta_2 \delta \theta_2 \hat{i} + r_{in} \cos \theta_2 \delta \theta_2 \hat{j} \tag{6.89}$$

and

$$\delta \vec{z}_{out} = -r_{out} \sin(\theta_2 + \beta) \delta \theta_2 \hat{i} + r_{out} \cos(\theta_2 + \beta) \delta \theta_2 \hat{j} \tag{6.90}$$

Step 5: The virtual work due to the forces are the dot product of the force vector in step 2 and the virtual displacement in step 4 (i.e., $\vec{F}_i \cdot \delta \vec{z}_i$), or

$$\delta W_{in} = \vec{F}_{in} \cdot \delta \vec{z}_{in} = F_{in} r_{in} \cos \theta_2 \delta \theta_2 \tag{6.91}$$

and

$$\delta W_{out} = \vec{F}_{out} \cdot \delta \vec{z}_{out} = F_{out} r_{out} \cos(\theta_2 + \beta) \delta \theta_2 \tag{6.92}$$

Step 6: The moments at the torsional springs can be taken into account by accounting for the work done by the moment or by the potential energy. To demonstrate both methods, the spring at the small-length flexural pivot will be taken into account as the work done by the moment, and the virtual work of the other torsional spring will be taken into account using the potential energy. The moment at the small-length flexural pivot is $\vec{T}_1 = -K_1(\theta_2 - \theta_{2_0})\hat{k}$.

Step 7: The moment in step 6 acts through an angle of $\vec{\Theta}_1 = (\theta_2 - \theta_{2_0})\hat{k}$.

Step 8: The virtual angular displacement is the derivative of the angular displacement of step 7 with respect to the generalized coordinate, or $\delta \vec{\Theta}_1 = \delta \theta_2 \hat{k}$.

Step 9: The virtual work caused by the moment is the dot product of the moment vector (step 6) and the virtual angular displacement (step 8), $\delta W_1 = \vec{M}_1 \delta \vec{\Theta}_1$, or

$$\delta W_1 = \vec{T}_1 \cdot \delta \vec{\Theta}_1 = -K_1(\theta_2 - \theta_{2_0}) \delta \theta_2 \tag{6.93}$$

Step 10: The torsional spring associated with the fixed–pinned member has not been taken into account in previous steps, and its potential energy is $V_2 = K_2 \Theta_2^3 / 2$ where $\Theta_2 = [(\theta_3 - \theta_2) - (\theta_{3_0} - \theta_{2_0})]$.

Step 11: The virtual work from the potential energy in step 10 is $-(dV_2/d\theta_2)\delta\theta_2$, or

$$\delta W_2 = -\frac{dV}{d\theta_2} \delta \theta_2 = -K_2 \Theta_2 \left(\frac{d\theta_3}{d\theta_2} - 1 \right) \delta \theta_2. \tag{6.94}$$

Spring Function for Fixed–Pinned Members

Step 12: The total virtual work is the sum of the virtual work terms in steps 5, 9, and 11, or

$$\delta W = F_{in} r_{in} \cos\theta_2 \delta\theta_2 + F_{out} r_{out} \cos(\theta_2 + \beta)\delta\theta_2 \\ - K_1(\theta_2 - \theta_{2_0})\delta\theta_2 - K_2|\theta_2|\left(\frac{d\theta_3}{d\theta_2} - 1\right)\delta\theta_2 \qquad (6.95)$$

Steps 13 and 14: The principle of virtual work is applied by setting the virtual work equal to zero. This equation is then solved for the unknown, F_{out}, as

$$F_{out} = \frac{K_1(\theta_2 - \theta_{2_0}) + K_2\Theta_2\left(\dfrac{d\theta_3}{d\theta_2} - 1\right) - F_{in} r_{in} \cos\theta_2}{r_{out} \cos(\theta_2 + \beta)} \qquad (6.96)$$

6.7 SPRING FUNCTION FOR FIXED–PINNED MEMBERS

The spring function for the pseudo-rigid-body model of a fixed–pinned member may be determined as another example of the use of the principle of virtual work. Consider the functionally binary fixed–pinned segment and its pseudo-rigid-body model shown in Figure 6.8. The spring function, m_k, may be determined from the principle of virtual work. Specifying $\vec{T}$ as the moment at the characteristic pivot, the virtual work for the pseudo-rigid-body model is

$$\delta W = \vec{F} \cdot \delta\vec{z} + \vec{T} \cdot \delta\vec{\Theta} \qquad (6.97)$$

where

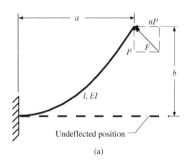
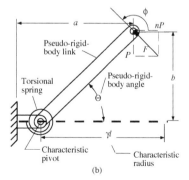

Figure 6.8. (a) Flexible cantilever beam with a force at its free end, and (b) its pseudo-rigid-body model.

$$\vec{F} = -nP\hat{i} + P\hat{j} \tag{6.98}$$

$$\vec{z} = [l(1-\gamma) + \gamma l\cos\Theta]\hat{i} + \gamma l\sin\Theta\hat{j} \tag{6.99}$$

Using Θ as the generalized coordinate, the virtual displacement is

$$\delta\vec{z} = -\gamma l\sin\Theta\delta\Theta\hat{i} + \gamma l\cos\Theta\hat{j} \tag{6.100}$$

The moment, $\vec{T}$, may be specified as a function of m_k:

$$\vec{T} = -m_k(\Theta)\hat{k} \tag{6.101}$$

The virtual work is found from equation (6.97) to be

$$\delta W = nP\gamma l\sin\Theta\delta\Theta + P\gamma l\cos\Theta\delta\Theta - m_k(\Theta)\delta\Theta \tag{6.102}$$

Applying the principle of virtual work ($\delta W = 0$) and rearranging yields

$$m_k(\Theta) = P\gamma l(n\sin\Theta + \cos\Theta) \tag{6.103}$$

Recalling that $F_t = P(n\sin\Theta + \cos\Theta)$ and $\alpha_t^2 = F_t l^2/EI$ and substituting into equation (6.103) yields

$$m_k(\Theta) = \alpha_t^2 \gamma \frac{EI}{l} \tag{6.104}$$

Recall that $\alpha_t^2 = K_\Theta \Theta$, where K_Θ is the stiffness coefficient. Substituting this into equation (6.104) results in

$$m_k(\Theta) = \gamma K_\Theta \frac{EI}{l} \Theta \tag{6.105}$$

or

$$m_k(\Theta) = K\Theta \tag{6.106}$$

where K is the value of the spring constant.

Table 6.1 summarizes the values for K and $m_k(\psi)$ for several types of flexible segments. Note that for all these cases

$$m_k(\psi) = K\psi \tag{6.107}$$

For the problem described in this section, $\psi = \Theta$.

TABLE 6.1. Spring functions for various types of flexible members

Type	K	$m_k(\psi)$
Small-length flexural pivot	$\dfrac{EI}{l}$	$\dfrac{EI}{l}\psi$
Fixed–pinned segment	$\gamma K_\Theta \dfrac{EI}{l}$	$\gamma K_\Theta \dfrac{EI}{l}\psi$
Fixed–fixed guided segment	$2\gamma K_\Theta \dfrac{EI}{l}$	$2\gamma K_\Theta \dfrac{EI}{l}\psi$
Initially curved fixed–pinned	$\rho K_\Theta \dfrac{EI}{l}$	$\rho K_\Theta \dfrac{EI}{l}\psi$

6.8 PSEUDO-RIGID-BODY FOUR-BAR MECHANISM

Rigid-body four-bar mechanisms are widely used because of their simplicity and versatility. Much rigid-body mechanism theory has been centered on four-bar mechanisms. Thus it is reasonable to assume that many useful compliant mechanisms will have pseudo-rigid-body mechanisms that are of four-bar mechanisms.

Consider a general pseudo-rigid-body four-bar mechanism with arbitrary externally applied force and moment loads on each link and torsional springs at each joint, as shown in Figure 6.9. The total virtual work of the system can be expressed as

$$\delta W = \sum_{i=2}^{4} \vec{F}_i \cdot \delta \vec{z}_i + \sum_{i=2}^{4} \vec{M}_i \cdot \delta \vec{\theta}_i + \sum_{i=1}^{4} \vec{T}_i \cdot \delta \vec{\psi}_i \qquad (6.108)$$

where $\vec{F}_i$ is the force applied to link i, which is expressed as

$$\vec{F}_i = X_i \hat{i} + Y_i \hat{j} \qquad (6.109)$$

$\vec{M}_i$ is the moment applied to link i, and $\vec{T}_i$ is the moment at characteristic pivot i.

The virtual displacements, $\delta \vec{z}_i$, are found by using the chain rule of differentiation on the displacement vectors, $\vec{z}_i$. For example,

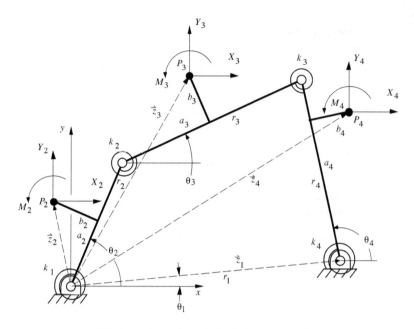

Figure 6.9. General pseudo-rigid-body four-bar mechanism.

$$\vec{z}_2 = (a_2 \cos\theta_2 - b_2 \sin\theta_2)\hat{i} + (a_2 \sin\theta_2 + b_2 \cos\theta_2)\hat{j} \qquad (6.110)$$

and

$$\delta\vec{z}_2 = (-a_2 \sin\theta_2 - b_2 \cos\theta_2)\delta\theta_2\hat{i} + (a_2 \cos\theta_2 - b_2 \sin\theta_2)\delta\theta_2\hat{j} \qquad (6.111)$$

The other virtual displacements are found in the same manner.

The virtual work due to the torsional spring at joint i may be determined from the moment at the joint, T_i, and an appropriate Lagrangian coordinate, ψ_i. For a general nonlinear torsional spring, the value of the moment is

$$T_i = -m_{k_i}(\psi_i) \qquad (6.112)$$

where m_k is the value of the moment as a function of the Lagrangian coordinate. For a pseudo-rigid-body model with a linear torsional spring constant, K_i,

$$T_i = -K_i\psi_i \qquad (6.113)$$

and values of K_i for various types of segments were listed in Table 6.1.

The Lagrangian coordinates for the joints are

Pseudo-Rigid-Body Four-Bar Mechanism

$$\psi_1 = \theta_2 - \theta_{2o} \tag{6.114}$$

$$\psi_2 = (\theta_2 - \theta_{2o}) - (\theta_3 - \theta_{3o}) \tag{6.115}$$

$$\psi_3 = (\theta_4 - \theta_{4o}) - (\theta_3 - \theta_{3o}) \tag{6.116}$$

$$\psi_4 = \theta_4 - \theta_{4o} \tag{6.117}$$

where the θ_{io} define the position of the mechanism when the springs are undeflected. Most compliant mechanisms have an initial position at which all the springs are undeflected, and $\theta_{io} = \theta_i$ at this initial position.

The $\delta\psi_i$ required are

$$\delta\psi_1 = \delta\theta_2 \tag{6.118}$$

$$\delta\psi_2 = \delta\theta_2 - \delta\theta_3 \tag{6.119}$$

$$\delta\psi_3 = \delta\theta_4 - \delta\theta_3 \tag{6.120}$$

$$\delta\psi_4 = \delta\theta_4 \tag{6.121}$$

The total virtual work for this system may now be found using equation (6.110). The result may be expressed as

$$\delta W = A\,\delta\theta_2 + B\,\delta\theta_3 + C\,\delta\theta_4 \tag{6.122}$$

where

$$\begin{aligned} A = &\ (-X_2 a_2 - Y_2 b_2 - r_2 X_3)\sin\theta_2 \\ &+ (-X_2 b_2 + Y_2 a_2 + r_2 Y_3)\cos\theta_2 \\ &+ M_2 + T_1 + T_2 \end{aligned} \tag{6.123}$$

$$\begin{aligned} B = &\ (-X_3 a_3 - Y_3 b_3)\sin\theta_3 \\ &+ (-X_3 b_3 + Y_3 a_3)\cos\theta_3 \\ &+ M_3 - T_2 - T_3 \end{aligned} \tag{6.124}$$

$$\begin{aligned} C = &\ (-X_4 a_4 - Y_4 b_4)\sin\theta_4 \\ &+ (-X_4 b_4 + Y_4 a_4)\cos\theta_4 \\ &+ M_4 + T_3 + T_4 \end{aligned} \tag{6.125}$$

r_i are the link lengths, and a_i and b_i are the axial and normal distances along the link to the load location, as shown in Figure 6.9.

Equation (6.122) is a general form of virtual work for any choice of a generalized coordinate. Once a generalized coordinate is specified, the equation may be simplified. For example, if θ_2 is the known mechanism input and is chosen as the generalized coordinate, equation (6.122) may be written as

$$\delta W = \left(A + B \frac{\delta \theta_3}{\delta \theta_2} + C \frac{\delta \theta_4}{\delta \theta_2} \right) \delta \theta_2 \tag{6.126}$$

Applying the principle of virtual work results in

$$\delta W = 0 \tag{6.127}$$

Combining equations (6.126) and (6.127) yields

$$A + B \frac{\delta \theta_3}{\delta \theta_2} + C \frac{\delta \theta_4}{\delta \theta_2} = 0 \tag{6.128}$$

Since $\delta \theta_3 / \delta \theta_2$ and $\delta \theta_4 / \delta \theta_2$ are kinematic coefficients, then h_{ij}, [121] has the values

$$\frac{\delta \theta_3}{\delta \theta_2} = h_{32} = \frac{r_2 \sin(\theta_4 - \theta_2)}{r_3 \sin(\theta_3 - \theta_4)} \tag{6.129}$$

and

$$\frac{\delta \theta_4}{\delta \theta_2} = h_{42} = \frac{r_2 \sin(\theta_3 - \theta_2)}{r_4 \sin(\theta_3 - \theta_4)} \tag{6.130}$$

Equation (6.126) may be rewritten as

$$\delta W = (A + B h_{32} + C h_{42}) \delta \theta_2 \tag{6.131}$$

If θ_3 or θ_4 is chosen as the generalized coordinate, equation (6.126) is expressed as

$$\delta W = (A h_{23} + B + C h_{43}) \delta \theta_3 \tag{6.132}$$

or

$$\delta W = (A h_{24} + B h_{34} + C) \delta \theta_4 \tag{6.133}$$

respectively, where

$$h_{43} = \frac{r_3 \sin(\theta_3 - \theta_2)}{r_4 \sin(\theta_4 - \theta_2)} \tag{6.134}$$

Pseudo-Rigid-Body Four-Bar Mechanism

and

$$h_{ij} = \frac{1}{h_{ji}} \tag{6.135}$$

Applying the principle of virtual work to equation (6.131)

$$A + Bh_{32} + Ch_{42} = 0 \text{ for } q = \theta_2 \tag{6.136}$$

or applied to equation (6.132),

$$Ah_{23} + B + Ch_{43} = 0 \text{ for } q = \theta_3 \tag{6.137}$$

or applied to equation (6.133),

$$Ah_{24} + Bh_{34} + C = 0 \text{ for } q = \theta_4 \tag{6.138}$$

These general equations may be used to study a variety of compliant mechanisms. Three examples follow.

Example: Bicycle Brakes Using Virtual Work. Recall the compliant bicycle brakes illustrated in Figure 6.3 and used in an example of the free-body diagram approach in Section 6.1. The general equations for a four-bar mechanism developed using the principle of virtual work may also be used to find the force–deflection relationships.

If θ is chosen as the generalized coordinate, equation (6.136) applies. The only externally applied load in the x direction is F_{in} on link 4, so

$$X_4 = F_{in} \tag{6.139}$$

and

$$X_2 = X_3 = 0 \tag{6.140}$$

There are no externally applied forces in the y direction, so

$$Y_2 = Y_3 = Y_4 = 0 \tag{6.141}$$

nor are there any externally applied moments, or

$$M_2 = M_3 = M_4 = 0 \tag{6.142}$$

There are no torsional springs at joints 3 and 4, so

$$T_3 = T_4 = 0 \tag{6.143}$$

Because of the special geometry of the parallelogram mechanism, $\theta = \theta_2 = \theta_4$ and $\theta_3 = 0$. The spring constants at joints 1 and 2 can be found from equations (6.113) to (6.115) as

$$T_1 = T_2 = -K(\theta - \theta_o) \tag{6.144}$$

where θ_o is the value of θ when the springs are undeflected and K_1 is the torsional spring constant described earlier as $K = 2\gamma K_\Theta EI/l$ for a fixed–guided segment.

Substituting the values above into equations (6.123) through (6.125) results in

$$A = T_1 + T_2 = 2T_1 \tag{6.145}$$

$$B = -T_2 \tag{6.146}$$

and

$$C = -F_{in} d \sin\theta \tag{6.147}$$

Substituting these values into equation (6.136) results in

$$2T_1 - T_1 h_{32} - F_{in} h_{42} d \sin\theta = 0 \tag{6.148}$$

The special geometry for the mechanism results in kinematic coefficients of

$$h_{32} = 0 \tag{6.149}$$

and

$$h_{42} = 1 \tag{6.150}$$

The input force may be found by substituting equations (6.144), (6.149), and (6.150) into equation (6.148) and rearranging. This results in

$$F_{in} = \frac{-2K(\theta - \theta_2)}{d \sin\theta} \tag{6.151}$$

Note that this is the same as equation (6.39) found using the free-body diagram approach.

Example: Fully Compliant Mechanism. Consider the fully compliant mechanism and its pseudo-rigid-body model shown in Figure 6.10. The flexible members are assumed to be small in length compared to the more rigid sections, and they may be modeled as turning pairs with torsional springs. The torsional spring function, m_{k_i}, is assumed to be linear in ψ_i:

Pseudo-Rigid-Body Four-Bar Mechanism

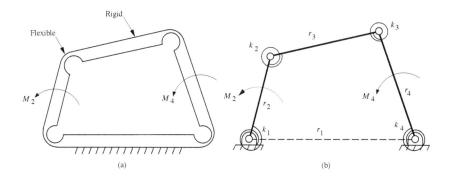

Figure 6.10. (a) Fully compliant mechanism, and (b) its pseudo-rigid-body model.

$$m_{k_i} = \frac{EI_i}{l_i}\psi_i \quad (6.152)$$

where E is the modulus of elasticity, I the moment of inertia, and l the length of the flexible section.

The general pseudo-rigid-body four-bar mechanism equations derived above are simplified using $X_i = Y_i = M_3 = 0$, to obtain

$$A = M_2 + T_1 + T_2 \quad (6.153)$$

$$B = -T_2 - T_3 \quad (6.154)$$

$$C = M_4 + T_3 + T_4 \quad (6.155)$$

Values for T_i are found using equations (6.112) through (6.117) and Table 6.1 as

$$T_1 = \frac{-EI_1}{l_1}(\theta_2 - \theta_{2o}) \quad (6.156)$$

$$T_2 = \frac{-EI_2}{l_2}[(\theta_2 - \theta_{2o}) - (\theta_3 - \theta_{3o})] \quad (6.157)$$

$$T_3 = \frac{-EI_3}{l_3}[(\theta_4 - \theta_{4o}) - (\theta_3 - \theta_{3o})] \quad (6.158)$$

$$T_4 = \frac{-EI_4}{l_4}(\theta_4 - \theta_{4o}) \tag{6.159}$$

If θ_2 is the generalized coordinate, the virtual work is given by equation (6.131). The principal of virtual work results in

$$M_2 + T_1 + T_2 - (T_2 + T_3)h_{32} + (M_4 + T_3 + T_4)h_{42} = 0 \tag{6.160}$$

These equations are consistent with the results obtained in [3] using Newtonian methods.

Example: Partially Compliant Mechanism. Consider the partially compliant mechanism in Figure 6.11a. A flexible, functionally binary, fixed–pinned segment, the "coupler" also has one small-length flexural pivot. The mechanism is made of polypropylene ($E = 200{,}000$ lb/in^2). The long flexible segment has a length $l_3 = 3.0$ in., width $b_3 = 0.5$ in., and thickness $h_3 = 0.04$ in. Constant values of $\gamma = 0.85$ and $K_\Theta = 2.65$ result in a pseudo-rigid coupler with length $r_3 = \gamma l_3 = 2.55$ in. The pseudo-rigid-body model is shown in Figure 6.10b; $r_1 = 3$ in., and $r_4 = 3$ in. The moments of inertia for the flexible segments are

$$I_3 = \frac{b_3 h_3^3}{12} = 2.7 \times 10^{-6} \text{ in.}^4 \tag{6.161}$$

$$I_4 = \frac{b_4 h_4^3}{12} = 3.3 \times 10^{-7} \text{ in.}^4 \tag{6.162}$$

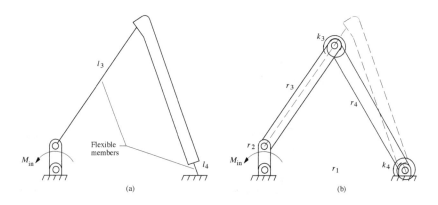

Figure 6.11. (a) Partially compliant mechanism, and (b) its pseudo-rigid-body model.

Pseudo-Rigid-Body Four-Bar Mechanism

The flexible segments are undeflected when $\theta_2 = 90°$. This calculates the values of θ_{3o} and θ_{4o} by using the closed-form equations presented in Chapter 4:

$$\theta_{3o} = 43.8° \tag{6.163}$$

$$\theta_{4o} = 112.8° \tag{6.164}$$

The general pseudo-rigid-body model four-bar mechanism equations are simplified using $X_i = Y_i = M_3 = M_4 = T_1 = T_2 = 0$, to obtain

$$A = M_2 = M_{in} \tag{6.165}$$

$$B = -T_3 \tag{6.166}$$

$$C = T_3 + T_4 \tag{6.167}$$

The value of T_3 is found by using equations (6.112) and (6.116) and Table 6.1:

$$T_3 = -\gamma K_\Theta \frac{EI}{l}[(\theta_4 - \theta_{4o}) - (\theta_3 - \theta_{3o})] \tag{6.168}$$

T_4 is the torque caused by the torsional spring modeling the small-length flexural pivot:

$$T_4 = -\left(\frac{EI}{l}\right)_4 (\theta_4 - \theta_{4o}) \tag{6.169}$$

If θ_2 is chosen as the generalized coordinate, the virtual work is given by equation (6.131). The principle of virtual work ($\delta w = 0$) results in

$$A + Bh_{32} + Ch_{42} = 0 \tag{6.170}$$

Substituting for each term in the equation above and solving for the input, M_{in}, as a function of θ_2 yields

$$M_{in} = \gamma K_\Theta \left(\frac{EI}{l}\right)_3 [(\theta_4 - \theta_{4o}) - (\theta_3 - \theta_{3o})] \left[\frac{r_2 \sin(\theta_3 - \theta_2)}{r_4 \sin(\theta_3 - \theta_4)} - \frac{r_2 \sin(\theta_4 - \theta_2)}{r_3 \sin(\theta_3 - \theta_4)}\right] + \left(\frac{EI}{l}\right)_4 (\theta_4 - \theta_{4o}) \frac{r_2 \sin(\theta_3 - \theta_2)}{r_4 \sin(\theta_3 - \theta_4)} \tag{6.171}$$

where θ_3 and θ_4 are functions of θ_2 and may be calculated using the closed-form equations in Chapter 4.

Figure 6.12 plots M_{in} versus θ_2. As expected, $M_{in} = 0$ at $\theta_2 = 90°$, where all flexible members are undeflected. M_{in} is equal to zero at two other positions. One of these is an unstable equilibrium position; the other is a stable equilibrium position. Bistable mechanisms are discussed in more detail in Chapter 11.

The two compliant mechanisms considered in these examples are very different from each other, yet both may be modeled using the general equations developed earlier.

6.9 PSEUDO-RIGID-BODY SLIDER MECHANISM

Consider the general slider mechanism of Figure 6.13. The slider mechanism is a four-bar mechanism with a prismatic pair (slider) that replaces a turning pair. Because of this relationship, the general equations for the slider mechanism may be obtained from those derived previously for the four-bar mechanism. Note that when a spring is added to the slider, the equation of virtual work becomes

$$\delta W = \sum_{i=2}^{4} \vec{F}_i \cdot \delta \vec{z}_i + \sum_{i=2}^{3} \vec{M}_i \cdot \delta \vec{\theta}_i + \sum_{i=1}^{3} \vec{T}_i \cdot \delta \vec{\psi}_i + \vec{F}_s \cdot \delta \vec{z}_4 \qquad (6.172)$$

where $\vec{F}_s = -f_k(\psi_4)\hat{i}$, and f_k is the spring force as a function of $\psi_4 = r_1 - r_{1o}$. The Lagrangian coordinates associated with $\vec{T}_i$ are

$$\psi_1 = \theta_2 - \theta_{2o} \qquad (6.173)$$

$$\psi_2 = (\theta_2 - \theta_{2o}) - (\theta_3 - \theta_{3o}) \qquad (6.174)$$

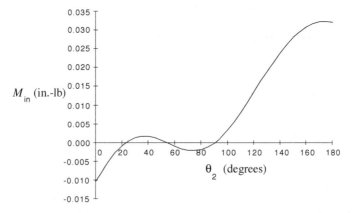

Figure 6.12. Required input moment, M_{in}, versus crank angle, θ_2.

Pseudo-Rigid-Body Slider Mechanism

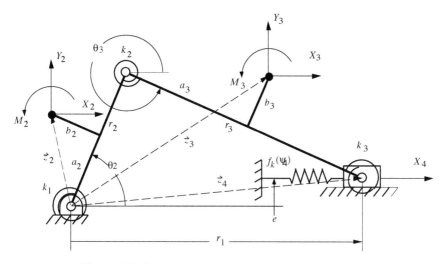

Figure 6.13. General pseudo-rigid-body slider mechanism.

$$\psi_3 = \theta_3 - \theta_{3o} \tag{6.175}$$

and those associated with $\delta\psi_i$ are

$$\delta\psi_1 = \delta\theta_2 \tag{6.176}$$

$$\delta\psi_2 = \delta\theta_2 - \delta\theta_3 \tag{6.177}$$

$$\delta\psi_3 = \delta\theta_3 \tag{6.178}$$

The virtual work may be calculated by using equation (6.172), resulting in

$$\delta W = A'\delta\theta_2 + B'\delta\theta_3 \tag{6.179}$$

where A' and B' are

$$\begin{aligned} A' = &[-X_2 a_2 - Y_2 b_2 - r_2(X_3 + X_4)]\sin\theta_2 \\ &+ (-X_2 b_2 + Y_2 a_2 + r_2 Y_3)\cos\theta_2 \\ &+ M_2 + T_1 + T_2 - F_s r_2 \sin\theta_2 \end{aligned} \tag{6.180}$$

$$\begin{aligned} B' = &(-X_3 a_3 - Y_3 b_3 - r_3 X_4)\sin\theta_3 \\ &+ (-X_3 b_3 + Y_3 a_3)\cos\theta_3 \\ &+ M_3 - T_2 + T_3 - F_s r_3 \sin\theta_3 \end{aligned} \tag{6.181}$$

The virtual work may be written for any generalized coordinate, q, as

$$\delta W = (g_{21}A' + g_{31}B')\delta r_1 \quad \text{for } q = r_1 \tag{6.182}$$

$$\delta W = (A' + g_{32}B')\delta \theta_2 \quad \text{for } q = \theta_2 \tag{6.183}$$

$$\delta W = (g_{23}A' + B')\delta \theta_3 \quad \text{for } q = \theta_3 \tag{6.184}$$

where

$$g_{ij} = \frac{C_i}{C_j} \tag{6.185}$$

and

$$C_1 = r_2 r_3 \sin(\theta_2 - \theta_3) \tag{6.186}$$

$$C_2 = -r_3 \cos \theta_3 \tag{6.187}$$

$$C_3 = r_2 \cos \theta_2 \tag{6.188}$$

Applying the principle of virtual work to equation (6.182) results in

$$g_{21}A' + g_{31}B' = 0 \quad \text{for } q = r_1 \tag{6.189}$$

Example: Compliant Slider Mechanism. Figure 6.14 shows a compliant slider mechanism. The flexible segment is steel ($E = 30 \times 10^6$ lb/in.^{2}), with length $l_2 = 24$ in., width $b = 1.25$ in., and thickness $1/32$ in., while the rigid link has length $r_3 = 17.6$ in. The mechanism is undeflected when $\theta_2 = 0°$ ($\theta_{2o} = 0$). Calculate the force required to deflect the slider 6 in.

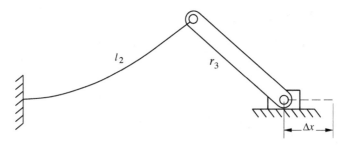

Figure 6.14. Example of a compliant slider mechanism.

Pseudo-Rigid-Body Slider Mechanism

Solution: Values of $K_\Theta = 2.6$ and $\gamma = 0.83$ apply to this loading. The length of the pseudo-rigid link, r_2, is

$$r_2 = \gamma l_2 = (0.83)(24 \text{ in.}) = 19.9 \text{ in.} \tag{6.190}$$

If a generalized coordinate of $q = r$ is chosen, the force–deflection relationship for this mechanism is found from equations (6.180) through (6.181) and (6.185) through (6.188):

$$\delta W = (g_{21} A' + g_{31} B') \delta r_1 \tag{6.191}$$

where

$$A' = -r_2 X_4 \sin \theta_2 + T_1 \tag{6.192}$$

$$B' = -r_3 X_4 \sin \theta_3 \tag{6.193}$$

$$g_{21} = \frac{-r_3 \cos \theta_3}{r_2 r_3 \sin(\theta_2 - \theta_3)} \tag{6.194}$$

$$g_{31} = \frac{r_2 \cos \theta_2}{r_2 r_3 \sin(\theta_2 - \theta_3)} \tag{6.195}$$

$$X_4 = -F \tag{6.196}$$

$$T_1 = -\gamma K_\Theta \frac{EI}{l_2} \theta_2 \tag{6.197}$$

Using the principal of virtual work ($\delta W = 0$ for equilibrium) gives

$$\frac{-r_3 \cos \theta_3}{r_2 r_3 \sin(\theta_2 - \theta_3)} \left(F r_2 \sin \theta_2 - \gamma K_\Theta \frac{EI}{l_2} \theta_2 \right) + \frac{r_2 \cos \theta_2}{r_2 r_3 \sin(\theta_2 - \theta_3)} F r_2 \sin \theta_3 = 0 \tag{6.198}$$

Solving for the force, F, results in

$$F = \frac{\gamma K_\Theta EI \theta_2 \cos \theta_3}{l_2 r_2 \sin(\theta_2 - \theta_3)} \tag{6.199}$$

The values of θ_2 and θ_3 when they are associated with an input slider displacement are found using geometry:

$$r_1 = r_2 + r_3 - \Delta x \tag{6.200}$$

$$\theta_2 = \text{acos} \frac{r_1^2 + r_2^2 - r_3^2}{2r_1 r_2} \qquad (6.201)$$

$$\theta_3 = \text{asin} \frac{-r_2 \sin \theta_2}{r_3} \qquad (6.202)$$

For the input of $\Delta x = 6$ in. and the link lengths specified,

$$\begin{aligned} r_1 &= 31.5 \text{ in.} \\ \theta_2 &= 0.54 \text{ rad} \\ \theta_3 &= -0.61 \text{ rad} \end{aligned} \qquad (6.203)$$

The force is found from equation (6.199):

$$\begin{aligned} F &= \frac{(0.83)(2.6)(30 \times 10^6 \text{ lb/in.}^2)\{[(1.25 \text{ in.})(1/32 \text{ in.})^3]/12\}(0.535) \cos(-0.614)}{(24 \text{ in.})(19.9 \text{ in.}) \sin(0.54 + 0.61)} \\ &= 0.21 \text{ lb} \end{aligned} \qquad (6.204)$$

Example: Equilibrium Position of a Slider Mechanism. Figure 6.15 shows a compliant slider mechanism and its pseudo-rigid-body model. The general equations given above are simplified by observing that

$$X_2 = X_3 = Y_2 = Y_3 = M_2 = M_3 = T_2 = f_k = 0 \qquad (6.205)$$

which results in

$$A' = (-r_2 X_4) \sin \theta_2 + T_1 \qquad (6.206)$$

and

$$B' = (-r_3 X_4) \sin \theta_3 + T_3 \qquad (6.207)$$

This particular mechanism does not have an orientation in which all members are unstrained, and the principle of virtual work may be used to determine the equilibrium position for the mechanism at rest. For this case $A' = T_1$ and $B' = T_3$. Choosing r_1 as the generalized coordinate, the principle of virtual work is applied using equation (6.182) as

$$\delta W = (g_{21} T_1 + g_{31} T_3) \delta r_1 = 0 \qquad (6.208)$$

or

Pseudo-Rigid-Body Slider Mechanism

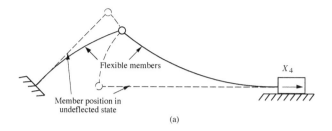

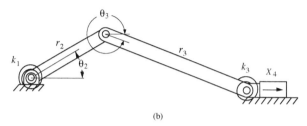

Figure 6.15. (a) Compliant slider mechanism, and (b) its pseudo-rigid-body model.

$$T_1 = \frac{-g_{31}}{g_{21}} T_3 = -g_{32} T_3 \qquad (6.209)$$

T_1 and T_3 are

$$T_1 = -\gamma_1 K_{\Theta_1} \frac{EI_1}{l_1} (\theta_2 - \theta_{2o}) \qquad (6.210)$$

and

$$T_3 = -\gamma_3 K_{\Theta_3} \frac{EI_3}{l_3} (\theta_3 - \theta_{3o}) \qquad (6.211)$$

Substituting into equation (6.209) yields

$$\gamma_1 K_{\Theta_1} \frac{EI_1}{l_1} (\theta_2 - \theta_{2o}) = \frac{r_2 \cos \theta_2}{r_3 \cos \theta_3} \gamma_3 K_{\Theta_3} \frac{EI_3}{l_3} (\theta_3 - \theta_{3o}) \qquad (6.212)$$

Because the reactive force at the pin joint is vertical, $n = 0$ for both flexible segments and $\gamma_1 K_{\Theta_1} = \gamma_3 K_{\Theta_3}$, resulting in

$$\frac{I_1}{l_1}(\theta_2 - \theta_{2o}) = \frac{r_2 \cos\theta_2}{r_3 \cos\theta_3} \frac{I_3}{l_3}(\theta_3 - \theta_{3o}) \tag{6.213}$$

This equation may be used with the kinematic equation that relates θ_2 to θ_3 to solve iteratively for the equilibrium position.

6.10 MULTI-DEGREE-OF-FREEDOM MECHANISMS

The previous examples have been single-degree-of-freedom mechanisms, but the principle of virtual work is easily applied to multi-degree-of-freedom mechanisms. The number of generalized coordinates must be equal to the number of degrees of freedom of the system. The number of equations which result from application of the principle of virtual work is also equal to the degrees of freedom.

For example, consider the open-link mechanism shown in Figure 6.16, which may have one, two, or three degrees of freedom, depending on the loading. If X_3 were the only active force, the mechanism would rotate about the joint at B. The compliance in the mechanism would not be active and the mechanism would have a

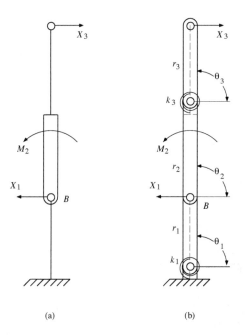

Figure 6.16. (a) Multi-degree-of-freedom open-link compliant mechanism, and (b) its pseudo-rigid-body model.

Multi-Degree-of-Freedom Mechanisms

single degree of freedom. In the following example, it is assumed that the loading is such that the compliance is active, and the three-degree-of-freedom pseudo-rigid-body model of Figure 6.16b is valid.

The virtual work for this system is

$$\delta W = \vec{F}_1 \cdot \delta\vec{z}_1 + \vec{F}_3 \cdot \delta\vec{z}_3 + \vec{M}_2 \cdot \delta\vec{\theta}_2 + \vec{T}_1 \cdot \delta\vec{\theta}_1 + \vec{T}_3 \cdot (\delta\vec{\theta}_3 - \delta\vec{\theta}_2) \qquad (6.214)$$

where

$$T_1 = -\left(\gamma K_\Theta \frac{EI}{l}\right)_1 (\theta_1 - \theta_{1o}) \qquad (6.215)$$

$$T_3 = -\left(\gamma K_\Theta \frac{EI}{l}\right)_3 [(\theta_3 - \theta_{3o}) - (\theta_2 - \theta_{2o})] \qquad (6.216)$$

$$\delta\vec{z}_j = \sum_{i=1}^{j} -r_i \sin\theta_i \delta\theta_i \hat{i} + \sum_{i=1}^{j} -r_i \cos\theta_i \delta\theta_i \hat{j} \qquad (6.217)$$

$$\vec{F}_1 = -X_1 \hat{i} \qquad (6.218)$$

$$\vec{F}_3 = X_3 \hat{i} \qquad (6.219)$$

where θ_{1o} is the value of θ_1 when k_1 is undeflected, and θ_{2o} and θ_{3o} are values of θ_2 and θ_3 that cause k_3 to be undeflected.

If θ_1, θ_2, and θ_3 are generalized coordinates, the virtual work may be written as

$$\delta W = (X_1 r_1 \sin\theta_1 + T_1)\delta\theta_1 + (M_2 - X_3 r_2 \sin\theta_2 - T_3)\delta\theta_2 \\ + (-X_3 r_3 \sin\theta_3 + T_3)\delta\theta_3 \qquad (6.220)$$

or

$$\delta W = \delta\vec{\theta}_1 \cdot \vec{Q}_1 + \delta\vec{\theta}_2 \cdot \vec{Q}_2 + \delta\vec{\theta}_3 \cdot \vec{Q}_3 \qquad (6.221)$$

where $\vec{Q}_i$ are known as the generalized forces. In applying the principle of virtual work, each of the generalized forces must be zero, or

$$Q_i = 0 \qquad (6.222)$$

This is consistent with previous results that show single-degree-of-freedom systems to have only one generalized force.

6.11 CONCLUSIONS

The principle of virtual work is well suited for use with the pseudo-rigid-body models of compliant mechanisms. The number of equations required to determine the force–deflection relationships is considerably less than that required for other methods, and several examples of compliant mechanisms have been presented to illustrate the use of the principle. General design equations have also been derived for pseudo-rigid-body four-bar mechanisms, including a slider mechanism. These equations are useful in analyzing many types of compliant mechanisms.

PROBLEMS

6.1 When would it be best to use the free-body diagram approach, and when would it be best to use the principle of virtual work?

6.2 What is an advantage of the principle of virtual work over other methods?

6.3 Refer to the mechanism illustrated in Figure 6.5.
(a) Derive the equation for the vertical force required to obtain an equilibrium position at an angle θ, using s_A as the generalized coordinate.
(b) Compare the equation from (a) to equation (6.68) and comment.

6.4 Use the general equations for a four-bar mechanism in Section 6.8 to derive the equation for the input force for the sample parallel mechanism in Figure 5.58.

6.5 Use the general equations for a slider mechanism in Section 6.9 to derive the equation for the input torque for the sample slider mechanism in Figure 5.56.

6.6 Use the general virtual work equations for a slider–crank developed in Section 6.9 to find the equation for the mechanism illustrated in Figure 6.1. Compare the answer to that obtained in Section using the free-body diagram approach.

6.7 Consider the mechanism shown in Figure P6.7.
(a) Draw the pseudo-rigid-body model of the mechanism. State the length of each pseudo-rigid link and the value of each torsional spring (in terms of the variables given and the appropriate E's and I's).
(b) Use Newtonian methods (free-body diagrams) to write an equation for T_{in}, assuming that the mechanism remains near the position shown (segment 3 is horizontal, and segments 2 and 4 are vertical).

6.8 The segment lengths and the location of the input torque are known for the mechanism shown in Figure P6.8.
(a) Draw the pseudo-rigid-body model of this mechanism. State symbolically the value of each pseudo-rigid-link length and spring constant.
(b) Develop the equation for the input torque, T_{in}, for the mechanism.
(c) Sketch on a diagram all variables (including link lengths and link angles) needed in the equation above.

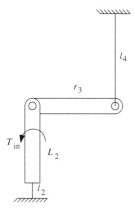

Figure P6.7. Figure for Problem 6.7.

6.9 Develop the equation for the input torque, T_{in}, of the crank–flapper mechanism shown in Figure P6.9 in terms of θ_2. The equation should be stated in terms of both the given variables (r_2, r_3, E, b, h, L, θ_2) and other known variables for which you must specify the numerical value (e.g., γ). Use each of the following methods and compare:
(a) The free-body diagram approach.
(b) The virtual work approach using the steps in Section 6.6.
(c) The virtual work approach using the general equations for a four-bar mechanism in Section 6.8.

6.10 Consider the parallel motion mechanism shown in Figure P6.10. The length of link 4 is such that its pseudo-rigid-link length is equal to r_2, and the pseudo-rigid-body model is a parallelogram mechanism (link 3 remains horizontal throughout the motion). Develop the equation for F_{out} as a function of T_{in} and θ_2 using the following methods:

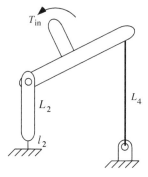

Figure P6.8. Figure for Problem 6.8.

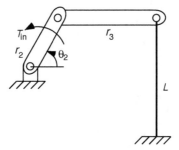

Figure P6.9. Crank–flapper mechanism (Problem 6.9).

(a) The free-body diagram approach.
(b) The virtual work approach using the steps in Section 6.6.
(c) The virtual work approach using the general equations for a four-bar mechanism in Section 6.8.

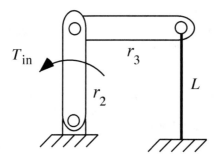

Figure P6.10. Figure for Problem 6.10.

CHAPTER 7

NUMERICAL METHODS

The deflection analysis methods described in Chapter 2 are useful in that they provide analytical solutions for a number of large- and small-deflection problems. The disadvantage of the methods, however, is that the derivations are complicated and solutions can be found only for relatively simple geometries and loadings. For this reason an alternative technique is needed to analyze more general flexible members.

Numerical methods such as finite element analysis and the chain algorithm are useful in determining the deflections and stresses in compliant mechanisms. There are two main reasons for using these methods. First, they are useful in validating or refining designs obtained using the pseudo-rigid-body model. This is not required but may be useful in critical applications or when a prototype is cost prohibitive. The pseudo-rigid-body model is particularly useful in the early design phases where many design iterations are required to design a mechanism that fulfills the design objectives specified. After the initial design is obtained, it may be further analyzed or refined using other methods, such as finite element analysis.

The second reason for using numerical methods is to analyze compliant mechanisms that have geometry that is not easily modeled using the pseudo-rigid-body model. For example, suppose that a mechanism has compliant segments with varying cross sections along their lengths, or long flexible segments that are neither straight or circular. It is very difficult to design such a mechanism for a very difficult task, but numerical methods may be used to assist in the analysis. For such mechanisms, it is still usually wise to use the pseudo-rigid-body model to make initial approximations in the design to obtain a general understanding of the mechanism characteristics, and then use other methods to improve the design.

7.1 FINITE ELEMENT ANALYSIS

The key to finite element analysis (FEA) is that a continuum is broken up into numerous smaller elements. These elements can be combined to model complex geometry and loading conditions that would be difficult, or impossible, to analyze using closed-form solutions. There are many different kinds of elements, including beam elements, plane elements (such as triangles and quads), and three-dimensional elements (such as bricks and tetrahedrals). The behavior of each element is well understood, and when combined together it is possible to create a *global stiffness matrix*, $[K]$. A large system of equations is developed from the stiffness matrix, the force vector, $\{f\}$, and the displacement vector, $\{u\}$, which can be expressed in matrix form as

$$\{f\} = [K]\{u\} \tag{7.1}$$

Because the resulting system of equations is very large, powerful computer algorithms and hardware are required to solve complex problems. A further challenge is introduced by geometric nonlinearities, such as those that are common in compliant mechanisms, because equation (7.1) becomes nonlinear and many iterations may be required to converge on a solution.

Many finite element analysis programs are available commercially, and many textbooks are available on the subject. Because information on finite element analysis is readily available from many sources, it will not be discussed in depth here. However, it is worthwhile to discuss a few issues that are more particular to the analysis of compliant mechanisms. A few of these are listed below.

- When selecting a commercial FEA program, ensure that it has the capability to perform nonlinear analysis. The large deflections associated with compliant mechanisms usually result in nonlinear deflection equations, and small-deflection linearization assumptions can lead to serious inaccuracies. Note that even when a program is capable of nonlinear analysis, it will need to be activated explicitly. Also beware that most programs scale the output plots so that small deflections are visible in the output. This can be misleading for large deflections, where you usually prefer to see the true deflections, and the scaling default should be changed.
- Make the model as simple as possible. A beam element model of the geometry can be very accurate, is easy to build, and can provide answers quickly. Trying to convert CAD files directly to an FEA model almost always results in extraneous information and difficulties that will take longer to get accurate answers in the long run. It is also more difficult to get complex nonlinear models to converge than for simple models.
- Use displacement loads rather than force loads as the input. For example, if the force–deflection relationship for a large-deflection beam is desired, the solution is much easier to obtain if the vertical displacement of the end is provided as an input and the resultant force is calculated. This is partic-

ularly true with more complex nonlinear analyses, such as the analysis of bistable mechanisms.

7.2 CHAIN ALGORITHM

The chain algorithm is similar to finite element analysis and lends itself well to compliant mechanism analysis. This method is discussed in more detail here because unlike FEA, information on this method is not readily available from many other sources.

A numerical chain algorithm uses the same basic beam element and stiffness matrix theory as those of conventional finite elements, but uses a different technique to combine and solve the resulting equations, making it computationally efficient in many applications [68]. Early references for the chain algorithm are found in [4], [42], [94], and [133].

The chain algorithm is so named because it requires discretization of the object being modeled into beam elements and analyzes each element in succession, as shown by the general compliant mechanism in Figure 7.1. Each element is treated as a beam cantilevered at the end of the previous element. Equivalent loads are found for each cantilevered element, and its deflections are calculated. Shooting methods are used to satisfy boundary conditions.

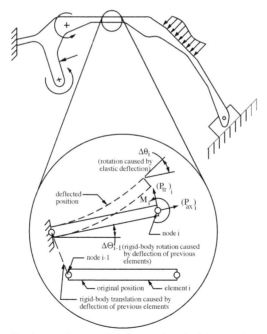

Figure 7.1. Generalized compliant mechanism and typical beam element as used by the chain algorithm.

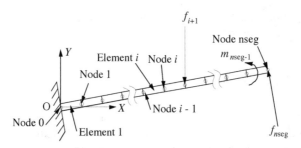

Figure 7.2. Flexible cantilever beam discretized into n_{seg} elements.

To illustrate the chain algorithm analysis procedure, consider a flexible cantilever beam (Figure 7.2), discretized into a number of beam elements. The first node (node 0) is considered fixed and located at the origin of the global coordinate system O–X–Y. The deflection of element 1 is calculated by treating it as a cantilever beam and loaded at node 1, as shown if Figure 7.3. The end loads are the internal axial, transverse and moment loads, calculated by use of the equations of static equilibrium, are

$$(P_{ax})_1 = \left[\sum_{i=1}^{n}(f_x)_i\right]\cos\theta_1 + \left[\sum_{i=1}^{n}(f_y)_i\right]\sin\theta_1 \qquad (7.2)$$

$$(P_{tr})_1 = -\left[\sum_{i=1}^{n}(f_x)_i\right]\sin\theta_1 + \left[\sum_{i=1}^{n}(f_y)_i\right]\cos\theta_1 \qquad (7.3)$$

$$M_1 = \sum_{i=1}^{n} m_i + \sum_{i=2}^{n}[(f_y)_i(x_i - x_1) - (f_x)_i(y_i - y_1)] \qquad (7.4)$$

where $(P_{ax})_1, (P_{tr})_1$ and M_1 are the internal axial load, transverse load, and moment at node 1, respectively, and $(f_x)_i, (f_y)_i$, and m_i are the externally applied loads in the global X and Y directions, and the externally applied moment at node i, respectively. These loads may now be used to calculate the deflections of the end of beam element 1:

$$\begin{bmatrix}\delta_{ax}\\ \delta_{tr}\\ \Delta\theta\end{bmatrix}_1 = [K]_1^{-1}\begin{bmatrix}P_{ax}\\ P_{tr}\\ M\end{bmatrix}_1 \qquad (7.5)$$

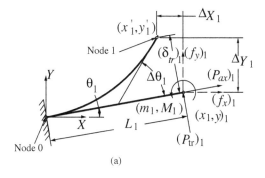

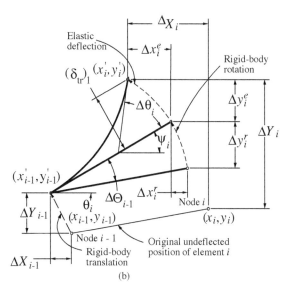

Figure 7.3. (a) Deflection of element 1, and (b) deflection of element i, as calculated by the chain algorithm.

where $(\delta_{ax})_1$, $(\delta_{tr})_1$, and $\Delta\theta_1$ are the elastic axial, transverse, and angular deflections, respectively, of element 1 at node 1, and $[K]_1^{-1}$ is the inverse of the stiffness matrix (flexibility matrix) for element 1. The axial deflection, δ_{ax}, is assumed to be negligible and the equations that follow reflect this assumption. The transverse deflection is easily transformed to global coordinates as follows:

$$\Delta X_1 = -(\delta_{tr})_1 \sin(\theta_1 + \Delta\theta_1) \tag{7.6}$$

$$\Delta Y_1 = (\delta_{tr})_1 \cos(\theta_1 + \Delta\theta_1) \tag{7.7}$$

The new coordinates of node 1 are found as

$$x'_1 = x_1 + \Delta X_1 \tag{7.8}$$

$$y'_1 = y_1 + \Delta Y_1 \tag{7.9}$$

With the new coordinates of node 1 known, the deflection of element 2 can be found. To ensure compatibility between elements, element 2 and the remaining elements go through a rigid-body rotation such that the angular deflection at the end of element 1 is the same as that at the beginning of element 2. Element 2 is now considered to be cantilevered at node 1, internal loads are found and applied at node 3, and the corresponding deflections are found. This process is continued for each segment in the chain until the last element is reached. In general, the calculations for the ith element are similar to the special case of element 1, except for the changes in moment arm calculations in equations (7.2) to (7.4) and rigid-body rotations of earlier elements. This is illustrated in Figure 7.3. The internal loads for the ith element may be written as

$$(P_{ax})_i = \left[\sum_{j=i}^{n} (f_x)_j\right] \cos \psi_i + \left[\sum_{j=i}^{n} (f_y)_j\right] \sin \psi_i \tag{7.10}$$

$$(P_{tr})_i = -\left[\sum_{j=i}^{n} (f_x)_j\right] \sin \psi_i + \left[\sum_{j=i}^{n} (f_y)_j\right] \cos \psi_i \tag{7.11}$$

$$M_i = \sum_{j=i}^{n} m_j + \sum_{j=i+1}^{n} [(f_y)_j \Delta \tilde{x}_{ji} - (f_x)_j \Delta \tilde{y}_{ji}] \tag{7.12}$$

where

$$\psi_i = \theta_i + \Delta \Theta_{i-1} \tag{7.13}$$

and $\Delta \Theta_{i-1}$ is the total angular displacement of the previous element; $\Delta \tilde{x}_{ji}$ and $\Delta \tilde{y}_{ji}$ are the distances from node i to node j using the most up-to-date coordinates [74] and are defined as

$$\Delta \tilde{x}_{ji} = (x_j - x_i) \cos \Delta \Theta_{i-1} - (y_j - y_i) \sin \Delta \Theta_{i-1} \tag{7.14}$$

$$\Delta \tilde{y}_{ji} = (x_j - x_i) \sin[\Delta \Theta_{i-1} + (y_j - y_i) \cos \Delta \Theta_{i-1}] \tag{7.15}$$

The deflection equations are similar in form to (7.5):

$$\begin{bmatrix} \delta_{ax} \\ \delta_{tr} \\ \Delta\theta \end{bmatrix}_i = [K]_i^{-1} \begin{bmatrix} P_{ax} \\ P_{tr} \\ M \end{bmatrix}_i \qquad (7.16)$$

The total displacements must now not only include the elastic displacements,

$$\Delta x_i^e = -(\delta_{tr})_i \sin \psi_i \qquad (7.17)$$

$$\Delta y_i^e = (\delta_{tr})_i \cos \psi_i \qquad (7.18)$$

but also the rigid-body displacements caused by the angular deflections of the previous elements

$$\Delta x_i^r = L_i(\cos \psi_i - \cos \theta_i) \qquad (7.19)$$

$$\Delta y_i^r = L_i(\sin \psi_i - \sin \theta_i) \qquad (7.20)$$

and the total x and y displacements of the previous elements. In other words:

$$\Delta X_i = \Delta X_{i-1} + \Delta x_i^r + \Delta x_i^e \qquad (7.21)$$

$$\Delta Y_i = \Delta Y_{i-1} + \Delta y_i^r + \Delta y_i^e \qquad (7.22)$$

$$\Delta \Theta_i = \Delta \Theta_{i-1} + \Delta \theta_i \qquad (7.23)$$

A considerable amount of error may be introduced into the formulation through inaccurate moment arm calculations in equations (7.10) through (7.12). This error stems from the fact that the calculations must be made from the latest available deflected positions rather than from the final deflected positions. This error can be reduced by using a load increment technique [4] and iteration [74].

The load increment technique works by applying the external load in increments. This means that some percentage of the load will be applied, the chain calculations performed, and the deflections found for this loading. The load is increased to the next percentage increment and the deflections are again calculated, but this time the moment arms are evaluated from the deflected position of the previous load increment calculations. The load increments may be written as

$$(f_x^n)_i = \frac{n}{n_{inc}} (f_x)_i \qquad (7.24)$$

$$(f_y^n)_i = \frac{n}{n_{inc}} (f_y)_i \qquad (7.25)$$

$$m_i^n = \frac{n}{n_{inc}} m_i \tag{7.26}$$

$$n = 1, 2, \ldots, n_{inc} \tag{7.27}$$

where n_{inc} is the number of load increments. The superscript n denotes the current load increment.

This method improves the accuracy of the chain calculations considerably, especially for relatively large deflections. As would be expected, the accuracy of the chain algorithm is increased with increasing number of load increments.

Iteration may be used with the load increment method to further improve accuracy. This is done by using the deflections found in the final load increment calculations to evaluate the new moment arms and new deflections for the full external load. This iteration may be continued until the desired accuracy is obtained.

The number of load increments and the number of iterations may be adjusted as needed to increase efficiency and accuracy. For instance, if a large number of load increments are used, a small number of iterations at the end are needed, and vice versa.

The flexibility matrix $\left[K\right]_i^{-1}$ used in the beam deflection calculations of equations (7.16) may be found by considering the six-degree-of-freedom beam element shown in Figure 7.4. The stiffness matrix for this element may be written as [134]

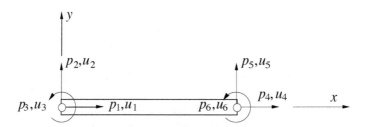

Figure 7.4. Six-degree-of-freedom beam element.

Chain Algorithm

$$\begin{bmatrix} p_1 \\ p_2 \\ p_3 \\ p_4 \\ p_5 \\ p_6 \end{bmatrix}_i = \frac{E_i}{L_i^3} \begin{bmatrix} AL^2 & 0 & 0 & -AL^2 & 0 & 0 \\ 0 & 12I & 6IL & 0 & -12I & 6IL \\ 0 & 6IL & 4IL^2 & 0 & -6IL & 2IL^2 \\ -AL^2 & 0 & 0 & AL^2 & 0 & 0 \\ 0 & -12I & -6IL & 0 & 12I & -6IL \\ 0 & 6IL & 2IL^2 & 0 & -6IL & 4IL^2 \end{bmatrix}_i \begin{bmatrix} u_1 \\ u_2 \\ u_3 \\ u_4 \\ u_5 \\ u_6 \end{bmatrix}_i$$

(7.28)

$$+ \frac{(P_{ax})_i}{30 L_i} \begin{bmatrix} 0 & 0 & 0 & 0 & 0 & 0 \\ 0 & 36 & 3L & 0 & -36 & 3L \\ 0 & 3L & 4L^2 & 0 & -3L & -L^2 \\ 0 & 0 & 0 & 0 & 0 & 0 \\ 0 & -36 & -3L & 0 & 36 & -3L \\ 0 & 3L & -L^2 & 0 & -3L & 4L^2 \end{bmatrix}_i \begin{bmatrix} u_1 \\ u_2 \\ u_3 \\ u_4 \\ u_5 \\ u_6 \end{bmatrix}_i$$

Using boundary conditions to reflect that each element in the chain algorithm is fixed to the previous element (i.e., $u_1 = u_2 = u_3 = 0$), the matrix equation above reduces to

$$\begin{bmatrix} P_{ax} \\ P_{tr} \\ M \end{bmatrix}_i = \frac{E}{L^3} \begin{bmatrix} AL^2 & 0 & 0 \\ 0 & 12I & -6IL \\ 0 & -6IL & 4IL^2 \end{bmatrix} + \frac{P_{ax}}{30L} \begin{bmatrix} 0 & 0 & 0 \\ 0 & 36 & -3L \\ 0 & -3L & 4L^2 \end{bmatrix} \begin{bmatrix} \delta_{ax} \\ \delta_{tr} \\ \Delta\theta \end{bmatrix}_i$$

(7.29)

The flexibility matrix, $[K]_i^{-1}$, may be obtained by inverting the stiffness (coefficient) matrix above to obtain

$$[K]_i^{-1} = \begin{bmatrix} \dfrac{L}{AE} & 0 & 0 \\ 0 & \dfrac{4L^2 Q_1 + 240 EIL^3}{3(Q_2 + 240 E^2 I^2)} & \dfrac{L Q_1 + 120 EIL^2}{Q_2 + 240 E^2 I^2} \\ 0 & \dfrac{L Q_1 + 120 EIL^2}{Q_2 + 240 E^2 I^2} & \dfrac{12 Q_1 + 240 EIL}{Q_2 + 240 E^2 I^2} \end{bmatrix}_i$$

(7.30)

where

$$(Q_1)_i = 2L_i^3(P_{ax})_i \tag{7.31}$$

$$(Q_2)_i = 3L_i^4(P_{ax})_i^2 + 104E_iI_iL_i^2(P_{ax})_i \tag{7.32}$$

An advantage of the flexibility matrix is that the deflection of each beam element is calculated individually, using the same equation. This eliminates the need to invert large stiffness matrices as required in conventional finite element analysis. As stated earlier, δ_{ax} in equation (7.29) is assumed negligible. Figure 7.5 summarizes the discussion above on the chain algorithm in flowchart form. This flowchart shows the general flow of calculations but does not show all programming details.

The deflection equations can be simplified even further by alternatively defining an equivalent internal transverse load to approximate the nonlinear effects of axial stiffening [180]. The transverse load $(P_{tr})_i$ in equations (7.3) helps yield

$$[P_{tr}]_{i,\,eq} = \frac{(P_{tr})_i}{1-\alpha_i} \tag{7.33}$$

where

$$\alpha_i = \frac{(P_{ax})_i(2L_i)^2}{E_iI_i\pi^2} \tag{7.34}$$

The beam deflections can now be calculated using linear beam theory and the equivalent transverse load:

$$(\delta_{tr})_i = \frac{1}{E_iI_i}\left[\frac{(P_{tr})_{i,\,eq}L_i^3}{3} + \frac{M_iL_i^2}{2}\right] \tag{7.35}$$

$$\Delta\theta_i = \frac{1}{E_iI_i}\left[\frac{(P_{tr})_{i,\,eq}L_i^2}{2} + M_iL_i\right] \tag{7.36}$$

The error in these approximations increases with increasing axial force $(P_{ax})_i$.

7.2.1 Shooting Method

The chain algorithm calculates the deflection of a flexible member fixed at one end. A shooting method may be used in conjunction with the chain algorithm when boundary conditions must be met. These boundary conditions are generally position and rotation constraints on a finite number of nodes. For example, consider the

Chain Algorithm

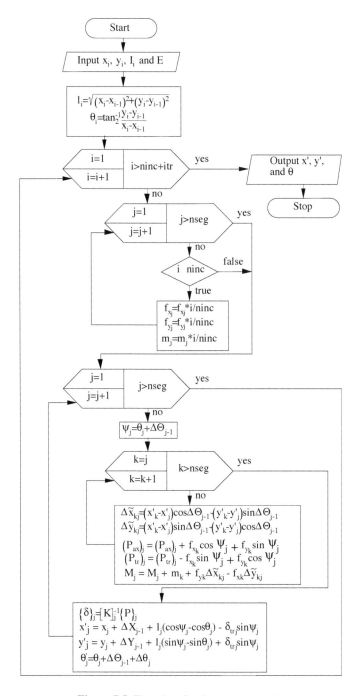

Figure 7.5. Flowchart for the chain algorithm.

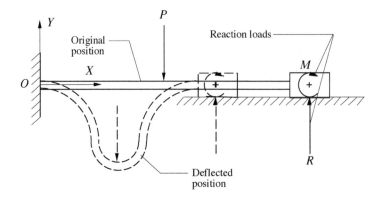

Figure 7.6. Compliant slider mechanism and its deflected configuration.

slider in Figure 7.6. The boundary conditions for this mechanism would be no rotation and no vertical deflection at the slider, and no translation or rotations at the fixed end. Since the chain algorithm assumes the first node to be fixed, the boundary condition at the fixed end are easily met. The other constraints are met by estimating the corresponding reaction loads and applying these as external loads in the chain algorithm and calculating the resulting deflections. In general, the error in the boundary conditions can then be calculated as

$$g_i[\vec{z}] = s_i[\vec{z}] - s_i^* \qquad i = 1, \ldots, n \qquad (7.37)$$

where $g_i[\vec{z}]$ is the boundary condition error (residual) function, $\vec{z}$ the reaction load vector, s_i^* the desired final value of $s_i[\vec{z}]$, and n the number of boundary conditions. An iterative technique, or shooting method, is needed to solve the set of n nonlinear error functions. Two such methods are the Newton–Raphson technique and unconstrained optimization. The Newton–Raphson technique finds a correction vector for $\vec{z}$, $\delta\vec{z}$, by truncating the Taylor's series of the error function by its Jacobian. The correction vector may then be found by

$$\begin{Bmatrix} \delta z_1 \\ \cdot \\ \cdot \\ \cdot \\ \delta z_n \end{Bmatrix} = \begin{bmatrix} \dfrac{\partial g_1}{\partial z_1} & \dfrac{\partial g_1}{\partial z_2} & \cdots & \dfrac{\partial g_1}{\partial z_n} \\ \cdot & \cdot & \cdots & \cdot \\ \cdot & \cdot & \cdots & \cdot \\ \cdot & \cdot & \cdots & \cdot \\ \dfrac{\partial g_n}{\partial z_1} & \dfrac{\partial g_n}{\partial z_2} & \cdots & \dfrac{\partial g_n}{\partial z_n} \end{bmatrix}_{\vec{z}}^{-1} \begin{Bmatrix} -g_1 \\ \cdot \\ \cdot \\ \cdot \\ -g_n \end{Bmatrix}_{\vec{z}} \qquad (7.38)$$

or

$$\{\delta z\} = [J]^{-1}\{-g\} \tag{7.39}$$

where $[J]^{-1}$ is the inverse of the Jacobian matrix. Since g_i cannot be expressed in closed form, the partial derivatives in $[J]$ must be evaluated numerically. The system of equations in (7.38) may be solved by any method used to solve simultaneous linear equations. Once the correction vector is found, a new guess for the reaction loads is obtained as

$$\{z\}^{new} = \{z\} + \{\delta z\} \tag{7.40}$$

This process is continued until the error (residual) functions are sufficiently close to zero.

An alternative to the Newton–Raphson method is the use of an unconstrained minimization algorithm such as the Davidon–Fletcher–Powell method or Powell's method [178]. Such a method would be used to find the reaction loads, $\grave{z}$, by minimizing the sum of the error functions squared, that is,

$$f(\grave{z}) = \sum_{i=1}^{n} [g_i(\grave{z})]^2 \tag{7.41}$$

The method has converged when the objective function, $f(\grave{z})$, is sufficiently close to zero.

Example: Analysis of a Compliant Mechanism. Compliant mechanisms gain at least some of their motion from the flexibility of their members. The member deflections resulting from such motions are often nonlinear and the chain algorithm is very useful in their analysis. As an example of its use in compliant mechanism analysis, consider the one-piece compliant hand tool shown in Figure 7.7. This mechanism was designed for high mechanical advantage and to have geometry and motion similar to a classical rigid-body four-bar mechanism. The relatively flexible sections, labeled flexural pivots, must deflect in order for the mechanism to go through the desired motion. The "passive pivot" was placed as shown since the rigid segments at the pivot are always in compression.

Figure 7.8 shows the compliant hand tool discretized into 22 beam elements. The chain algorithm assumes node 0 to be fixed; therefore, the boundary conditions of no translation or rotation at node 0 are automatically met. The passive pivot at node 22 is treated as a pin joint, requiring two additional boundary conditions of no translation in the x and y directions at that node. Another boundary condition may be required if a determined amount of deflection is desired at the output port. This is done by meeting the condition that the distance between nodes 6 and 12 is $\delta_0 - \delta$, where δ_0 is the original undeflected distance and δ is the desired deflection (Figure

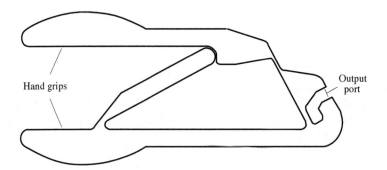

Figure 7.7. Compliant hand tool design.

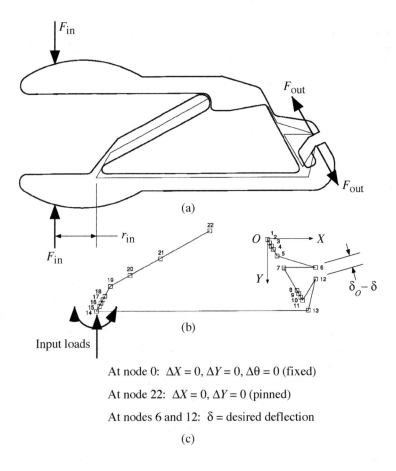

At node 0: $\Delta X = 0, \Delta Y = 0, \Delta \theta = 0$ (fixed)

At node 22: $\Delta X = 0, \Delta Y = 0$ (pinned)

At nodes 6 and 12: δ = desired deflection

(c)

Figure 7.8. (a) Compliant mechanism with an outline of the fully compliant model, (b) discretization of the model, and (c) boundary conditions.

Chain Algorithm

7.8). This condition will be used when a workpiece is placed in the output port. For instance, if a rigid workpiece with a height of δ_0 is placed in the output port, a deflection of $\delta = 0$ must be maintained at the output port. The mechanism's deformed configurations for different input loads and no workpiece are shown in Figure 7.9a. Figure 7.9b shows how the mechanism deforms under different input loads when gripping a rigid workpiece at the output port.

Performance analysis is another important application of the chain algorithm in compliant mechanism analysis. This is particularly important since the input/output characteristics are not easily calculated. For instance, the mechanical advantage of a compliant mechanism may vary as a function of the energy stored in the flexural pivots, the initial and final orientations, the method of loading, and the flexible and rigid segment dimensions.

Figures 7.10a and b show variations of the output force, F_{out}, and mechanical advantage, $ma = F_{out}/F_{in}$, respectively, with increasing input force, F_{in}, for the example design. The three graphs in each figure represent the effects of using different-sized workpieces, where δ is the amount the output port deflects before gripping the workpiece. In comparison, a rigid-body mechanism working on a rigid workpiece would have a constant mechanical advantage with F_{in} for each given

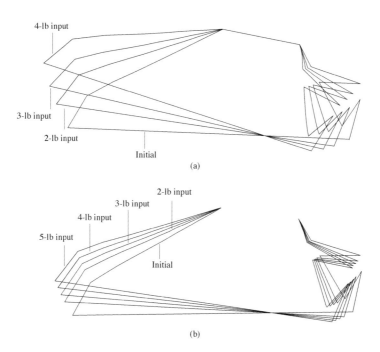

Figure 7.9. (a) Mechanism deflection with no workpiece, and (b) mechanism deflection with rigid workpiece.

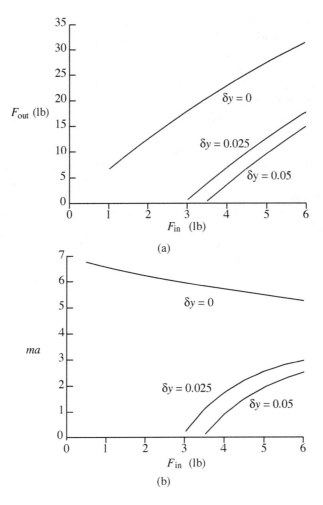

Figure 7.10. Input/output characteristic of compliant hand tool: (a) output force versus input force, and (b) mechanical advantage versus input force, for various workpieces.

workpiece. The varying graphs for different workpieces illustrate the loss of energy to member deflection in compliant mechanisms. The increase in mechanical advantage, *ma* (Figure 7.10b), after initial contact with the workpiece, for $\delta = 0.025$ and $\delta = 0.050$, is due to the compliant mechanism's ability to deflect toward its toggle position. The toggle position describes the mechanism orientation when the two links connected by the passive pivots are collinear.

CHAPTER 8

COMPLIANT MECHANISM SYNTHESIS

Compliant mechanisms pose several challenges that are not found in rigid-body mechanisms. The motion of a compliant mechanism depends on the location, direction, and magnitude of applied forces. For example, a compliant mechanism that carries an object from one place to another may follow a different path depending on the weight of the object being carried. The practical limits on the geometry of a compliant mechanism are often more restrictive. For instance, flexural pivots cannot fully rotate, and crossing links are often not acceptable because of the truly planar nature of many compliant components. Stress and fatigue are greater concerns in compliant mechanism design than in rigid-body mechanisms; for a compliant mechanism to undergo any amount of motion, some member(s) must deflect, and stresses are therefore induced.

Compliant mechanism synthesis will be divided into two major classes: rigid-body replacement synthesis, and synthesis with compliance. This chapter discusses how the pseudo-rigid-body model can be used in these two classes of synthesis. The next chapter discusses the use of optimization methods to perform synthesis with compliance. Subsequent chapters on special-purpose mechanisms also provide design examples.

8.1 RIGID-BODY REPLACEMENT (KINEMATIC) SYNTHESIS

In its simplest form, compliant mechanism synthesis is accomplished by obtaining a pseudo-rigid-body model for a compliant mechanism, assuming constant link lengths, and directly applying rigid-body kinematics equations. This approach is useful when a compliant mechanism is to be used to perform a traditional rigid-body mechanism task, such as path or motion generation, without concern for the energy storage in the flexible members. Once the kinematic geometry of the mechanism is determined, the structural properties of the flexible members may be chosen according to allowable stresses and input requirements. This class of synthesis problems in which rigid-body equations are applied directly to the pseudo-rigid-body model will hereafter be referred to as *rigid-body replacement synthesis*. Because only the kinematics of the mechanism is an issue, it may also be called *kinematic synthesis*. The examples that could be cited as applications of this type of synthesis problem are endless. Since compliant mechanisms could be designed to perform many functions currently accomplished by rigid-body mechanisms, many rigid-body synthesis problems are also examples of rigid-body replacement synthesis.

Given that rigid-body equations may be applied to a model directly, the major challenge of this class of synthesis problems lies in determining the appropriate pseudo-rigid-body model for a compliant mechanism and evaluating the feasibility of the resulting model. Evaluating the design is a significant step because the rigid-body mechanism synthesis may yield configurations that are adequate for rigid-link mechanisms but are not acceptable for compliant mechanisms. An example of an unacceptable design is a compliant segment, such as a flexural pivot, that must deflect through a complete rotation. Compliant mechanism synthesis may require more iterations on the free choices before an acceptable design is obtained. Many of these concerns may be accounted by using an optimization routine with appropriate constraints. Several examples of rigid-body replacement synthesis examples illustrate the approach.

Often in compliant mechanism analysis, a pseudo-rigid-body model is obtained from the compliant mechanism. However, in rigid-body replacement synthesis, a pseudo-rigid-body model is equivalent to a rigid-body mechanism model, and the resulting compliant mechanism is determined from these models. Many different compliant mechanisms can be made from one pseudo-rigid-body model.

Example: Hoeken Straight-Line Mechanism. A Hoeken straight-line mechanism has a coupler point, P, that follows an almost straight line over much of its path. The link lengths of this mechanism may be specified as functions of the crank length, r_2:

$$\begin{aligned} r_1 &= 2r_2 \\ r_3 &= 2.5r_2 \\ r_4 &= 2.5r_2 \\ a_3 &= 5r_2 \\ b_3 &= 0 \end{aligned} \quad (8.1)$$

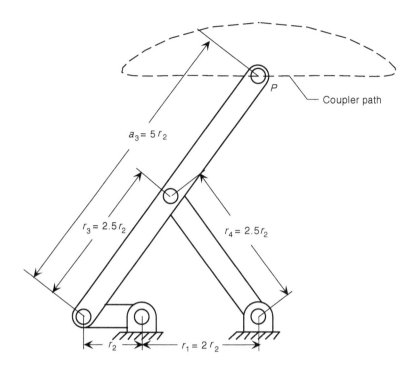

Figure 8.1. Rigid-link Hoeken straight-line mechanism.

The path for a mechanism with $r_2 = 1$ is shown in Figure 8.1. The midpoint of the straight path is $\theta_2 = 180°$, a good undeflected position for a compliant mechanism.

The rigid-body skeleton diagram of the mechanism is like a pseudo-rigid-body model of the desired compliant mechanism. A fully compliant mechanism may be designed by replacing each pin joint with flexural pivots, as shown in Figure 8.2a. The pseudo-rigid-body model of this mechanism is the same as that shown in Figure 8.1, with torsional springs at all joints to represent the strain energy associated with the deflection of the flexible segments. This mechanism does not have full rotation, but can move over much of the range of the straight-line path. Full rotation could be obtained by using a partially compliant mechanism with pin joints that connect both ends of the crank and using small-length flexural pivots elsewhere (Figure 8.2b). The small-length flexural pivots are located such that their centers are at the location of the pin joint of the pseudo-rigid-body model.

Another possible configuration uses two fixed–pinned segments, as illustrated in Figure 8.3. For the characteristic pivots to be in the proper locations,

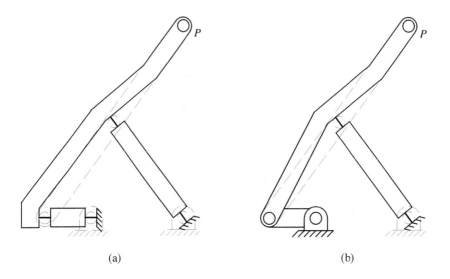

Figure 8.2. Compliant straight-line mechanisms with (a) four and (b) two small-length flexural pivots.

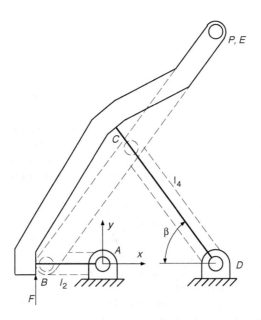

Figure 8.3. Compliant straight-line mechanism with two fixed–pinned segments.

Rigid-Body Replacement (Kinematic) Synthesis

$$l_2 = \frac{r_2}{\gamma} \tag{8.2}$$

$$l_4 = 2.5\frac{r_2}{\gamma} = 2.5l_2 \tag{8.3}$$

The angle β has the same value for this mechanism as for the rigid-body mechanism, and

$$\cos\beta = \frac{1.5}{2.5} = 0.6 \tag{8.4}$$

$$\sin\beta = \frac{2}{2.5} = 0.8 \tag{8.5}$$

These values may be used to calculate the initial coordinates of points A through E as listed in Table 8.1, and as shown in Figure 8.3. For example, for point C

$$x_c = 2r_2 - l_4 \cos\beta = 2r_2 - 1.5\frac{r_2}{\gamma} \tag{8.6}$$

$$y_c = l_4 \sin\beta = 2\frac{r_2}{\gamma} \tag{8.7}$$

If a force, F, is applied at point B as shown in Figure 8.3, the principle of virtual work may be used to determine the force–deflection relationships for the mechanism:

$$F = \frac{T_3 h_{42} - (T_2 + T_3)h_{32} + T_2}{r_2 \cos\theta_2 - b_3 \sin\theta_3 h_{32}} \tag{8.8}$$

where, for this specific geometry,

TABLE 8.1. Coordinates for a compliant straight-line mechanism such as that shown in Figure 8.3

Point	x	y
A	0	0
B	$-r_2/\gamma$	0
C	$2r_2 - 1.5r_2/\gamma$	$2r_2/\gamma$
D	$2r_2$	0
E	$2r_2$	$4r_2$

$$b_3 = l_2(1 - \gamma) \tag{8.9}$$

$$h_{32} = \frac{\sin(\theta_4 - \theta_2)}{2.5 \sin(\theta_3 - \theta_4)} \tag{8.10}$$

$$h_{42} = \frac{\sin(\theta_3 - \theta_2)}{2.5 \sin(\theta_3 - \theta_4)} \tag{8.11}$$

$$T_2 = -\gamma K_\Theta \frac{EI_2}{l_2}(\theta_2 - \theta_{2o}) \tag{8.12}$$

$$T_3 = -\gamma K_\Theta \frac{EI_4}{l_4}[(\theta_4 - \theta_{4o}) - (\theta_3 - \theta_{3o})] \tag{8.13}$$

If the undeflected position is like that shown in Figure 8.3, then

$$\theta_{2o} = \pi \tag{8.14}$$

$$\theta_{3o} = \mathrm{acos}\frac{1.5}{2.5} \tag{8.15}$$

$$\theta_{4o} = \mathrm{acos}\frac{-1.5}{2.5} \tag{8.16}$$

This example illustrates an approach to rigid-body replacement in which a known rigid-body mechanism is replaced directly by a compliant mechanism. Another approach uses rigid-body synthesis equations to design a compliant mechanism. Several examples illustrate this concept by using loop closure equations.

8.1.1 Loop Closure Equations

Loop closure equations for rigid-body mechanisms were derived in Chapter 4. The vector loop for a four-bar is shown again in Figure 8.4. The standard form equations of dyadic construction associated with this mechanism are

$$\vec{Z}_2(e^{i\phi j} - 1) + \vec{Z}_3(e^{i\gamma j} - 1) = \vec{\delta}_j \tag{8.17}$$

$$\vec{Z}_5(e^{i\gamma j} - 1) + \vec{Z}_4(e^{i\psi j} - 1) = \vec{\delta}_j \tag{8.18}$$

where $\vec{\delta}_j$ is the displacement vector of point P from position 1 to position j, and ϕ_j, γ_j, and ψ_j are the angular rotations of links 2, 3, and 4 between positions 1 and j.

Rigid-Body Replacement (Kinematic) Synthesis

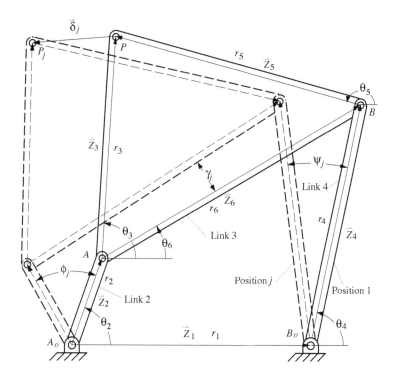

Figure 8.4. Vector loop for a four-bar mechanism with coupler point *P*.

These equations can be used in conjunction with the pseudo-rigid-body model to design compliant mechanisms for function, path, and motion generation.

Example: Function Generation. A compliant mechanism is to be synthesized to have the following side-link angles: $\phi_2 = 20°$, $\phi_3 = 40°$, $\psi_2 = 30°$, $\psi_3 = 50°$. The pseudo-rigid-body model of the mechanism should be a four-bar mechanism.

Neither pseudo-rigid-body mechanism side link should represent a flexible segment because the pseudo-rigid-body model would show misleading results. If the side links were to represent flexible segments, the required relationship between the input and output would be maintained for the pseudo-rigid-body model, but the side link would actually be a flexible beam, not a rigid link with the specified angular rotation. Figure 8.5a shows a compliant function generator, and Figure 8.5b shows its pseudo-rigid-body model.

One method of solving this problem is to choose values for $\vec{Z}_2$, γ_2, and γ_3 and find the other values by solving equation (4.74). Although this is an acceptable approach, another method will be used here to illustrate a useful property in linear synthesis.

Compliant Mechanism Synthesis

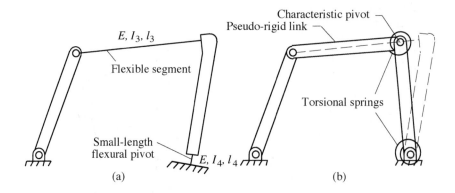

Figure 8.5. (a) Compliant mechanism for function generation, and (b) its pseudo-rigid-body model.

Since several free choices can be made in this problem, it is possible to prescribe the deflections of an imaginary coupler point as the free choices. Equations (8.17) and (8.18) would then be used to find the remaining unknowns. The disadvantage of this method is that the number of linear, scalar equations that must be solved doubles from four to eight. The advantage, however, is that the same equations and solution methods may be applied to function, path, and motion generation.

Employing this method with the free choices of $\gamma_2 = 10°$, $\gamma_3 = 15°$, $\vec{\delta}_2 = (1, 1)$, and $\vec{\delta}_3 = (2, 2)$ yields

$$\begin{aligned}
\vec{Z}_1 &= -5.20 - 6.26i = 8.14e^{-129.7°} \\
\vec{Z}_2 &= -0.30 - 3.95i = 3.96e^{-94.4°} \\
\vec{Z}_3 &= 5.13 - 1.67i = 5.39e^{18.0°} \\
\vec{Z}_4 &= -3.66 - 4.32i = 5.66e^{-130.3°} \\
\vec{Z}_5 &= 13.68 + 8.30i = 16.00e^{31.3°} \\
\vec{Z}_6 &= -8.56 - 6.64i = 10.83e^{-142.2°}
\end{aligned} \quad (8.19)$$

Figure 8.6 shows a skeleton drawing of the pseudo-rigid-body model. Since no constraints have been placed on energy storage, the undeflected position of the flexible segments may be chosen arbitrarily. If the characteristic radius factor is $\gamma = 0.85$, the length of the flexible member is

$$\text{length} = \frac{|\vec{Z}_6|}{0.85} = 12.74 \quad (8.20)$$

Rigid-Body Replacement (Kinematic) Synthesis

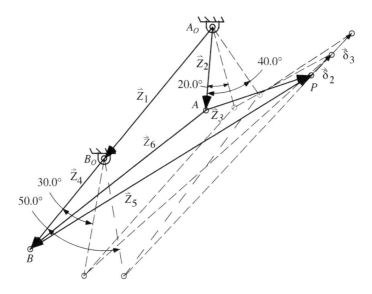

Figure 8.6. Pseudo-rigid-body skeleton drawing of a compliant mechanism for function generation.

Once a potential mechanism has been identified, it must be evaluated to ensure that it is acceptable. Possible limitations include those associated with both compliant mechanisms and rigid-body mechanisms. The mechanism described previously requires deflections for flexible segments that are reasonable for very flexible members. The order of the precision points is correct, and the mechanism moves continuously between the required positions.

Example: Path Generation with Prescribed Timing. A fully compliant (one-piece) mechanism is to be designed for three-precision-point path generation with prescribed timing. The coupler point is to have the displacements of $\vec{\delta}_2 = -5 + 3i$ for $\phi_2 = 10°$ and $\vec{\delta}_3 = -8 + 10i$ for $\phi_3 = 25°$. The general form of the mechanism is shown in Figure 8.7a; and its pseudo-rigid-body model is shown in Figure 8.7b. If the remaining unknown angles are chosen as free choices, the solution is linear. Equations (8.17) and (8.18) may be rearranged as

$$\begin{bmatrix} e^{i\phi_2} - 1 & e^{i\gamma_2} - 1 \\ e^{i\phi_3} - 1 & e^{i\gamma_3} - 1 \end{bmatrix} \begin{pmatrix} \vec{Z}_2 \\ \vec{Z}_3 \end{pmatrix} = \begin{pmatrix} \vec{\delta}_2 \\ \vec{\delta}_3 \end{pmatrix} \qquad (8.21)$$

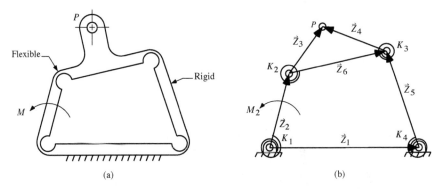

Figure 8.7. (a) Fully compliant mechanism for path generation, and (b) its pseudo-rigid-body model.

$$\begin{bmatrix} e^{i\psi_2} - 1 & e^{i\gamma_2} - 1 \\ e^{i\psi_3} - 1 & e^{i\gamma_3} - 1 \end{bmatrix} \begin{pmatrix} \vec{Z}_4 \\ \vec{Z}_5 \end{pmatrix} = \begin{pmatrix} \vec{\delta}_2 \\ \vec{\delta}_3 \end{pmatrix} \qquad (8.22)$$

With the free choices $\gamma_2 = -8°$, ($\gamma_3 = -25°$), $\psi_2 = 10°$, and $\psi_3 = 15°$, equations (8.21) and (8.22) may be solved to obtain

$$\begin{aligned}
\vec{Z}_1 &= -45.75 + 15.26i = 48.23e^{161.6°} \\
\vec{Z}_2 &= -19.48 + 29.85i = 35.65e^{123.1°} \\
\vec{Z}_3 &= -48.82 + -4.22i = 49.01e^{-175.1°} \\
\vec{Z}_4 &= 0.25 + 21.30i = 21.30e^{89.3°} \\
\vec{Z}_5 &= -22.80 - 10.92i = 25.28e^{154.4°} \\
\vec{Z}_6 &= -26.02 + 6.70i = 26.87e^{165.6°}
\end{aligned} \qquad (8.23)$$

Figure 8.8 shows a pseudo-rigid-body skeleton drawing of this mechanism.

Example: Motion Generation–Nonlinear Solution. A compliant mechanism and its pseudo-rigid-body model, similar to those shown in Figure 8.9, are to be synthesized for three-precision-point motion generation. The preceding examples used free choices that allowed linear solutions, but in this problem, because the value of $\vec{Z}_2$ is controlled, not enough free choices remain to allow a linear solution. The prescribed motion is described by $\vec{\delta}_2 = -10 + 5i$, $\vec{\delta}_3 = -15 - 2i$, $\gamma_2 = -20°$, $\gamma_3 = -10°$, and $\vec{Z}_2 = 10e^{60°}$. It can be reasoned from equations (8.17) and (8.21)

Rigid-Body Replacement (Kinematic) Synthesis

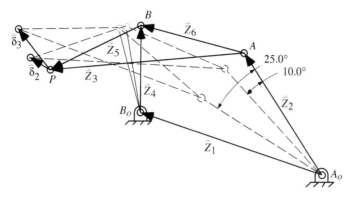

Figure 8.8. Pseudo-rigid-body skeleton drawing of a compliant mechanism for path generation.

that the values of ϕ_j cannot also be used as free choices because the problem would be overconstrained. Choosing values of ψ_j in equation (8.18), however, allows equation (8.22) to be solved linearly. The remaining nonlinear equations may be expressed using equation (8.21):

$$\vec{Z}_2(e^{i\phi_2} - 1) + \vec{Z}_3(e^{i\gamma_2} - 1) - \vec{\delta}_2 = 0 \qquad (8.24)$$

$$\vec{Z}_2(e^{i\phi_3} - 1) + \vec{Z}_3(e^{i\gamma_3} - 1) - \vec{\delta}_3 = 0 \qquad (8.25)$$

These equations are then solved simultaneously. The Newton–Raphson method is commonly used to solve systems of nonlinear equations. An alternative method is to use an unconstrained optimization routine to minimize the objective function

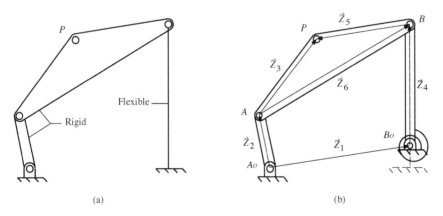

Figure 8.9. (a) Compliant mechanism for motion generation, and (b) its pseudo-rigid-body model.

$$f = f_1^2 + f_2^2 + \cdots + f_n^2 \tag{8.26}$$

where f_i is the value of scalar equation i and the design variables are the problem unknowns. The objective function, f, is minimized until it is sufficiently close to zero, and the values of the design variables are then assumed to be the solution. For this example, the optimization problem could be stated as follows: Find the values of ϕ_2, ϕ_3, and the real and imaginary parts of $\vec{Z}_3$ that minimize

$$f = f_1^2 + f_2^2 + f_3^2 + f_4^2 \tag{8.27}$$

where

$$f_1 = \text{real part of equation (8.24)} \tag{8.28}$$

$$f_2 = \text{imaginary part of equation (8.24)} \tag{8.29}$$

$$f_3 = \text{real part of equation (8.25)} \tag{8.30}$$

$$f_4 = \text{imaginary part of equation (8.25)} \tag{8.31}$$

The values of the free choices are $\psi_2 = 15°$, $\psi_3 = 30°$, and $\vec{Z}_2 = 5.00 + 8.66i = 10.00e^{60.0°}$. Initial guesses of $\phi_2 = 50°$, $\phi_3 = 75°$, and $\vec{Z}_3 = 1.87 + 3.26i$ were used, resulting in the following values:

$$\begin{aligned}
\vec{Z}_1 &= 15.09 - 14.10i = 20.65e^{-43.1°} \\
\vec{Z}_3 &= -10.73 - 7.65i = 13.17e^{-144.5°} \\
\vec{Z}_4 &= -3.20 + 27.20i = 27.38e^{-96.7°} \\
\vec{Z}_5 &= -17.6i - 12.08i = 21.36e^{145.6°} \\
\vec{Z}_6 &= 6.89 + 4.43i = 8.20e^{32.8°} \\
\phi_2 &= 47.7° \\
\phi_3 &= 92.1°
\end{aligned} \tag{8.32}$$

Figure 8.10 shows a pseudo-rigid-body skeleton drawing of this mechanism.

8.2 SYNTHESIS WITH COMPLIANCE: KINETOSTATIC SYNTHESIS

The other class of compliant mechanism synthesis accounts for energy storage. With this type of synthesis, the unique characteristics inherent in compliant mechanisms may be used to design mechanisms with specified energy storage characteris-

Synthesis with Compliance: Kinetostatic Synthesis

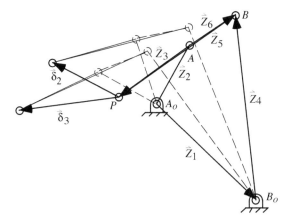

Figure 8.10. Pseudo-rigid-body skeleton drawing of a compliant mechanism for motion generation.

tics. The synthesis equations include not only the rigid-body loop closure equations from the pseudo-rigid-body model, but also include equations dealing with the desired energy-storage relationships. These introduce unknowns that are not common to rigid-body synthesis. For instance, the values of the spring constants of the pseudo-rigid-body model may be either unknowns or free choices. This type of synthesis is called *synthesis with compliance*, and since both the kinematic and static force characteristics are considered, it can be considered *kinetostatic synthesis*.

An example of synthesis with compliance is a mechanism designed for path generation with a prescribed input torque. The input torque and the displacement path of a particular point is specified at each precision point. Another example is a compliant constant-force mechanism that produces the same output force for various input displacements.

In compliant mechanisms, energy is stored as strain energy in the flexible members. One method of accounting for this energy storage is to include springs, with appropriate stiffnesses and locations, on the pseudo-rigid-body model. The resulting model may be used to determine the performance of the mechanism. One powerful analysis method is the principle of virtual work. With this method, only the necessary forces need to be taken into account. This method was described in some detail in Chapter 6 and will be used in the work that follows.

8.2.1 Additional Equations and Unknowns

In rigid-body replacement synthesis, only rigid-body equations are required. This results in the number of equations and unknowns for various tasks as listed in Table 4.2. In synthesis with energy consideration, another equation is added for each precision point. This additional equation is necessary in order to describe the energy

TABLE 8.2. Prescribed and unknown variables for synthesis with energy considerations

Task	Prescribed	Unknown
Function	E_k, ϕ_j, ψ_j	$\vec{Z}_2, \vec{Z}_4, \vec{Z}_6, \gamma_j, k_i, \theta_{o_i}$
Path without prescribed timing	$k_k, \vec{\delta}_j$	$\vec{Z}_2, \vec{Z}_3, \vec{Z}_4, \vec{Z}_5, \phi_j, \gamma_j, \psi_j, k_i, \theta_{o_i}$
Path with prescribed timing	$E_k, \vec{\delta}_j, \phi_j$	$\vec{Z}_2, \vec{Z}_3, \vec{Z}_4, \vec{Z}_5, \gamma_j, \psi_j, k_i, \theta_{o_i}$
Motion	$E_k, \vec{\delta}_j, \gamma_j$	$\vec{Z}_2, \vec{Z}_3, \vec{Z}_4, \vec{Z}_5, \phi_j, \psi_j, k_i, \theta_{o_i}$

where $i = 1, \ldots, m$; $j = 2, \ldots, n-1$; and $k = 1, \ldots, n$

storage or the relationship between the input and output and may represent any number of attributes of the system. Three common examples are the total amount of energy stored in the system as a function of the input, the required input force or torque, and the required input and output force or torque (mechanical advantage or torque gain) at each precision point. The prescribed value of the energy equation at precision point j will be expressed as E_j.

The consideration of energy in the design adds new unknowns to the problem. The value of the spring stiffness and the undeflected position for each flexible segment are now included as unknowns. In the discussion that follows, these quantities will be expressed as k_i and θ_{o_i}, where $i = 1, \ldots, m$, and m is the number of flexible segments in the mechanism. The spring stiffness, k_i, may take different forms, depending on the nature of the flexible segment. For example, a small-length flexural pivot has a stiffness value $k = EI/l$, but for a functionally binary fixed–pinned segment, $k = K_\Theta \gamma EI/l$. The specific form of θ_{o_i} also varies, depending on the specific geometry of the mechanism. The prescribed and unknown variables for various tasks are listed in Table 8.2. Note that $\vec{Z}$ and $\vec{\delta}$ are complex values, and each quantity therefore represents two scalar variables. The numbers of unknowns, scalar equations, and free choices for various tasks are listed in Table 8.3, where n is the number of precision points and m is the number of flexible segments (or the number of springs in the pseudo-rigid-body model).

8.2.2 Coupling of Equations

Recall the standard form of the dyadic construction of equations (8.17) and (8.18). These result in two sets of $2(n-1)$ scalar equations for a problem with n precision points. Example 2 showed that these equations could be solved as two individual

Synthesis with Compliance: Kinetostatic Synthesis

TABLE 8.3. Numbers of unknowns, equations, and free choices for various synthesis problems

Task	No. of unknowns	No. of equations	No. of free choices
Function	$5 + n + 2m$	$3n - 2$	$7 - 2n + 2m$
Path without prescribed timing	$5 + 3n + 2m$	$5n - 4$	$9 - 2n + 2m$
Path with prescribed timing	$6 + 2n + 2m$	$5n - 4$	$10 - 3n + 2m$
Motion	$6 + 2n + 2m$	$5n - 4$	$10 - 3n + 2m$

sets of equations, (8.21) and (8.22), without one affecting the other. In such a case, equations (8.17) and (8.18) are said to be *uncoupled*. This occurs when there are no unknown variables common to the two sets of equations. In the case of the function generation example of Section 8.1.1, this means that the problem could be solved as two sets of second-order matrix equations instead of one set of fourth-order matrix equations. This can significantly reduce the effort required to obtain the solution, particularly when the equations are nonlinear.

The introduction of input-output or energy-storage equations to a synthesis problem adds n equations to the system of equations to be solved. If these equations could be decoupled from the kinematic equations, or if the effects of the coupling could be minimized, the kinematic synthesis and energy synthesis equations could be solved individually. This can be very useful because the energy equations are usually nonlinear in the kinematic variables.

The addition of energy considerations in an n-precision-point problem adds up to n equations and $2m$ unknowns. Typically, the additional equations also include unknown kinematic variables that couple the equations, and in such cases, it is useful to attempt to render the system *weakly coupled*. In a weakly coupled system, the kinematic synthesis equations can be solved without regard to the energy equations. Once all kinematic variables are known, the energy equations are then solved for the remaining unknowns. The system can be made weakly coupled only if

$$2m \geq n \tag{8.33}$$

If, however, more equations than unknowns are introduced into the system, kinematic variables that were previously treated as free choices must now be used as unknowns. The system then becomes *strongly coupled*, and the kinematic and energy equations must be solved simultaneously.

There are some special cases in which the kinematic and energy synthesis equations can be completely uncoupled, and one set may be solved independent from the other. One example is a two-precision-point synthesis with a specified input torque, which uses a flexural pivot between the ground and the input link.

8.2.3 Design Constraints

The same constraints used to determine acceptable designs in rigid-body synthesis and rigid-body replacement synthesis apply to a synthesis with energy considerations. However, some additional constraints must be taken into account. If a spring stiffness is unknown, the resulting spring constants must be positive. The results for the spring constants may also suggest unreasonably large or small thicknesses for the flexible members. When such cases occur, new values for the free choices should be used to obtain a new design.

Following are examples of function, path, and motion generation with energy considerations. These examples include weakly coupled, uncoupled, and strongly coupled systems of equations.

Example: Function Generation with Prescribed Input Torque. Recall the compliant mechanism synthesized for function generation in Section 8.1. This mechanism is to be synthesized for the motion characteristics of the earlier example, as well as for the prescribed input torque. The input torque, M_2, at the precision points is specified as $M_{2_1} = -500$ lb-in., $M_{2_2} = 60$ lb-in., and $M_{2_3} = 200$ lb-in.

The principle of virtual work may be used to determine the input torque, M_2. From the formulation for the general pseudo-rigid-body four-bar mechanism that was presented previously, M_2 at precision point i is

$$M_{2_i} = \frac{k_3}{r_6}[(\theta_{4_i} - \theta_{4_o}) - (\theta_{6_i} - \theta_{6_o})](h_{62_i} - h_{42_i}) + k_4(\theta_{4_i} - \theta_{4_o})h_{42_i} \qquad (8.34)$$

where

$$k_3 = \gamma K_\Theta \frac{EI_3}{l_3} \qquad (8.35)$$

$$k_4 = \frac{EI_4}{l_4} \qquad (8.36)$$

$$h_{62_i} = \frac{r_2 \sin(\theta_{4_i} - \theta_{2_i})}{r_6 \sin(\theta_{6_i} - \theta_{4_i})} \qquad (8.37)$$

Synthesis with Compliance: Kinetostatic Synthesis

$$h_{42_i} = \frac{r_2 \sin(\theta_{6_i} - \theta_{2_i})}{r_4 \sin(\theta_{4_i} - \theta_{6_i})} \quad (8.38)$$

and γ is the characteristic radius factor of the flexible coupler, E is the modulus of elasticity, I is the moment of inertia of the flexible segments, and θ_i and r_i are as defined in Figure 6.9.

Introducing energy considerations to the problem adds three equations ($M_{2_i}; i = 1, 2, 3$) and four new unknowns ($k_3, k_4, \theta_{4_o}, \theta_{6_o}$). Since there are more new unknowns than equations, one unknown may be used as a free choice. Electing to use θ_{4_o} as the free choice and expressing the input torque equations in linear form results in

$$\begin{bmatrix} -(\theta_{4_1} - \theta_{6_1} - \theta_{4_o})(h_{62_1} - h_{42_1}) & (\theta_{5_1} - \theta_{5_o})h_{42_1} & h_{62_1} - h_{42_1} \\ -(\theta_{4_2} - \theta_{6_2} - \theta_{4_o})(h_{62_2} - h_{42_2}) & (\theta_{5_2} - \theta_{5_o})h_{42_2} & h_{62_2} - h_{42_2} \\ -(\theta_{4_3} - \theta_{6_3} - \theta_{4_o})(h_{62_3} - h_{42_3}) & (\theta_{5_3} - \theta_{5_o})h_{42_3} & h_{62_3} - h_{42_3} \end{bmatrix} \begin{pmatrix} k_3 \\ k_3 \theta_{6_o} \\ k_4 \end{pmatrix}$$

$$= \begin{pmatrix} M_{2_1} \\ M_{2_2} \\ M_{2_3} \end{pmatrix} \quad (8.39)$$

The unknowns of the position analysis are present in equation (8.39), and the analysis of the kinematics and input torque is therefore coupled. The system is weakly coupled, however, because the kinematic synthesis equations may be solved independent from the input torque equations. Once the kinematic variables are known, equation (8.39) may be solved as a set of linear equations to find the values of the remaining unknowns. If unreasonable results are obtained, the free choices may be modified until acceptable values are found.

The motion synthesis was accomplished in the function generation example of Section 8.1, and with a value of $\theta_{4_o} = \theta_{4_2} = \theta_{4_1} + \phi_2 = 79.75°$, equation (8.39) yields

$$k_3 = 14.66$$
$$\theta_{6_o} = -154.90° \quad (8.40)$$
$$k_4 = 458.23$$

Once the values of k_3 and k_4 are known, the specific dimensions of the flexible segments may be determined. The unit of length is assumed to be inches, and $K_\Theta = 2.65$, and $\gamma_{ps} = 0.85$. The material is steel with $E = 30 \times 10^6$ lb/in.^{2}, and the length of the small-length flexural pivot is 0.2 in. The resulting moments of

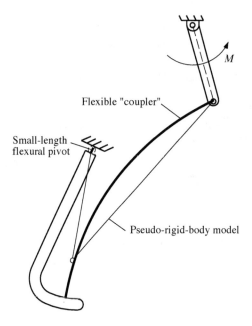

Figure 8.11. Compliant mechanism for function generation with prescribed input torque.

inertia for the two flexible segments are $I_3 = 2.76 \times 10^{-6}$ in.^{4} and $I_4 = 3.05 \times 10^{-6}$ in.^{4} Using a rectangular cross section with a thickness of 1/32 in., the widths of the segments are $b_3 = 1.09$ in. and $b_4 = 1.20$ in. The resulting mechanism is shown in Figure 8.11.

8.2.4 Special Case of $\theta_o = \theta_j$

The undeflected position of the flexible members of a compliant mechanism can be specified to be at a precision point. In such a case, $\theta_o = \theta_j$. If all flexible members have their undeflected position at the same precision point, the energy stored at that position is zero, and $E_j = 0$.

Because $E_j = 0$ and the undeflected position of the mechanism is prescribed, one less energy equation and m fewer unknowns exist. The effects of this on the number of free choices available in a synthesis problem are listed in Table 8.4. This special case is significant because it is possible to prescribe the location of stable equilibrium positions. Since the energy at this position is minimal, the mechanism will tend toward it when it is in a nearby position. This is the position the mechanism adopts when it is at rest, and it is vital in many compliant mechanism designs.

Synthesis with Compliance: Kinetostatic Synthesis

TABLE 8.4. Numbers of unknowns, equations, and free choices when $\theta_o = \theta_j$

Task	No. of unknowns	No. of equations	No. of free choices
Function	$5 + n + m$	$3n - 3$	$8 - 2n + m$
Path without prescribed timing	$5 + 3n + m$	$5n - 5$	$10 - 2n + m$
Path with prescribed timing	$6 + 2n + m$	$5n - 5$	$11 - 3n + m$
Motion	$6 + 2n + m$	$5n - 5$	$11 - 3n + m$

Example: Path Generation with Prescribed Potential Energy. Consider the path generation with prescribed timing example in Section 8.1. This mechanism is now to be designed so that a stable equilibrium position will occur at the first precision point ($\theta_{o_i} = \theta_{i_1}$; $i = 1, 2, 6, 5$), and the total potential energy, V, at each precision point will be

$$V_1 = 0 \text{ in.-lb}$$
$$V_2 = 5 \text{ in.-lb} \tag{8.41}$$
$$V_3 = 30 \text{ in.-lb}$$

The torsional spring constant for a flexural pivot is approximated as EI/l, and the potential energy associated with it is

$$V = \frac{EI}{2l}(\theta - \theta_o)^2 \tag{8.42}$$

The total potential energy for the system at point j is

$$V_j = \frac{1}{2}k_1(\theta_{2_j} - \theta_{2_o})^2 + k_2[(\theta_{2_j} - \theta_{2_o}) - (\theta_{6_j} - \theta_{6_o})]^2$$
$$+ k_3[(\theta_{5_j} - \theta_{5_o}) - (\theta_{6_j} - \theta_{6_o})]^2 + k_4(\theta_{5_j} - \theta_{5_o})^2 \tag{8.43}$$

However, since $\theta_{i_o} = \theta_{i_1}$, the system may be simplified:

$$V_1 = 0$$
$$V_2 = \frac{1}{2}[k_1\phi_2^2 + k_2(\phi_2 - \gamma_2)^2 + k_3(\psi_2 - \gamma_2)^2 + k_4\psi_2^2]$$
$$V_3 = \frac{1}{2}[k_1\phi_3^2 + k_2(\phi_3 - \gamma_3)^2 + k_3(\psi_3 - \gamma_3)^2 + k_4\psi_3^2]$$
(8.44)

Since the only unknowns in the equations above are k_i; $i = 1, 2, 6, 5$, the energy and kinematic synthesis equations are uncoupled. This means that once the free choices for γ_j and ψ_j are made, either the energy or the kinematic equations may be solved independent from the other set of equations.

The energy equations involve four unknowns and two equations, resulting in two free choices. The flexible members are assumed to be strips of spring steel with $E = 30 \times 10^6$ lb/in.² The free choices are k_2 and k_3; the member thickness, width, and length are taken to be 1/32 in., 1 in., and 2 in., respectively. This results in $k_2 = k_3 = 38.147$ in.-lb. The values of k_1 and k_4 may be found by rearranging equation (8.44) in linear form as

$$\begin{bmatrix} \phi_2^2 & \psi_2^2 \\ \phi_3^2 & \psi_3^2 \end{bmatrix} \begin{pmatrix} k_1 \\ k_4 \end{pmatrix} = \begin{bmatrix} 2V_2 - k_2(\phi_2 - \gamma_2)^2 - k_3(\psi_2 - \gamma_2)^2 \\ 2V_3 - k_2(\phi_3 - \gamma_3)^2 - k_3(\psi_3 - \gamma_3)^2 \end{bmatrix}$$
(8.45)

Solving the equation yields

$$k_1 = 55.802 \text{ in.-lb}$$
$$k_4 = 25.286 \text{ in.-lb}$$
(8.46)

If the thickness is 1/32 in. and the width is 1 in., the lengths are calculated from k_3 and k_4 to be $l_3 = 1.37$ in. and $l_4 = 3.02$ in. The mechanism has now been synthesized for prescribed potential energy, specified equilibrium position, and path generation with prescribed timing.

Example: Motion Generation with Prescribed Input Torque. The mechanism discussed in the motion generation example in Section 8.1 is to be synthesized for motion generation and prescribed input torque. The input torque, M_{2_j}, at each precision position is given as

$$M_{2_1} = 0.5 \text{ in.-lb}$$
$$M_{2_2} = 0.5 \text{ in.-lb}$$
$$M_{2_3} = 2.0 \text{ in.-lb}$$
(8.47)

The input torque may be calculated using the principle of virtual work:

Synthesis with Compliance: Kinetostatic Synthesis

$$M_j = k_4(\theta_{4_j} - \theta_{4_o}) \frac{r_2 \sin(\theta_{6_j} - \theta_{2_j})}{r_4 \sin(\theta_{4_j} - \theta_{6_j})} \qquad j = 1, 2, 3 \qquad (8.48)$$

This adds three equations but only two unknowns, k_4 and θ_{4_o}, and one additional kinematic variable must therefore be used as an unknown. Three unknowns are then common between the kinematic and input torque equations, resulting in a strongly coupled system. Recall that due to the choice of unknown variables, the motion synthesis required a nonlinear solution. The coupled system here requires the simultaneous solution of 11 nonlinear equations with 11 unknowns.

In the previous motion analysis, ψ_2, ψ_3, and $\vec{Z}_2$ were chosen as free variables. An additional unknown is required for this analysis, resulting in only one free choice. Values of $\psi_2 = 15°$ and $\vec{Z}_2 = 10e^{60°}$ are chosen, and ψ_3 is now an unknown. The eleven nonlinear equations are solved to obtain

$$\vec{Z}_1 = 6.48 + 2.64i = 7.00e^{22.2°}$$
$$\vec{Z}_3 = -10.74 - 7.64i = 13.17^{-144.6°}$$
$$\vec{Z}_4 = 0.46 + 16.97i = 16.98e^{88.4°}$$
$$\vec{Z}_5 = -12.68 - 18.59i = 22.50e^{-124.3°}$$
$$\vec{Z}_6 = 1.95 + 10.95i = 11.12e^{79.9°} \qquad (8.49)$$
$$\phi_2 = 47.7°$$
$$\phi_3 = 92.1°$$
$$\psi_3 = 44.2°$$
$$k_4 = 4.43 \text{ in.-lb}$$
$$\theta_{4_o} = 273.2°$$

where

$$k_4 = \gamma K_\Theta \frac{EI}{l} \qquad (8.50)$$

Values of $K_\Theta = 2.58$, $\gamma = 0.83$, and $E = 30 \times 10^6$ lb/in.² are assumed. The length of the flexible member is $l = r_5/\gamma_{ps} = 20.45$ in. A rectangular cross section with a thickness of 1/32 in., and a width of 0.554 in. is used. Figure 8.12 shows this mechanism in its initial, undeflected position.

Once a design has been obtained, it should be analyzed to ensure that the design constraints are met. Three different analysis methods show that they give consistent results. The first method is the most straightforward. Constant values of γ and K_Θ are used in the pseudo-rigid-body model when the rigid-body mechanism analysis is performed. Since identical values of γ and K_Θ are used for both the analysis and synthesis, this should result in the exact path specified. The second method uses

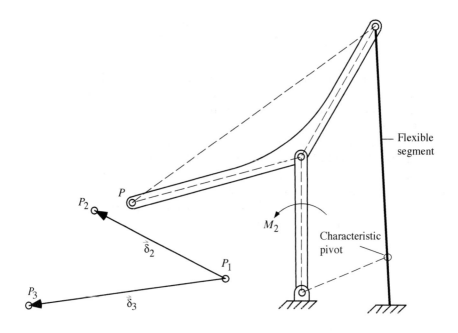

Figure 8.12. Compliant mechanism in its undeflected position.

updated values of γ and K_Θ for the pseudo-rigid-body model. The updated values change as the load direction changes for different mechanism positions. Since γ and K_Θ values change only slightly, the results from this method should be similar to those obtained using constant values of γ and K_Θ. The third analysis method is nonlinear finite element analysis. Although this method is not practical in the synthesis phase, it is very useful in verifying the design once an initial design is obtained.

The displacement vectors of point P, $\vec{\delta}_j$, at different values of the crank angle (precision point positions) are listed in Table 8.5. The 3% relative error in the y displacement at precision point 3, between finite element analysis and pseudo-rigid-body model results, is the largest difference calculated.

TABLE 8.5. Coupler point displacement comparison

	Finite element analysis	Pseudo-rigid-body model	
θ_2		Updated	Constant
60.000	$0.000 + 0.000i$	$0.000 + 0.000i$	$0.000 + 0.000i$
107.674	$-10.006 + 5.009i$	$-10.006 + 5.008i$	$-10.000 + 5.000i$
152.095	$-15.030 - 1.941i$	$-15.011 - 1.982i$	$-15.000 - 2.000i$

Other Synthesis Methods

Table 8.6 lists the angles of the coupler link, θ_6, at the precision positions. The input torque, M_2, at the precision points are listed in Table 8.7. A maximum relative error of 2.2% occurs between the updated pseudo-rigid-body model and finite element analysis results.

8.3 OTHER SYNTHESIS METHODS

The equations and methods described thus far are general; they are intended to apply to a large number of synthesis problems rather than specific cases. The equations and examples in this chapter have concentrated on planar, three-finitely-separated-precision-point function, motion, and path generation, but they are not limited to such cases. A number of other synthesis methods applicable to compliant mechanisms are described briefly below.

8.3.1 Burmester Theory for Finite Displacements

Burmester theory is often used to design rigid-body mechanisms in four-precision-point synthesis. A nonlinear solution of the general equations derived above will also provide solutions to the four-precision-point problem. An advantage of Burmester theory is that the resulting equations are linear, and convergence is not a concern. It is also advantageous because it provides the designer with a graphical representation of the infinite number of solutions available. The center- and circle-

TABLE 8.6. Coupler angle comparison

θ_2	Finite element analysis	Pseudo-rigid-body model	
		Updated	Constant
60.000	79.963	79.966	79.919
107.674	59.920	59.925	59.919
152.095	69.681	69.873	69.919

TABLE 8.7. Input torque comparison

θ_2	Finite element analysis	Pseudo-rigid-body model	
		Updated	Constant
60.000	0.47916	0.47743	0.50000
107.674	0.48319	0.50304	0.50000
152.095	1.96510	2.00758	2.00000

point curves show the possible ground pivot and coupler pivot locations for the specified design requirements.

The application of Burmester theory to compliant mechanisms is similar to that described for the general problem [56]. The simplest case involves rigid-body replacement. In this type of synthesis, a pseudo-rigid-body mechanism is designed in the same manner as a rigid-body mechanism. The resulting design is evaluated, taking into account the constraints associated with compliant mechanisms. A compliant mechanism is then designed from an acceptable pseudo-rigid-body mechanism. The more challenging problem is synthesis for compliance. Again, the kinematic and compliance equations may be weakly coupled when $m \geq 2$ and there is an appropriate choice of free variables.

The intersections of two sets of circle- and center-point curves may be used to obtain solutions for five-precision-point problems. Burmester theory has been extended to five- and six-bar mechanisms, and may also be extended to compliant mechanisms.

8.3.2 Infinitesimal Displacements

Path curvature theory is based on infinitesimal displacements. Many of the concepts of curvature theory have proven useful in kinematic synthesis; for example, the inflection circle is the locus of points on the moving plane (of which the coupler link is a part) which, at a given mechanism orientation, move in straight lines. Burmester's circle- and center-point curves for four infinitesimally close positions, or the cubic of stationary curvature, are the locus of points in a moving plane which, at a given mechanism orientation, have stationary curvature. Ball's point is the intersection of the cubic of stationary curvature and the inflection circle, representing a point with a displacement path that is particularly rectilinear for a considerable distance. These important concepts, and many others from path curvature theory, may be applied to compliant mechanisms by using the pseudo-rigid-body model.

8.3.3 Optimization of Pseudo-Rigid-Body Model

The closed-form synthesis methods discussed in this work are used to design mechanisms for which the pseudo-rigid-body model has no structural error at the precision points specified. However, the methods do not control the structural error between these points. Often, it is desired that the error be maintained within a certain range for a specified motion. There may be points that have no structural error, but their locations are not specified. Optimization is a useful tool in solving such design problems. A design optimization problem may be specified that minimizes the deviation from the desired behavior while maintaining specified constraints.

8.3.4 Optimization

Another approach to the design of compliant mechanism design is the use of optimization to design mechanisms that are outside the design space of the pseudo-rigid-body model. The advantage of a structural optimization approach is that it is possible to investigate designs that are outside the experience base of the designer or of mechanical design in general. The disadvantages include the complexity of the methods, the difficulty in including nonlinear deflection analysis. These methods are discussed in Chapter 9.

8.4 PROBLEMS

8.1 What is the difference between weakly coupled and strongly coupled equations in compliant mechanism synthesis?

8.2 Design a compliant mechanism to replace an existing rigid-body mechanism. Follow the steps given below.
   (a) Evaluate a number of rigid-body mechanisms and choose one that is well suited to be replaced by a compliant mechanism.
   (b) Specify the design requirements that are necessary for the mechanism to adequately perform its function.
   (c) Use rigid-body replacement synthesis to design a compliant mechanism.
   (d) Analyze the compliant mechanism to ensure that the design constraints are met.

8.3 A Roberts straight-line mechanism is shown in Figure P8.3. Use rigid-body replacement synthesis to design compliant mechanisms for the following conditions:
   (a) Use four small length flexural pivots.
   (b) Use two fixed-pinned segments pinned to ground.
   (c) Sketch other possible compliant configurations of the mechanism.

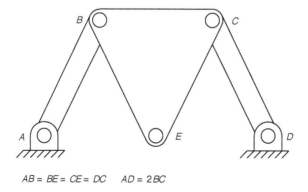

$AB = BE = CE = DC \quad AD = 2BC$

Figure P8.3. Rigid-body Roberts approximate-straight-line mechanism.

CHAPTER 9

OPTIMAL SYNTHESIS WITH CONTINUUM MODELS

G. K. Ananthasuresh
University of Pennsylvania

Mary I. Frecker
Pennsylvania State University

9.1 INTRODUCTION

Fully compliant mechanisms can be viewed as flexible continua and can be treated as such in their analysis and synthesis. For instance, small and large deformations of a flexible cantilever beam can be modeled in the beam's actual continuum form instead of pseudo-rigid-body models. Thus in this chapter, compliant mechanisms are modeled using the methods of continuum solid mechanics rather than rigid-body kinematics. Consequently, structural optimization techniques can be adapted for compliant mechanism design. The geometric structure of a compliant mechanism and the properties of the material it is made of determine its force and motion transmission capability. Design of compliant mechanisms then becomes an inverse problem wherein the geometry of the flexible material continuum for a given material is to be obtained for prescribed force and displacement specifications. Continuous optimization methods are employed in solving this inverse problem.

Figure 9.1 depicts the overall view of this chapter using the example of a certain kind of compliant pliers. As shown in Figure 9.1a, the desired behavior of the pliers is specified by indicating where the force is applied, and where and in which direction the output displacement is desired. The region in which the mechanism to be designed must fit is also specified. This region is called the *design domain*. A material is also chosen and its properties are specified. The information shown in Figure 9.1a is the only input given to the systematic optimal design method, which then generates the solution shown in Figure 9.1b. Shown in Figure 9.1c is the finite element analysis solution, in which it can be seen that the behavior desired is indeed

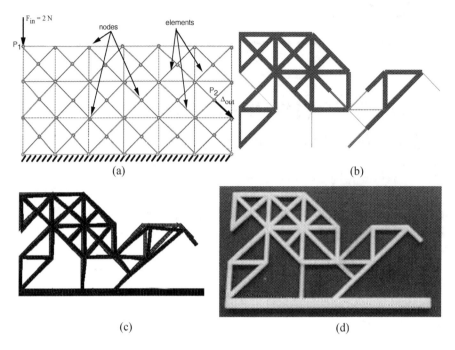

Figure 9.1. Compliant pliers from problem specifications to prototype—a systematic method for the design of compliant mechanisms using continuum mechanics models and optimization. (a) Problem specifications. Optimization method determines the out-of-plane thickness of each beam element in the super structure. Elements that reach the lowest limit are removed. (b) Optimum solution. The figure shows the relative out-of-plane thickness of each element. (c) Verification of the solution with the continuum model using the finite element analysis. (d) Manufactured prototype using ABS plastic. *Note*: Segments have different out-of-plane thickness following the optimal solution in (b).

obtained. Figure 9.1d shows the photograph of the fabricated laboratory prototype using ABS plastic material. The unique feature of this design method is that optimal solutions of compliant mechanisms can be generated automatically to obtain desired force–deflection behavior. As can be seen in this example, no decisions are made about the physical form of the compliant mechanism at the outset except specifying the space in which it should fit. The optimization algorithm generates the best solution for a given problem specification. The solutions obtained in this manner have adequate detail to generate the manufacturing information automatically in the form of computer numerically controlled machine code for macro-devices or the photolithographic mask layouts for micromachined mechanical devices. The purpose of this chapter is to give an overview of the underlying principles, hypotheses, methods, and numerical techniques that make this possible. The following topics are covered in this chapter:

Introduction

- Elastostatic analysis using the finite element analysis
- Structural optimization
- Conditions for a minimum in unconstrained and constrained minimization
- Measures of stiffness and flexibility
- Formulation of the optimal synthesis problem for the design of compliant mechanisms
- Size, shape, and topology optimizations
- Computational aspects: numerical optimization techniques and sensitivity analysis
- Optimality criteria methods

Since only a brief overview is given on each of these topics, interested readers should consult the references cited herein for more information.

9.1.1 Distributed Compliance

The elastic deformation in a compliant mechanism can be limited only to a small portion as in mechanisms with flexural or notch hinges. Such mechanisms are called *lumped compliant* mechanisms and they can be analyzed and designed using rigid-body kinematic techniques. On the other hand, if a large portion of the structure deforms as shown in Figure 9.2, it is called *distributed compliance*. The methods described in this chapter pertain to the distributed-type compliant mechanisms.

9.1.2 Continuum Models

If the deforming structure has a regular geometry such as beams with constant cross-section area or a diaphragm of uniform thickness, analysis and design can be accomplished with simple analytical models such as the Euler beam theory model and pseudo-rigid-body model. If the geometry of compliant mechanisms is irregular, the deformation and stress analysis can be done by solving the partial differen-

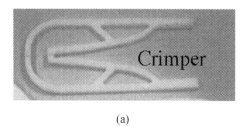
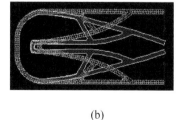

(a)   (b)

Figure 9.2. (a) Compliant crimping mechanism with distributed compliance, and (b) deformed model using finite element analysis.

tial equations governing the elastic deflection behavior. In general, closed-form analytical solutions are not possible for irregular geometries. Therefore, these equations are solved using numerical methods such as finite element method [135], finite difference method [136, 137], boundary element method [138], chain algorithm [4, 91], and so on. In this chapter we use the finite element method when we consider the design of compliant mechanisms.

9.1.3 Elastostatic Analysis Using the Finite Element Method

The term *elastostatic analysis* implies static analysis of elastic structures. The finite element method is used extensively for elastostatic structural analysis to determine deflections, stresses, and strains of a structure with any geometry and boundary conditions. The boundary conditions specify where the structure is supported or fixed and where the loads are applied on the boundary of the structure. The basis for the finite element method is discretization of the structure into small elements and interpolation of scalar field variables within the element. The interpolation is done in terms of the values of the variables at the vertices of the element using what are known as *shape functions*. The vertices of a finite element are called *nodes*. The scalar field variables include components of deflections, stresses, and strains in orthogonal directions at each node in the discretized finite element model. For linear elastic analysis, the equilibrium equation for a finite element model of a flexible continuum of any shape is written in the following matrix equation form:

$$[K]\{u\} = \{f\} \tag{9.1}$$

where $\{u\}$ is a $N \times 1$ vector of displacement degrees of freedom, $\{f\}$ is a $N \times 1$ vector of external forces applied on the degrees of freedom, $[K]$ is a $N \times N$ symmetric matrix called the *stiffness matrix* of the flexible continuum, and N is the total number of displacement degrees of freedom. Equation (9.1) has the same form as the equilibrium equation of a simple linear spring: $kx = f$. Thus the finite element method renders a linear elastic system of any geometry and loading conditions to a discretized linear spring model. Elastostatic analysis entails the determination of the unknown displacement vector $\{u\}$ for specified geometry, material properties, and force vector $\{f\}$. This can be done by solving N linear equations in equation (9.1). As an example, consider the continuous beam in a plane and its finite element representation using only 3 two-noded elements, as shown in Figure 9.3a and b. It has a total of 12 displacement degrees of freedom, 3 at each node. The three degrees of freedom at each node for a planar beam element are axial and transverse displacements and the slope. As another example, consider a planar continuum discretized using 9 four-noded plane-stress elements with 18 nodes and a total of 36 degrees of freedom (Figure 9.3c and d). In a plane-stress element, each node has two orthogonal displacements as its degrees of freedom.

The stiffness matrix depends on the geometry and material properties that make up the flexible continuum under consideration. In the design problem, which is the

Introduction

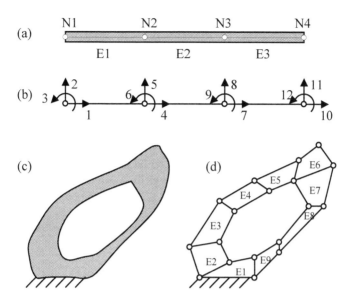

Figure 9.3. (a) Continuous beam discretized into three beam elements, (b) degrees of freedom in the discretized beam model, (c) two-dimensional continuum, and (d) discretized finite element model of the two-dimensional continuum.

inverse of the analysis problem, the stiffness matrix is unknown because the geometry of the continuum is unknown. Thus, the design variables in the optimization are in the stiffness matrix. We use the linear elastic model shown in equation (9.1) later in this chapter.

9.1.4 Structural Optimization

If we look around closely at various structures, ancient and modern, we can observe that they have certain shape that might appear peculiar at first sight. For example, bridges have arches of certain shape and grillages of certain pattern—they have that shape because the engineer who designed it thought that for that situation that shape was optimal. It is often argued that the Eiffel tower and many bridges designed by Eiffel, the great French engineer, are optimal [139, 140]. An optimum can be defined on the basis of many criteria. One can design a structure for maximum stiffness, minimum weight, maximum strength, suitable flexibility, maximum crashworthiness, suitable dynamic characteristics such as natural frequency and normal mode shapes, and many other objectives. There may also be constraints on the amount of material used, type of material used, space available to fit the structure, maximum permissible stress, maximum permissible deformation, buckling load, and so on. In fact, the objectives and constraints can be interchanged depending on which characteristic is more crucial. The most important characteristic is often chosen as the objective function, and the characteristics of secondary importance are formulated as constraints. In the distant past, the "optimum" structures were designed by engineers who had great intuition for design and perhaps sometimes by trial and error. But today, the advances in mathematics (optimization theory [141] and variational calculus [142]), computational mechanics (finite element method [135] and boundary element method [138]), and computer technology have given rise to the field of structural optimization. Structural optimization techniques enable us to design optimum structures systematically. The essence of structural optimization is the economical and efficient use of material in designing a structural component to meet certain objectives under some constraints. Structural optimization is a broad topic in itself. Interested readers may want to consult one of several books on structural optimization [143].

Much of the literature in the structural optimization field is concerned with the design of the stiffest structures with the least weight. Compliant mechanisms, on the other hand, are structures that are intended to be sufficiently flexible. However, the methods of structural optimization can be extended and applied to compliant mechanism design if continuum models are used. To this end, new formulations must be developed and some challenges in the numerical optimization must be addressed. The remainder of this chapter is a brief overview of structural optimization and its application to compliant mechanism design.

9.2 FORMULATION OF THE OPTIMIZATION PROBLEM

9.2.1 Objective Function, Constraints, and Design Variables

A general *constrained optimization problem* with an *objective function* $f(x_1, x_2, ..., x_n)$, *p equality constraints*, and *m inequality constraints* in n variables $(x_1, x_2, ..., x_n)$ is stated as shown below. Here we want to find the optimal set of values for the variables such that the objective function has the least value while satisfying the equality and inequality constraints.

$$\text{Minimize } f(x_1, x_2, ..., x_n)$$
$$\text{subject to}$$
$$h_i(x_1, x_2, ..., x_n) = 0 \quad i = 1, 2, ..., p \quad \text{[P 9.1]}$$
$$g_j(x_1, x_2, ..., x_n) \leq 0 \quad j = 1, 2, ..., m$$

If there are no constraints, it will be called an *unconstrained optimization problem*. In that case, the necessary conditions for the minimum of f are

$$\frac{\partial f}{\partial x_i} = 0 \quad i = 1, 2, ..., n \tag{9.2}$$

Formulation of the Optimization Problem

By solving the n equations for n variables above, the optimum for the unconstrained problem can be determined.

The necessary conditions for the constrained optimization problem, also known as *Karush–Kuhn–Tucker conditions*, are given by

$$\frac{\partial f}{\partial x_k} + \sum_{i=1}^{p} \mu_i \frac{\partial h_i}{\partial x_k} + \sum_{j=1}^{m} \lambda_j \frac{\partial g_j}{\partial x_k} = 0 \quad k = 1, 2, \ldots, n$$

$$h_i = 0 \quad i = 1, 2, \ldots, p \text{ and } g_j \leq 0 \quad j = 1, 2, \ldots, m \tag{9.3}$$

$$\lambda_j g_j = 0 \text{ (switching condition)}$$

$$\lambda_j \geq 0$$

The equations and inequalities above can be solved to obtain the optimal values of the n variables and the multipliers μ_i ($i = 1, 2, \ldots, p$) and λ_j ($j = 1, 2, \ldots, m$). These multipliers are called *Lagrange multipliers*. Using these multipliers as additional unknown variables, the constrained problem of [P 9.1] can be posed as an unconstrained problem by defining the *Lagrangian*, L, as follows:

minimize

$$L = f(x_1, x_2, \ldots, x_n) + \sum_{i=1}^{p} \mu_i h_i(x_1, x_2, \ldots, x_n) + \sum_{j=1}^{m} \lambda_j g_j(x_1, x_2, \ldots, x_n) \quad [\text{P 9.2}]$$

The solution to the "unconstrained" problem above is the same as equation (9.3) [Apply equation (9.2) to [P 9.2] and get equation (9.3)].

It should be noted that conditions in equations (9.2) and (9.3) are only necessary, but not sufficient, for a minimum. Therefore, we also need to check if the resulting solution is a minimum or a maximum using second-order sufficient conditions. Interested readers may refer to one of several books on optimization [143, 144] to learn more about necessary and sufficient conditions and numerical algorithms to solve them.

In the context of optimal design of compliant mechanisms, the objective function is formulated to quantify a function- or performance-related requirement. Quantitative measures of flexibility, stiffness, strength, weight, mechanical efficiency, maximum stress, buckling load, natural frequency, normal mode shapes, and so on, are of importance in the design of compliant mechanisms. One or more of these are used to define the objective function f, depending on the primary requirements of the design task at hand. The remaining criteria are included as equality or inequality constraints h_i ($i = 1, 2, \ldots, p$) and g_j ($j = 1, 2, \ldots, m$). For instance, we may wish to minimize the weight of the structure while satisfying a flexibility requirement as an equality constraint. Minimization of the weight often leads to the simplest design that uses the material most economically. The design variables x_k ($k = 1, 2, \ldots, n$) are chosen such that by varying these values, the geometry and structural form of the continuum model of the compliant mechanism can be modified to minimize the objective function and thus obtain the optimal design. The choice of the design variables, the objective function, and the constraints constitute the optimization problem for compliant mechanism design.

9.2.2 Measures of Stiffness and Flexibility

Mean compliance, defined as the work done by the applied external forces, is often used as a measure of the stiffness of a structure. Mathematically, it is given by

$$\text{mean compliance} = \int_V fu \, dV + \int_S tu \, dS + \sum_i F_i u_i \tag{9.4}$$

where u is the displacement field, f is the distributed body force (e.g., gravity force), t is distributed surface force (e.g., fluid pressure), F_i is ith point force, and u_i is ith displacement degree of freedom. V denotes the volume of the continuum, and S is the surface area of the continuum. The smaller the mean compliance, the stiffer the structure, and vice versa. As can be seen in equation (9.4), for specified forces, minimizing mean compliance implies that the displacements at the points of application of the forces are minimized as a weighted sum. Minimum displacement at the points of application of forces implies that minimum work is done by the forces on the structure, thus resulting in the least elastic energy stored in the structure. The stiffer the structure, the less elastic energy is stored in it. Therefore, a more intuitive measure of stiffness is the *strain energy* (SE) stored in a structure when it deforms. At equilibrium, the strain energy is equal to half the mean compliance. Thus the smaller the strain energy, the stiffer the structure, and vice versa. The strain energy of a single spring of stiffness constant k and displacement u is $(ku^2/2)$. The strain energy of a general three-dimensional continuum is given by

$$\text{strain energy} = \int_V \frac{1}{2}\sigma\varepsilon \, dV \tag{9.5}$$

The integrand in equation (9.5) is called the *strain energy density* (i.e., the strain energy stored in unit volume of the structure). For a discretized finite element model of a continuum

$$\text{strain energy} = \frac{1}{2}\{u\}^T[K]\{u\} \tag{9.6}$$

Expression (9.6) is the multi-degree-of-freedom analogue of the strain energy of a single spring. In general, in the design of compliant mechanisms we want one or more points, called *output ports*, to displace by a large amount to generate motion. Just as minimizing the displacements at the points of application of forces is equivalent to maximizing the stiffness of a structure, maximizing displacements at the

Formulation of the Optimization Problem

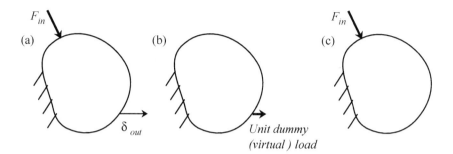

Figure 9.4. Illustration of the unit dummy (virtual) load method. (a) $\delta_{out} = \int_V \sigma_d \varepsilon \, dV$. (b) Displacement field, V; stress, σ_d; strain, ε_d. (c) Displacement field, U; stress, σ; and strain, ε.

output ports is equivalent to maximizing the flexibility. Consider the general situation, shown in Figure 9.4a, of a flexible continuum of arbitrary shape with a fixed portion, an applied load F_{in} at the input port, and an output port with a desired direction of maximum displacement. The expression for the displacement at the output port, δ_{out}, in the specified direction under the applied force F_{in} is given by the *mutual strain energy* (MSE) [145], defined as

$$\delta_{out} = \text{mutual strain energy} = \int_V \sigma_d \varepsilon \, dV \qquad (9.7)$$

where σ_d is the stress field when only a unit dummy (virtual) load is applied in the direction of the output displacement at the output port (Figure 9.4b), and ε is the strain field when only the actual load F_{in} is applied at the input port (Figure 9.4c).

For a discretized finite element model, MSE is given by

$$\text{mutual strain energy} = \{v\}^T [K]\{u\} \qquad (9.8)$$

where $\{v\}$ is the displacement vector when only the unit dummy load is applied, and $\{u\}$ is the displacement vector when only the actual load F_{in} is applied. Thus $\{v\}$ and $\{u\}$ are the solutions of the following equilibrium equations in matrix form [see equation (9.1)]:

$$\begin{aligned} [K]\{v\} &= \{f\}_{\text{unit dummy load}} \\ [K]\{u\} &= \{f\}_{\text{actual}} \end{aligned} \qquad (9.9)$$

In this chapter SE and MSE are used as measures of stiffness and flexibility, respectively.

9.2.3 Multicriteria Formulations

The desired motion specified in the form of an output port displacing in a prescribed direction when an input force is applied in a given direction at the input port constitutes the kinematic requirements for a compliant mechanism. Therefore, sufficient flexibility is necessary to generate the displacement required at the output port. However, if the continuum is too flexible, it will not be able to support any output loads. Therefore, adequate stiffness is also needed to be able to support applied loads or resisting the force created by an external object at the output port. Thus there are two objectives to be met simultaneously when designing a compliant mechanism. That is, we seek an optimal continuum (1) flexible enough to satisfy the kinematic requirements and (2) stiff enough to support external loads. The two objectives mentioned above are in conflict with each other, and that provides the scope for optimization.

Whenever more than one objective criteria is involved, it becomes a multicriteria formulation. The two objectives can be combined in several ways. A weighted linear combination formulation gives rise to an objective function defined as

$$\text{minimize } f = \{-(\alpha)\text{MSE} + (1-\alpha)\text{SE}\} \quad 0 \leq \alpha \leq 1 \qquad [\text{P 9.3}]$$

The weighting factor α is used to assign relative importance to normalized values of MSE and SE. Since there is a negative sign in front of MSE and a positive sign for SE, it means that when we minimize f, we maximize MSE and minimize SE. The resulting solutions will be different for different values of α. Alternatively, MSE and SE can be combined in a ratio-type formulation as

$$\text{minimize } f = \left\{\frac{-\text{MSE}}{\text{SE}}\right\} \qquad [\text{P 9.4}]$$

Here again MSE is maximized and SE is minimized to minimize f, but there is no need to choose any other parameter such as the α used in problem [P 9.3].

As a third alternative, MSE and SE can be combined to maximize one type of *mechanical efficiency* of a compliant mechanism. Mechanical efficiency is defined here as the ratio of the useful work done at the output port to the work done by the input forces. It should be noted that in compliant mechanisms all the work done by the input forces is not available for use at the output port because a portion of that energy is used to deform the structure. This portion, called the *lost energy*, should be minimized. Consider the flexible continuum of arbitrary shape shown in Figure 9.5 with a spring attached at the output port. This spring, with a linear stiffness constant k_s, represents an elastic object acted upon by the compliant mechanism. If δ_{out} is the output port displacement, the useful work done on the elastic object is $k_s \delta_{\text{out}}^2 / 2$. Note that this quantity is less than the total work done by the input

Formulation of the Optimization Problem

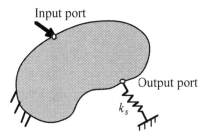

Figure 9.5. Flexible continuum with a linear spring at the output port to model the workpiece.

forces. As noted above, δ_{out} is nothing but MSE. Furthermore, at equilibrium the work done by the input forces is twice the strain energy, SE. Our objective is to minimize the energy lost as elastic energy and maximize the energy available at the output port. Therefore, the multicriteria objective function is given by

$$\text{maximize} \left\{ f = \frac{\text{mechanical}}{\text{efficiency}} = \frac{\text{output work}}{\text{input work}} = \frac{(1/2)k_s \text{MSE}^2}{2\text{SE}} \right\} \qquad [\text{P 9.5}]$$

Since the scalar multipliers in the objective function do not affect the optimum solution, they can be removed from the objective function. Converting the maximization formulation to a minimization, we get

$$\text{minimize} \left\{ f = \frac{-\text{sign}(\text{MSE})\text{MSE}^2}{\text{SE}} \right\} \qquad [\text{P 9.6}]$$

In the formulation above, sign(MSE) is introduced to restore the direction of MSE (i.e., δ_{out}), which is lost in squaring MSE. This is important because the output port displacing in a direction opposite to the intended direction is not useful in the compliant mechanism design.

In fact, the formulations above can be expressed in a general way by using any monotonically increasing functions of MSE and SE, denoted, respectively, by $\phi(\text{MSE})$ and $\psi(\text{SE})$, and combining them in the weighted linear combination form or the ratio form. The three formulations above then become special cases of the following general formulations [146]:

$$\text{minimize} \left\{ f = -\alpha\, \phi(\text{MSE}) + (1-\alpha)\psi(\text{SE}) \right\} \qquad [\text{P 9.7}]$$

$$\text{minimize} \left\{ f = \frac{-\phi(\text{MSE})}{\psi(\text{SE})} \right\} \qquad [\text{P 9.8}]$$

Compliant mechanism design objectives can also be posed as geometric advantage and mechanical advantage [147], and mechanical efficiency defined in a certain way [148]. Other types of formulations can be found in [149] through [151].

9.3 SIZE, SHAPE, AND TOPOLOGY OPTIMIZATION

The structural form of a flexible continuum can be interpreted hierarchically in terms of *topology, shape,* and *size*. At the highest level, the *topology* of a flexible continuum determines the material connectivity among various portions of the compliant mechanism (i.e., how the fixed portion, input ports, and output ports are connected to each other). The number of holes in the continuum is also a part of topology definition. Deciding a suitable topology of a continuum is equivalent to the *type synthesis* of rigid-link mechanisms. In other words, topology design is the same as conceptual design. At the next level, we have the *shape* of individual segments that connect different portions of the compliant mechanism and also the shape and location of the holes in a segment. At the lowest level, we consider the *sizes* of various segments. In the optimal design problem, if the design variables describe various possible topologies, it is called a *topology optimization* problem. For a selected topology, if the design variables describe different shapes, it is called *shape optimization*. Figure 9.6 conceptually illustrates the difference between topology and shape of a flexible continuum. And finally, when a topology and shapes of individual segments are selected, if the design variables describe sizes, it is called *size optimization*. Of the three, topology optimization is the most general and the design space searched is the largest and holds the promise for the best optimal solution with no a priori assumptions. The choice of design variables, called *design parameterization*, determines the type of optimization. Specific examples of each type are considered in this section.

9.3.1 Size Optimization

In size optimization the cross section and thickness dimensions of beam and truss-like segments, thickness of plate- and shell-like elements, size of a hole, and so on are the design variables. Figure 9.7a shows a distributed compliant crimping mechanism with curved beam segments. If we want to optimize the width profile of the beam segments to minimize an objective function, we can discretize the finite element model (Figure 9.7b) and let the width of the segment in each element be an optimization design variable (note that in this case the width of the segment is actually the thickness of the beam finite element as the deformation takes place in the plane of the mechanism). If we have N beam elements, we will have an optimization problem in N variables. We can then solve for these variables using numerical optimization techniques reviewed briefly later in this chapter. Let us now consider a

Size, Shape, and Topology Optimization 313

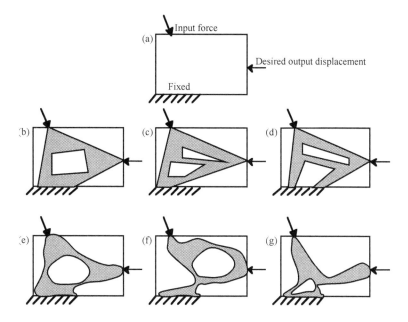

Figure 9.6. Difference between the topology and shape of a flexible continuum. (a) Specification for the compliant mechanism design problem. (b) through (d) Different topologies that connect the input and output ports, and the fixed portions. (e) through (g) Different shapes possible for the same topology shown in (b). *Note:* The topologies and shapes shown in the figure are not necessarily the solutions for these specifications; they are only conceptual examples which illustrate the different possibilities in topology and shape.

simple size optimization problem which can be solved in continuous form analytically without resorting to discretization and numerical techniques.

Figure 9.8 shows a simply supported compliant beam with a distributed transverse force $p(x)$. We want to determine the width profile of this beam so that the volume of the beam is minimized subject to constraints on the deflection at a point and the overall strain energy of the beam. Mathematically, the problem is stated as

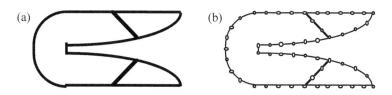

Figure 9.7. (a) Compliant mechanism with curved beam segments, and (b) discretized finite element beam model for size optimization. The in-plane widths of the elements are the design variables.

Optimal Synthesis with Continuum Models

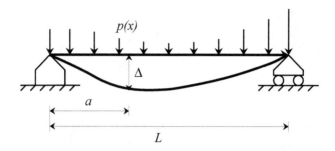

Figure 9.8. Specifications for the size optimization of a simply supported beam for desired deflection, strain energy and weight.

$$\text{minimize } \left\{ \text{volume} = \int_0^L dV \right\}$$

subject to [P 9.9]

at $x = a$, deflection $= \Delta$

(strain energy $-$ SE*) ≤ 0

SE* is the upper bound on the strain energy. Assuming that t is the constant thickness of the beam, and the width is given by the unknown function $w(x)$, area and moment of inertia of the beam are given by

$$A(x) = tw(x)$$
$$I(x) = \frac{t^3}{12}w(x) \qquad (9.10)$$

The strain energy in a beam can be written as [refer to a book on strength of materials and use equation (9.5)]

$$\text{strain energy} = \int_0^L \frac{1}{2} \frac{M^2(x)}{EI(x)} dx \qquad (9.11)$$

where E is Young's modulus of material, and $M(x)$ is the bending moment in the beam under the applied load $p(x)$. The deflection at $x = a$ can be expressed mathematically as follows using the concept of MSE [152].

$$\text{at } x = a, \text{ deflection} = \int_0^L \frac{M(x)m(x)}{EI(x)} dx \qquad (9.12)$$

where $m(x)$ is the bending moment in the beam when only the transverse unit dummy (virtual) load is applied at $x = a$ in the direction of the desired displacement. Since this is a statically determinate beam, the two bending moments $M(x)$ and $m(x)$ do not depend on the width of the beam $w(x)$. Using the expressions above, problem [P 9.9] can be written as

$$\text{minimize} \left\{ \text{volume} = \int_0^L tw(x)\,dx \right\}$$

subject to

$$\int_0^L 12\frac{M(x)m(x)}{Et^3 w(x)}\,dx - \Delta = 0 \qquad \text{[P 9.10]}$$

$$\int_0^L 6\frac{M^2(x)}{Et^3 w(x)}\,dx - SE \le 0$$

Unlike the multiple scalar variable optimization problem [P 9.1], in the problem above the unknown design variable is a function. The optimality conditions for such a problem, which come from the *calculus of variations*, are reviewed briefly next.

Euler–Lagrange Variational Calculus Equation. Calculus of variations is a counterpart of multivariable differential calculus when the variables are continuous functions. Several books are available on this fascinating subject. Interested readers should consult [142] or another book, because what follows is only a brief mention of one the main results of variational calculus. A general problem with an objective function involving an unknown function $y(x)$ and its derivatives $y' = dy/dx$ and $y'' = d^2y/dx^2$ has the solution given by the Euler–Lagrange equation:

Problem: $\text{minimize} \left\{ f = \int_{x_1}^{x_2} F(y, y', y'', x)\,dx \right\}$ (9.13)

Solution: $\dfrac{d^2}{dx^2}\left(\dfrac{\partial F}{\partial y''}\right) - \dfrac{d}{dx}\left(\dfrac{\partial F}{\partial y'}\right) + \dfrac{\partial F}{\partial y} = 0$ and boundary conditions (9.14)

The objective function in the problem above is called a *functional* as it is a function of a function. In problem [P 9.10], the functionals representing the objective function and constraints depend only on the unknown function $w(x)$ and not its derivatives. So only the third term in equation (9.14) is applicable. But this is not the case in general.

Solution of Problem [P 9.10]. As in equation [P 9.2], a constrained variational calculus problem can be transformed into an unconstrained problem by introducing two Lagrange multipliers, Λ and Γ, associated with the two constraints.

$$\text{Min } L = \int_0^L tw(x)\,dx + \Lambda\left\{\int_0^L \frac{12M(x)m(x)}{Et^3 w(x)}\,dx - \Delta\right\} + \Gamma\left\{\int_0^L \frac{6M^2(x)}{Et^3 w(x)}\,dx - SE^*\right\} \quad [\text{P 9.11}]$$

Using the Euler–Lagrange equation [equation (9.14)], the optimal width profile of the beam can be obtained as

$$w^*(x) = \sqrt{\frac{6}{Et^4}[2\Lambda M(x)m(x) + \Gamma M^2(x)]} \quad (9.15)$$

The values of unknown multipliers Λ and Γ are determined by substituting optimal width profile $w^*(x)$ into the two constraint equations in problem [P 9.10]. Therefore, the prescribed values of Δ and SE^* determine the values of the Lagrange multipliers and thus the optimal width profile.

Example: Size Optimization. The side view of a simply supported beam of rectangular cross section with two point forces P_1 and P_2 acting at distances a and c from the left support is shown in Figure 9.9a. A transverse deflection of Δ is desired at a distance of b from the left support. The thickness of the beam is t. Obtain the optimal width profile $w(x)$ of the beam to minimize the volume of the beam to achieve the desired deflection at b. The strain energy should not exceed a prespecified value SE^*. Use the following numerical data:

Young's modulus of the material = $E = 0.7$ GPa

thickness of the beam = $t = 5$ mm

length of the beam = $L = 100$ mm

$a = 30$ mm; $b = 50$ mm; and $c = 60$ mm

$P_1 = 30$ N; $P_2 = 13$ N

$\Delta = 10$ mm

$SE^* = 0.078$ N-m

Solution: The solution is given by equation (9.15). Therefore, first we need to compute $M(x)$ and $m(x)$. The bending moment in the beam due to the applied loads P_1 and P_2 and the supports shown in Figure 9.9a is given by

Size, Shape, and Topology Optimization

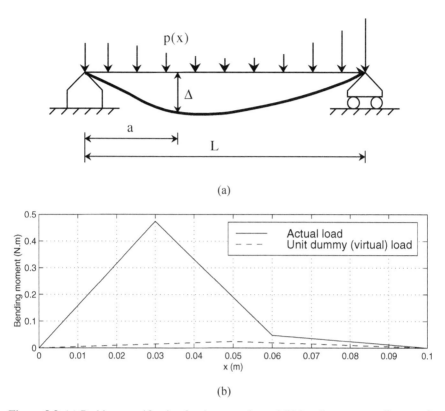

Figure 9.9. (a) Problem specification for the example, and (b) bending moment diagrams for the applied and unit dummy (virtual) loads.

$$M(x) = \begin{cases} \left(P_1 - P_2 - \dfrac{P_1 b - P_2 c}{L}\right)x & \text{for } 0 \le x \le b \\ \left(P_1 - P_2 - \dfrac{P_1 b - P_2 c}{L}\right)x - P_1(x-b) & \text{for } b \le x \le c \\ \left(P_1 - P_2 - \dfrac{P_1 b - P_2 c}{L}\right)x - P_1(x-b) + P_2(x-c) & \text{for } c \le x \le a \end{cases}$$

The bending moment due to the unit dummy (virtual) load applied at b is given by

$$m(x) = \begin{cases} \dfrac{(L-a)x}{L} & \text{for } 0 \le x \le a \\ \dfrac{(L-a)x}{L} - x + a & \text{for } a \le x \le L \end{cases} \quad (9.16)$$

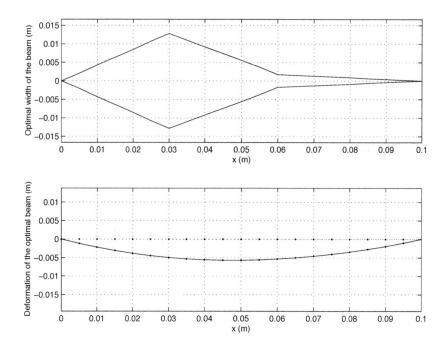

Figure 9.10. Top: Optimal width profile (top view of the beam). Bottom: Deformation of the optimal beam (side view of the beam).

Figure 9.9b shows the plot of the two bending moments. To evaluate the values of the two Lagrange multipliers Λ and Γ, we solve the following two equations, which are nothing but the constraints in problem statement [P 9.10]:

$$\int_0^L 12 \frac{M(x)m(x)}{Et^3 w^*(x)} dx - \Delta = 0$$

$$\int_0^L 6 \frac{M^2(x)}{Et^3 w^*(x)} dx - \text{SE}^* \leq 0$$

After substituting the expression for $w^*(x)$ and the numerical data in the two equations above, we get two nonlinear equations in Λ and Γ. They can be solved numerically to yield the following solution:

$$\Lambda = 5.0750 \times 10^{-5} \qquad \Gamma = 8.300 \times 10^{-6}$$

Figure 9.10a shows the resulting optimal width profile, and Figure 9.10b shows the deflection of the optimal beam.

Size, Shape, and Topology Optimization

9.3.2 Shape Optimization

In general, shape optimization is more difficult than size optimization. The difficulties arise in three ways in the implementation of shape optimization. First, we need to identify appropriate design variables that define the shape of the compliant mechanism so that many different types of shapes can be included in the design search space. The design search space is defined as the set of all possible shapes. Control points of a Bezier or spline interpolation curves are one type of design variable for shape optimization. The coordinates of the nodes in the finite element mesh is another way to vary shape in the optimization process. In these and other shape parameterization schemes, the second difficulty arises from the need for revising the finite element mesh. Whenever shape changes substantially, resulting in highly distorted mesh, the accuracy of the finite element model will diminish without appropriate remeshing. Thus for automated shape optimization, monitoring the accuracy of the mesh and automating the remeshing is necessary. The third difficulty is analytical computation of the effects of shape design variables on the objective and constraint functions. This is called *sensitivity analysis* and is discussed in Section 9.4.2. For more details on the shape optimization method applied to structures, interested readers are referred to a survey paper [153]. Those methods can readily be adapted to compliant mechanisms [154].

9.3.3 Topology Optimization

As stated above, topology optimization is at the highest level in structural optimization and is the most general. When we have a design domain in which the compliant mechanism has to fit and we want to use only a limited amount of material to create the mechanism, we should consider all possible ways of distributing the material within the domain. Like type synthesis techniques of rigid-link mechanisms, topology optimization procedures bring a multitude of possible designs into the hands of a designer. The designer does not have to commit to any particular topology; rather, the optimization algorithm determines the optimal topology for the problem at hand. Pioneering work on topology optimization was done by A. G. M. Michell [155], who arrived at optimal network of structural elements, now known as *Michell continua*. The developments in the last decade have resulted in very efficient design parameterizations that lead not only to optimal topology but also to shape and size simultaneously. Practical implementation of these topology optimization schemes is easier than shape optimization. Two such approaches are described next. A review article [156] is recommended for further reading on topology optimization methods.

Ground Structure Parameterization. The easiest topology optimization design parameterization is to use an exhaustive set of truss or beam/frame elements in the design domain to approximate the continuum domain, and to vary their individual cross-section dimensions by defining them as design variables. For a truss element, the area of cross section is the appropriate optimization design variable. When the area of cross section of an element goes to zero, that element is removed. Thus after the optimization procedure converges, some elements will be removed from the original exhaustive set. The remaining elements will define the topology and shape for the compliant mechanism. The exhaustive set of structural elements is known as the *ground structure* [157] or *structural universe* [158] or *super-structure*. Figure 9.11 shows the ground structure using truss elements and a possible topology after removing some elements. By increasing the resolution of the ground structure, a more refined topology can be obtained. That also increases the number of elements in the ground structure, and hence the optimization problem becomes large. In the practical implementation, we don't really let any element vanish completely from the finite element model. Instead, we use a very small value for the lower bound on the variables so that no numerical difficulties arise. Similarly, an upper bound is also used in some cases. The variables that do not reach either the lower or the upper bound will have values in between defining the size-related features of the segments in the compliant mechanism. Thus this procedure gives not only the optimal topology but also the optimal shape and size. There is no need for remeshing in this procedure because the mesh remains unchanged and undistorted. Examples that illustrate this approach are given in Section 9.5.3.

Example: Topology Optimization—Compliant Gripper Mechanism. Using a ground structure of frame elements, we would like to design the optimal topology for a compliant gripper mechanism which, when subject to a single applied force F_{in}, will result in the deflections Δ_{out} to grasp a workpiece at the output ports. The design domain is constructed as shown in Figure 9.12a, where space is allowed for application of the force and for a workpiece at the output ports. Note that because of the symmetry of the problem, only half the design domain is needed for optimization, as shown in Figure 9.12b. The point of application of F_A is restricted from moving in the vertical direction, and the constrained nodes provide support on the left-hand edge of the design domain. The starting point for topology

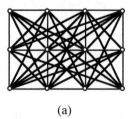
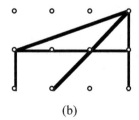

(a) (b)

Figure 9.11. (a) Truss element ground structure for topology optimization, and (b) possible topology after removing some elements.

Size, Shape, and Topology Optimization

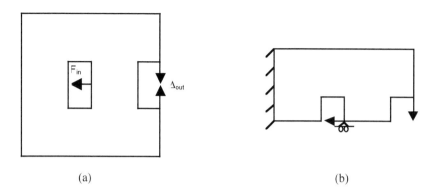

Figure 9.12. (a) Compliant gripper design problem, and (b) symmetric half view.

optimization is the full ground structure of frame elements shown in Figure 9.13a, where each design variable is given an initial value between the upper and lower bound constraints. The solution obtained using the sequential linear programming (SLP) optimization algorithm is shown in Figure 9.13b, where the optimal values of the design variables are illustrated in gray scale—black denotes the upper bound constraint; white denotes the lower bound constraint, and gray denotes intermediate values. Figure 9.13c shows the undeformed and deformed configurations: the solid lines indicate the undeformed configuration; the dashed lines indicate the deformed configuration. A physical prototype of the compliant gripper mechanism was fabricated, as shown in Figure 9.14. This device is a three-dimensional version of the optimal compliant gripper topology and was fabricated using nylon and a rapid prototyping technique called *fused deposition modeling* [159].

Continuous Material Density Parameterization. The discrete nature of the ground structure approach and the manner in which we define the exhaustive set of elements determine the type of solution we get. A more general approach is to vary the *artificial material density* at every point in the design domain [160]. Consider

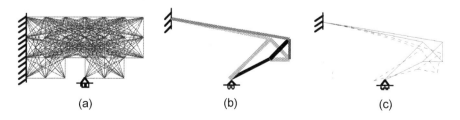

Figure 9.13. (a) Initial design, (b) optimal solution, and (c) verification.

Figure 9.14. Nylon prototype of a three-dimensional compliant gripper.

the design domain discretized with square elements as shown in Figure 9.15a and define an artificial density variable ρ_i for element i $(i = 1, 2, 3, ...)$ such that the effective Young's modulus of the element is given by

$$E_i = E_o \rho_i \qquad (9.17)$$

where E_0 is the actual Young's modulus of the material. When ρ_i reaches the very small value of the lower bound, it implies that the element is made of an artificially very soft material and thus making it virtually absent from the structure. This creates a small void. If a group of adjacent elements reach the lower limit, a large hole will be created. If ρ_i reaches the upper bound, that element forms the solid portion of the resulting optimal compliant mechanism. If ρ_i takes a value in between the two limits, those elements form the transition region. Since ρ_i of each element can take any value, this makes possible a variety of topologies and shapes. One possibility is shown as a gray scale in Figure 9.15b, where black represents solid elements; white, void elements; and gray, the remaining elements. Since manufacturing with intermediate artificial density is expensive, numerical algorithms are developed to push the variables to either of the two limits. A detailed compliant design method based on this approach is described in Section 9.4.1.

The artificial material density approach was developed by researchers on the basis of a more rigorous method called the *homogenization method*, which was

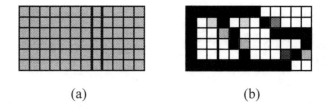

(a)   (b)

Figure 9.15. (a) Discretization for artificial material density distribution. The density of each cell is a design variable. (b) Possible topology with black (solid), white (void), and gray (intermediate density) cells.

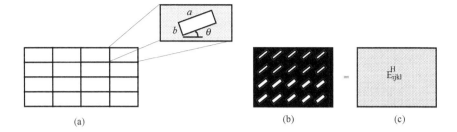

Figure 9.16. (a) Design parameterization in the homogenization method. Each element has three parameters that determine the size of the hole and its orientation. (b) A continuum with holes is approximated as a homogeneous material with homogenized (averaged) material properties. E_{ijkl}^H is the homogenized elasticity tensor (material properties) that relates the stress-strain.

developed by Bendsøe and Kikuchi in 1988 [161]. The homogenization method treats the material inside the entire design domain as a composite material made of a microstructure consisting of material and void. The *material* is like the matrix and *void* is the inclusion in this "composite material." For each element in the discretized design domain, its void is defined using three parameters: a, b, and θ, as shown in Figure 9.16a. Each element is then interpreted as a new composite material with very small microstructural inclusions with those three parameters. The *homogenization theory* provides the *homogenized* or *averaged* material properties of the composite material as if it is a homogeneous material (see Figures 9.16b and c). The homogenized effective material properties are then used for each element. The optimization algorithm will determine the values of the three parameters for each element. If the values of a and b are such that the void is as large as the element, that leaves a hole in the final design. On the other hand, if a and b are both zero, we get a solid element resulting in the final shape. Like the artificial density approach, this method also searches a multitude of designs before arriving at the optimal topology. The θ variable introduces anisotropy in the homogenized material properties, making this method more general. Details of this method can be found in [161] and [162] and its application to compliant mechanisms in [83], [84], and [163].

9.4 COMPUTATIONAL ASPECTS

An important practical aspect of optimal design problems is solving them numerically. As described above, the problem formulation consists of a large number of variables. If it is a topology optimization problem and we desire a well-defined shape with a large number of elements, the variables can be in the hundreds or more. Finding the optimal set of values for these variables necessitates simultaneous solution of the nonlinear equations in equation (9.3). Since these equations are in general too complex to solve directly, numerical optimization algorithms use a different technique. Interested readers should consult [143] and [164], since the following is a very brief overview.

9.4.1 Optimization Algorithms

Instead of solving the nonlinear equations corresponding to the necessary conditions for an optimum [equation (9.3)] directly, numerical optimization algorithms *search* the design space to find the minimum of the objective function. *Feasibility* (i.e., satisfying all the constrains) is also maintained throughout. They all begin with a starting design, called an *initial guess*, given by the user and search the design space for better designs until an optimum is found. At every design point a search direction is identified and the step length along that direction is computed. Determination of the search direction distinguishes one algorithm from another. These types of optimization algorithms are called methods of *mathematical programming*.

On the basis of the information used to search the design space, mathematical programming algorithms can be divided into three categories: zeroth order, first order, and second order. *Zeroth-order methods* use only function information—the numerical values of the objective and constraint functions at the current design—to move to a better design. Local polynomial approximation, golden section search, Fibonacci search, and Powell's conjugate direction method are some examples of the zeroth-order methods. In the *first-order methods*, in addition to function information, we also need *gradient*—the first derivative—information. The gradients of the objective and constraint functions with respect to the design variables are used in this case to search for a better design. Computation of the gradients is called *sensitivity analysis;* it is explained in the next section. Cauchy's steepest descent method is an example of the first-order method. In the *second-order methods*, the second derivative, called the *Hessian*, is also used. There are also many other methods for problems involving constraints, such as generalized reduced gradient (GRG) method, quasi-Newton methods of BFGS and DFP, and so on. In general, higher-order methods have better convergence properties than lower order methods, but at each step in the optimization search, called an *iteration*, more computations are needed in higher-order methods to evaluate not only the functions but also their derivatives.

In the context of optimization of structures and compliant mechanisms, a class of local approximation techniques are widely used: sequential linear programming (SLP), sequential quadratic programming (SQP), convex linearization (CONLIN), method of moving asymptotes (MMA), generalized convex approximation (GCA), and others. As the name implies, in SLP [165] the function and constraints are approximated locally at the current design point using first-order (linear) terms in the Taylor's series expansion. Then a linear programming problem is solved to find the optimum for that approximation. At the new design point the procedure is repeated and continued until the final optimum for the original problem is reached.

SQP [166], CONLIN [167], MMA [168], and GCA [169] use better (higher-order) local approximations and yield better results. An easy-to-use evolutionary structural optimization (ESO) has also been used for solving large structural optimization problems [170].

A good understanding of the numerical optimization algorithms is necessary to implement the optimal design methods discussed in this chapter. Good sources for more information are [143] and [164]. Wherever applicable, commercial optimization software can be used to solve well-formulated optimization problems.

9.4.2 Sensitivity Analysis

While formulating the optimization problem statement and choosing the most appropriate design variables are the most important aspects of optimal design problems, solving it numerically is the crucial aspect from the practical viewpoint. If the chosen numerical algorithm requires the evaluation of gradients and Hessian, it is necessary to develop computationally efficient methods for the sensitivity analysis. Sensitivity analysis indicates the effects on the objective and constraint functions due to changes in the design variables. Sometimes this also helps in reformulating the problem statement. It should be noted that usually a function evaluation in structural optimization problems requires one finite element analysis; and in compliant mechanism design problems, two finite element analyses, one to calculate the displacements due to the actual loads and the other for calculating the displacements due to the dummy (virtual) load. For large problems, function evaluation alone takes a few minutes of CPU time. Since this should be done at every iteration, it adds significantly to the total time taken for optimization. It is therefore necessary to minimize the computations involved in sensitivity analysis, which must also be done at every iteration if the algorithm chosen is not of zeroth order. An excellent source of reference on sensitivity analysis is [171]. We briefly mention two approaches below.

Finite Difference Method. For a function $\varphi(x)$ three types of finite difference computations for the approximate first derivative are defined as:

$$\frac{d\varphi}{dx} = \begin{cases} \dfrac{\varphi(x+\Delta x) - \varphi(x)}{\Delta x} & \text{forward difference} \\ \dfrac{\varphi(x) - \varphi(x-\Delta x)}{\Delta x} & \text{backward difference} \\ \dfrac{\varphi(x+\Delta x) - \varphi(x-\Delta x)}{2\Delta x} & \text{central difference} \end{cases} \quad (9.18)$$

where Δx is a small perturbation from x.

As can be seen above, this method of computing the derivatives requires additional function evaluations. This means additional finite element analyses. For example, if there are N variables, in order to compute the first derivative with respect to each variable, at each iteration, we will need to perform $2N$ additional function evaluations if a central difference scheme is used. The finite difference method is not preferred in structural optimization because of these large computations.

Analytical Gradients. If the gradients and Hessians can be computed analytically, the total number of computations will decrease and the performance of the optimization algorithm will improve because now the derivatives are exact. Fortunately, for many measures used in structural optimization, analytical gradients are easy to obtain. For example, the volume measure depends on the design variables, and since this dependence relationship is known explicitly, computing the derivative analytically is straightforward. Most other measures involve displacement at different points in the flexible continuum. So we will describe how we can compute the first derivative of displacement at a point. We begin with the discretized finite element equilibrium equation [same as equation (9.1)]

$$[K]\{u\} = \{f\} \tag{9.19}$$

Differentiating the equation above with respect to the design variable x_i,

$$\frac{\partial [K]}{\partial x_i}\{u\} + [K]\frac{\partial \{u\}}{\partial x_i} = 0 \tag{9.20}$$

because $\{f\}$ is usually independent of the design variables. If $\{f\}$ involves body forces (e.g., gravity), design variables do influence $\{f\}$. Then the right-hand side of equation (9.20) will not be zero, but it can be computed analytically. Similarly, the partial derivative of $[K]$ with respect to x_i can also be computed analytically for most choices of design variables. Now, using equation (9.20), we can obtain the derivative of $\{u\}$ with respect to x_i analytically as follows:

$$\frac{\partial \{u\}}{\partial x_i} = -[K]^{-1}\frac{\partial [K]}{\partial x_i}\{u\} \tag{9.21}$$

Thus, without any additional finite element analyses, we can compute the first derivative. This is called the *direct method*. There is also the *adjoint method*, which is more efficient when few constraints in large number of variables are involved. Interested readers should see [143] and [171] for more details.

9.5 OPTIMALITY CRITERIA METHODS

Unlike the mathematical programming methods discussed in Section 9.4, *optimality criteria methods* [158] find the optimum solution directly by solving the equations resulting from the necessary and/or sufficient conditions [see equation (9.3)]. These conditions are *the optimality criteria* that the final solution must satisfy. Such optimality criteria are derived rigorously using the Karush–Kuhn–Tucker conditions or the Euler–Lagrange variational calculus equations. Sometimes these criteria could also be intuitive stipulations based on past experience. The *fully stressed design method* is used often in designing the stiffest structures. In this, the design variables are adjusted so that the stress has the same maximum permissible value at all points. The *uniform strain energy density method* is another example. In optimality criteria methods, we search the design space indirectly by using simple recurrence formulas derived from the optimality criteria. Optimality criteria methods are deemed very efficient when the optimization problem has few constraints and many variables. In this section we derive an optimality criterion for compliant mechanisms and solve some topology optimization problems.

9.5.1 Derivation of the Optimality Criterion

We will derive the optimality criterion for the simplest formulation given in problem [P 9.4]. The necessary conditions for optimality [equation (9.2)] require that the first derivative of the objective function be zero at the optimum. Differentiating the objective function in problem [P 9.4] with respect to ith design variable x_i and equating it to zero yields

$$\frac{\partial(\text{MSE})/\partial x_i}{\partial(\text{SE})/\partial x_i} = \frac{\text{MSE}}{\text{SE}} \tag{9.22}$$

By differentiating the expression for MSE from equation (9.8), we get

$$\frac{\partial(\text{MSE})}{\partial x_i} = \frac{\partial \{v\}^T}{\partial x_i}[K]\{u\} + \{v\}^T \frac{\partial([K]\{u\})}{\partial x_i} \tag{9.23}$$

In the absence of body forces and forces that depend on the design variables, $[K]\{u\} = \{f\}$ is independent of the design variables, and therefore the second term in equation (9.23) is zero. Furthermore, since the unit dummy load force vector $\{f\}_{\text{unit dummy load}}$ is independent of the design variables, the differentiation of the first equation of equation (9.9) gives

$$\frac{\partial \{v\}^T}{\partial x_i}[K] = -\{v\}^T \frac{\partial [K]}{\partial x_i} \tag{9.24}$$

Combining equations (9.23) and (9.24) yields

$$\frac{\partial(\text{MSE})}{\partial x_i} = -\{v\}^T \frac{\partial[K]}{\partial x_i}\{u\} \tag{9.25}$$

If we assume that the stiffness matrix $[K]$ is linear in design variable x_i, the following equation is true:

$$\frac{\partial[K]}{\partial x_i} = \frac{[K]}{x_i} \tag{9.26}$$

As *i*th design variable occurs only in the *i*th element stiffness matrix $\{k\}_i$, equations (9.25) and (9.26) can be combined and simplified as

$$\frac{\partial(\text{MSE})}{\partial x_i} = -\{v\}_i^T \frac{\partial\{k\}_i}{\partial x_i}\{u\}_i = -\{v\}_i^T \frac{\{k\}_i}{x_i}\{u\}_i = -\frac{\text{MSE}_i}{x_i} \tag{9.27}$$

where $\{u\}_i$ and $\{v\}_i$ are *i*th element displacement vectors due to actual and dummy loads, respectively. MPE$_i$ is the mutual potential energy of the element *i*. Similarly, it can be shown that

$$\frac{\partial(\text{SE})}{\partial k_i} = -\{u\}_i^T \frac{\{k\}_i}{x_i}\{u\}_i = -\frac{\text{SE}_i}{x_i} \tag{9.28}$$

Now, by substituting equations (9.27) and (9.28) into equation (9.22), we get

$$\frac{\text{MSE}_i}{\text{SE}_i} = \frac{\text{MSE}}{\text{SE}} = R \tag{9.29}$$

where R is a constant. Equation (9.28) reveals an interesting criterion for compliant topologies which becomes even more meaningful if the element size in the finite element discretized model is decreased to an infinitesimally small value. The ratio of element mutual potential energy and strain energy can then be interpreted as the ratio of *mutual potential energy density* to the *strain energy density* at the point. Therefore, the optimality criterion can be stated as: *For an optimal compliant topology that satisfies both flexibility and stiffness requirements, the ratio of the mutual potential energy density and the strain energy density is uniform throughout the continuum provided that the stiffness is linear in the design variables.* The validity of this property is subject to another condition—that the design variables do not reach their upper or lower limits. Further discussion of this criterion can be found in [146] and [172].

9.5.2 Solution Procedure

We will now discuss a solution algorithm using the optimality criterion above. In equation (9.29), if we divide the numerator and denominator on the left-hand side by x_i and rewrite the equation as a recurrence relationship, we get

$$x_i^{k+1} = \frac{\text{MSE}_i}{\text{SE}_i} \frac{1}{R} x_i^k \tag{9.30}$$

where the superscript refers to the iteration number. Starting with an arbitrary initial guess, we can update all the design variables using equation (9.30) until convergence is reached. At the converged solution, when we apply equation (9.30) the design variables do not change within the specified tolerance. Then all the elements satisfy the optimality criterion. However, if an element reaches its upper or lower limit, it will not satisfy the optimality property. This is because the property above was derived without assuming any constraints on the design variables. But in practice, the design variables have physical significance (e.g., cross section of truss element) and therefore cannot be negative or very large. Further discussion on this and a detailed algorithm can be found in [146].

9.5.3 Examples

Example: Maximize Deflection. By implementing the design variable update equation shown in equation (9.30) in conjunction with a one-variable search along a minimizing direction, a practical example problem shown in Figure 9.17 was solved. The design intent in this problem is to obtain a compliant mechanism such that the output displacement Δ at the output port is maximized when an input force F is applied at the input port. This mechanism was used in a micromachined accelerometer [146]. The design domain is a rectangle with a superstructure of frame elements. The design variables are the out-of-plane thicknesses of the elements. The mechanism is to be pivoted only at the four corners. Figure 9.17a depicts all the specifications. The following numerical data used in this problem are furnished in Table 9.1.

Figure 9.17b is the optimal solution generated by the algorithm. In Figure 9.17c, the deflection behavior of the solution mechanism is shown. Finite element analysis using two-dimensional plane-stress elements was used to obtain the deflected configuration shown in this figure. This device has an amplification ratio of 7 (i.e., the output port displacement is seven times the input port displacement). The macro ABS plastic prototype shown in Figure 9.17d is a scaled-up version of size 12 cm × 56 cm.

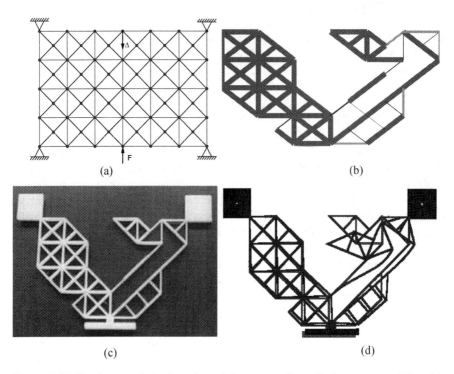

Figure 9.17. Topology optimization of a miniature compliant displacement amplifier: (a) problems specifications, (b) optimal solution using beam elements, (c) validation of the solution using finite element analysis using plane-stress elements, and (d) scaled-up prototype in ABS plastic.

TABLE 9.1. Numerical data for example

Problem specifications	
Design domain	200 μm × 100 μm
Input force	500 μN
In-plane width	0.053 μm
Upper limit on out-of-plane thickness	5 μm
Lower limit on out-in-plane thickness	0.0001 μm
Young's modulus	160 GPa
Poisson's ratio	0.29
Solution	
Output port displacement	0.29 μm
Strain energy	0.1 N-μm
Ratio of output port displacement to input port displacement = amplification ratio	7

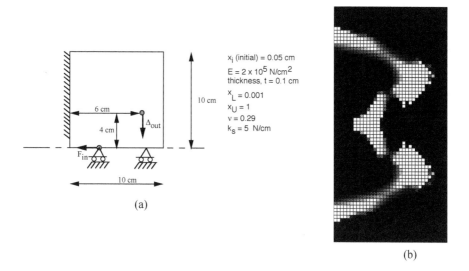

Figure 9.18. Topology optimization of a compliant gripper with plane-stress elements: (a) problem specification for the symmetric half of a compliant gripper, and (b) optimal solution using four-noded plane-stress finite elements.

Example: Compliant Gripper. This example is concerned with a compliant gripper. Since the two jaws of the gripper are assumed to be identical in shape, only the symmetric half is considered in synthesis. Figure 9.18a shows the specifications for this problem. Symmetric boundary conditions are imposed on the line of symmetry. A linear spring of spring constant $k_s = 0.5$ N/mm was included to simulate the effect of an elastic object grasped by the gripper. The numerical data used in this problem are also shown in the figure. Figure 9.18b shows the resulting optimum solution after reflecting about the axis of symmetry. In this example, four-noded bilinear plane-stress finite elements were used. The artificial material density design parameterization [see equation (9.17)] was used for topology optimization. It can be seen that the solution obtained in this manner not only gives the optimal topology, but also the shape of the individual segments and their sizes. It is also interesting to see that the optimal topology for the gripper has a flexure pivot. Similarities to a slider–crank mechanism can also be seen if we look at the symmetric half of the optimal gripper. Thus the solution methods described in this chapter appear to be promising for the systematic-type synthesis of mechanisms in a much broader context than is considered traditionally where nonrigid body motions are not included.

9.6 CONCLUSION

If we treat fully compliant mechanisms as flexible continua, we can apply the methods of structural optimization to design them. Since arbitrary topologies and shapes are permitted by these methods, many types of solutions are searched by the optimization algorithm before arriving at the optimum solution. The design procedure requires nominal specifications from the user and can be fully automated. Many other types of objective functions, design parameterizations, and solution algorithms are possible in this setup. Better formulations, large deformations, unconventional actuations, and dynamic objectives are some of the many ongoing studies related to compliant mechanisms. In this chapter we could only touch on various aspects of optimal synthesis with continuum models. Elastostatic analysis using the finite element method, optimization theory, calculus of variations, mathematical programming-type numerical optimization algorithms, optimality criteria methods, sensitivity analysis, and finally, application of all of these to compliant mechanisms are all big topics individually. In this chapter, a brief overview of each of these was provided. Interested readers should consult the references to learn more about these subjects.

9.7 ACKNOWLEDGMENTS

The chapter authors would like to thank their Ph.D. advisors Sridhar Kota and Noboru Kikuchi, both at the University of Michigan, Ann Arbor; and a graduate student, Anupam Saxena (University of Pennsylvania, Philadelphia), with whom a part of the research described in this chapter was performed.

PROBLEMS

9.1 *Unconstrained optimization problem.* The static equilibrium analysis of an elastic system can be posed as an optimization problem wherein the potential energy expressed as a function of displacements is minimized to obtain the displaced stable equilibrium configuration. This is known as the minimum potential energy principle. For the two-bar truss shown in Figure P9.1, the potential energy is given by

$$f(u, v) = \frac{EA}{s}\left(\frac{l}{2s}\right)^2 u^2 + \frac{EA}{s}\left(\frac{h}{s}\right)^2 v^2 - Fu\cos\theta - Fv\sin\theta$$

where E is Young's modulus, A the cross-sectional area of each bar, s the length of each bar, and u and v the displacements in the horizontal and verti-

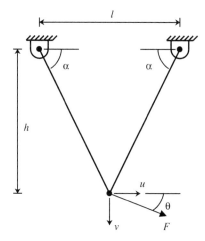

Figure P9.1. Two-bar truss (Problem 9.1).

cal directions, respectively. Find the values of the displacements when $E = 2$ GPa, $A = 10^{-5}$ m^2, $l = 1.5$ m, $h = 4$ m, $P = 10^4$ N, and $\theta = 30°$.

9.2 *Two-bar truss.* Figure P9.2 shows a two-bar truss with a vertical force at the free end. If a deflection u is desired in the direction, compute the strain energy and mutual strain energy of the two elements. Use the following numerical data: $l = 750$ mm, $h = 500$ mm, area of cross section of the horizontal element = 1000 mm^2, area of cross section of the inclined element = 1250 mm^2, and the Young's modulus of both elements = 200×10^3 N/mm^2.

9.3 *Cantilever with a prescribed deflection and minimum volume.* To minimize the volume of the material used, determine the width profile of a cantilever beam of length l and thickness t with a vertical load of P at the tip if a vertical tip deflection of Δ is desired at the midpoint of the beam. What is wrong with this solution, and how can it be reformulated to realize a solution that makes sense in practice?

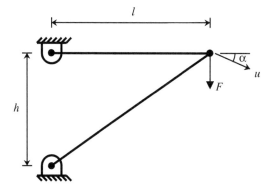

Figure P9.2. Two-bar truss (Problem 9.2).

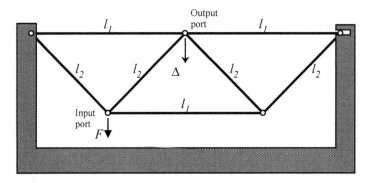

Figure P9.4. Seven-bar truss (Problem 9.4).

9.4 *Barnett's problem.* For the seven-bar truss shown in Figure P9.4, a force F is applied at the input port and a downward deflection Δ is desired at the output port. This deflection is given by

$$\Delta = \sum_{i=1}^{7} \frac{p_i P_i l_i}{A_i E}$$

where
 p_i = interforce ith bar due to unit dummy (virtual) load applied at the output port in the downward direction
 P_i = interforce ith bar due to F applied at the input port
 A_i = area of cross section of bar i
 l_i = length of bar i
Use the following numerical data: $l_1 = 0.5$ m, $l_2 = 0.6$ m, $E = 200$ GPa, $F = 100$ N, and $\Delta = 20$ mm.
(a) Find the optimal cross-section areas of all seven elements that give the desired deflection to minimize the total volume of the truss. (*Hint:* Pose it as a constrained optimization problem and solve the necessary conditions for a minimum to obtain the areas of cross sections.)
(b) Repeat the problem if we desire the deflection Δ in upward direction when the force is applied in the same (i.e., downward) direction.

9.5 *Displacement inverter.* Shown in Figure P9.5 is a *superstructure* made of 28 beam elements. The four corners are pinned to a fixed frame. For the force F applied as shown, we desire Δ displacement in the direction shown. The resulting compliant mechanism will serve as a *displacement inverter*. Apply the optimality criteria–based topology optimization method discussed in Section 9.5 to obtain a solution. For the geometrical and material properties, assume any numbers that are appropriate, as we are only interested in a solu-

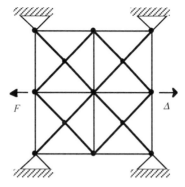

Figure P9.5. Beam element superstructure for compliant displacement inverter of Problem 9.5.

tion that is qualitatively correct (i.e., the output port moves in the right direction when the force is applied as shown). (*Note:* You would need a finite element program that uses beam elements.)

CHAPTER 10

SPECIAL-PURPOSE MECHANISMS

Compliant mechanisms have properties that make them well suited for some unique applications. The ability of a compliant member to store energy is one of the most important of these properties. This energy storage capability may be exploited to create mechanisms that are capable of difficult tasks. Bistable mechanisms and constant-force mechanisms are examples. In addition to energy storage, part reduction and the ability to miniaturize compliant mechanisms offer other unique design possibilities.

This and the following chapter discuss three types of special-purpose compliant mechanisms: bistable mechanisms, constant-force mechanisms, and parallel guiding mechanisms. These represent only a few of the many possibilities, but they are explored because they have many possible applications and offer a good cross section of mechanism types. They also offer examples to further illustrate the methods described in earlier chapters. Constant-force mechanisms and parallel-guiding mechanisms are described in this chapter, and bistable mechanisms in Chapter 11.

10.1 COMPLIANT CONSTANT-FORCE MECHANISMS*

Constant-force mechanisms produce a reaction force at the output port that does not change for a large range of input motion. Rigid-link constant-force mechanisms have been designed [173–175]. Constant-force springs [176] produce a constant force as they are extended and are used in a wide variety of applications. More

---

*See [60].

recently, compliant constant-force mechanisms have been proposed [49, 67, 177]. Type synthesis was used to identify 28 possible configurations for compliant constant-force slider mechanisms [62, 67].

Constant-force mechanisms may be useful in many applications. A few of these applications include the following: a gripping device to hold delicate parts of varying size; electronic connectors that maintain a constant contact force regardless of part tolerances; and wear testing, where a constant force needs to be applied to a surface even as the surface is worn down.

10.1.1 Pseudo-Rigid-Body Model of Compliant Slider Mechanisms

Consider the compliant slider mechanism with three small-length flexural pivots illustrated in Figure 10.1a. The pseudo-rigid-body model for the mechanism is as shown in Figure 10.1b. This mechanism is readily analyzed using common equations for rigid-link slider mechanisms. Assuming that the slider deflection, Δx, is the input,

$$r_1 = r_2 + r_3 - \Delta x \tag{10.1}$$

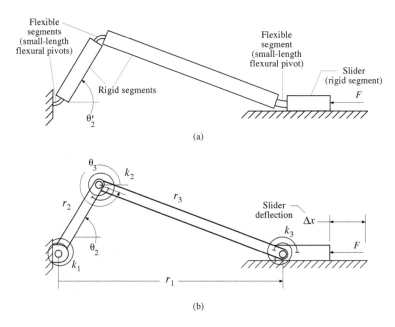

Figure 10.1. (a) Compliant slider mechanism, and (b) its pseudo-rigid-body model.

Compliant Constant-Force Mechanisms

$$\theta_2 = \text{acos} \frac{r_1^2 + r_2^2 - r_3^2}{2r_1 r_2} \tag{10.2}$$

$$\theta_3 = \text{asin} \frac{-r_2 \sin \theta_2}{r_3} \tag{10.3}$$

The force at the slider may be found using the principle of virtual work as

$$F = \frac{r_3 \cos \theta_3 [k_1 \theta_2 + k_2(2\pi + \theta_2 - \theta_3)]}{r_2 r_3 \sin(\theta_2 - \theta_3)}$$
$$+ \frac{r_2 \cos \theta_2 [k_2(2\pi + \theta_2 - \theta_3) + k_3(2\pi - \theta_3)]}{r_2 r_3 \sin(\theta_2 - \theta_3)} \tag{10.4}$$

10.1.2 Dimensional Synthesis

The mechanisms shown in Figure 10.2 represent a number of configurations of compliant constant-force mechanisms that may be analyzed using the equations presented in Section 10.1.1. The five main classes of mechanisms to be analyzed are shown. Class 1A consists of mechanisms with a compliant segment at an end. Class 1B are mechanisms with pin joints at the ends and a compliant segment for motion of the middle joint. Mechanisms with two flexible segments, one at one end and the other at the center belong to class 2A, while those with two flexible segments at the ends belong to class 2B, and class 3A are those with three flexible segments. The significance of these classes is that one pseudo-rigid-body model may be used to model all mechanisms within a particular class. This will be described in more detail later. Note that the equations from Section 10.1.1 apply to all configurations, and some classes have values of k that are zero. Mechanisms with long flexible segments are not shown in class 2A, and a mechanism with two long flexible members is not shown in class 2B. The reason for this is that each has a member with a loading that causes it to take on a deflection shape with a point of inflection (the undulating elastica, or an S shape), and they behave differently than the flexible segments analyzed here.

The dimensions of the pseudo-rigid-body model that result in a constant-force mechanism may be found by minimizing the variation in the output force over the given input displacement. The optimization problem may be stated formally [178] as follows:

$$\text{find } X = \begin{Bmatrix} K_1 \\ K_2 \\ R \end{Bmatrix} = \begin{Bmatrix} k_2/k_1 \\ k_3/k_1 \\ r_3/r_2 \end{Bmatrix} \tag{10.5}$$

Special-Purpose Mechanisms

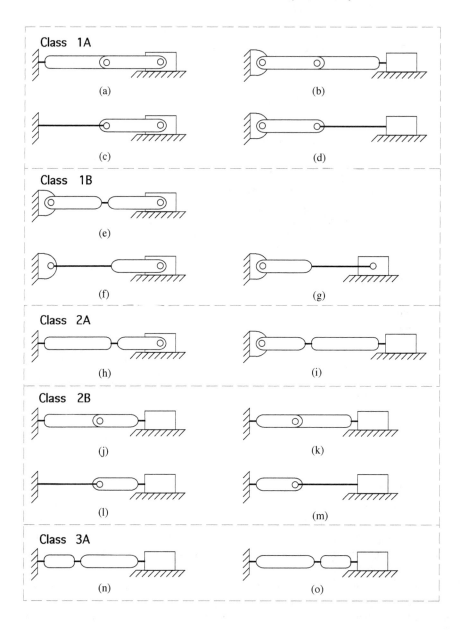

Figure 10.2. Fifteen configurations of the constant-force mechanisms for which results are presented. The mechanisms on the right are the same as those on the left, except the subscripts are switched (i.e., r_2 for r_3, r_3 for r_2, and k_3 for k_1).

Compliant Constant-Force Mechanisms

which minimizes

$$f(X) = \frac{|F|_{max}}{|F|_{min}} = \Xi \qquad (10.6)$$

subject to the constraints

$$k_i \geq 0 \qquad (10.7)$$

$$r_1 > 0 \qquad (10.8)$$

$$k_1 \neq 0 \qquad (10.9)$$

and $|F|_{min}$ and $|F|_{max}$ are the minimum and maximum values of the output force over the slider displacement, Δx. The number of design variables is reduced for mechanisms with fewer flexural joints. The variables are easily rearranged for cases where $k_i = 0$.

The optimization was performed for two cases for each mechanism class: (1) $0 \leq \Delta x \leq 0.16(r_2 + r_3)$ (i.e., a slider deflection that goes from 0 to 16% of the length of the pseudo-rigid-body mechanism), and (2) $0 \leq \Delta x \leq 0.40(r_2 + r_3)$ (a slider deflection 40% of length of pseudo-rigid-body mechanism). The value of the output force was calculated at 50 evenly spaced points along the displacement range to find the maximum and minimum force values. The results for each class are shown in Table 10.1. The value of Ξ is shown for each class, where $\Xi = 1$ represents a mechanism with a perfectly constant force over the entire range of deflection. The deviation from unity is the percentage difference between the maximum and minimum forces [i.e., % difference $= (\Xi - 1) \times 100\%$]. The results are listed in terms of the following nondimensionalized parameters:

$$R = \frac{r_3}{r_2} \qquad (10.10)$$

$$K_1 = \frac{k_2}{k_1} \qquad (10.11)$$

$$K_2 = \frac{k_3}{k_1} \qquad (10.12)$$

The results in Table 10.1 are for the mechanisms shown in the left column of Figure 10.2. The subscripts are switched (i.e., r_2 for r_3, r_3 for r_2, and k_3 for k_1) for the mechanisms in the right column.

10.1.3 Determination of Force Magnitude

The dimensional synthesis above may be used to determine the dimensions of a constant-force mechanism, but it does not provide the magnitude of the constant force. Since the force is not perfectly constant over the input motion, an average force is calculated.

The equation for the output force is given in equation (10.4). Rearranging this equation to include the nondimensionalized parameters R, K_1, and K_2, results in

$$F = \frac{k_1}{r_2}\Phi \qquad (10.13)$$

where

$$\Phi = \frac{R\cos\theta_3[\theta_2 + K_1(2\pi + \theta_2 - \theta_3)] + \cos\theta_2[K_1(2\pi + \theta_2 - \theta_3) + K_2(2\pi - \theta_3)]}{R\sin(\theta_2 - \theta_3)} \qquad (10.14)$$

The mechanisms of class 1B, where $k_1 = 0$, are an exception to the above equation. The following equation can be used with the values of Φ from Table 10.1 [but not equation (10.14)] to estimate the magnitude of the constant force:

$$F = \frac{k_2}{r_2}\Phi \quad \text{(for class 1B)} \qquad (10.15)$$

For a mechanism with a near-constant output force, F, the nondimensionalized force term, Φ, is nearly constant. Each class of constant-force mechanism has a value of Φ associated with it. The values listed in Table 10.1 were found by calcu-

TABLE 10.1. Dimensional synthesis results written in nondimensionalized form

Class	$\Delta x/(r_2+r_3)$	R	K_1	K_2	Ξ	Φ
1A	0.16	0.8274	—	—	1.0030	0.4537
	0.40	0.8853	—	—	1.0241	0.4773
1B	0.16	1.0000	—	—	1.0564	2.0563
	0.40	1.0000	—	—	1.1576	2.1513
2A	0.16	0.3945	0.1906	—	1.0015	0.9575
	0.40	0.4323	0.2237	—	1.0058	1.0466
2B	0.16	0.7591	—	0.1208	1.0721	1.2259
	0.40	0.8441	—	0.1208	1.1914	1.2154
3A	0.16	2.6633	1.0000	12.6704	1.0002	3.4016
	0.40	2.0821	1.0000	9.3816	1.0049	3.6286

lating the average of Φ for 50 equally spaced points along the input motion. The contributions of geometry and stiffness to the force magnitude are taken into account by multiplying Φ by k_1/r_2.

10.1.4 Examples

The following three examples are given to illustrate the use of the results above in the design of compliant constant-force mechanisms.

Example: Class 1A Mechanism—Short-Length Flexural Pivot. Consider the mechanisms of class 1A that have a slider displacement of $\Delta x/(r_2+r_3) = 0.40$. Table 1 shows $R = r_3/r_2 = 0.8853$, and $\Phi = 0.4773$. If $r_2 = 78.74$ mm (3.10 in.), then $r_3 = r_2 R = 69.70$ mm (2.744 in.). If a small-length flexural pivot is used as shown in Figure 10.2a, then $k_1 = EI/l$, where l is small compared to the rigid segment lengths.

If the material is polypropylene with a modulus of elasticity of $E = 1655$ MPa (240,000 lb/in.^{2}), and the flexural pivot has length $l = 5.08$ mm (0.20 in.), width $w = 12.70$ mm (0.50 in.), and thickness $h = 0.76$ mm (0.030 in.), then

$$k_1 = \frac{EI}{l} = 151 \frac{\text{N-mm}}{\text{rad}} \quad \left(1.35 \frac{\text{in.-lb}}{\text{rad}}\right) \tag{10.16}$$

The magnitude of the constant force is therefore

$$F = \frac{k_1}{r_2}\Phi = 0.92 \text{ N} \quad (0.21 \text{ lb}) \tag{10.17}$$

For the mechanism in Figure 10.2b, the results are the same except the subscripts are switched (i.e., r_2 for r_3, r_3 for r_2, and k_3 for k_1). This procedure is the same for all mechanisms shown on the right side of Figure 10.2.

For the mechanism above and other mechanisms with only one flexural pivot, the stiffness of the torsional spring only affects the magnitude of the force but does not affect whether or not it is a constant-force mechanism. This is not the case when more than one joint is represented by a flexible segment.

Example: Class 1A Mechanism—Fixed–Pin Segment. Suppose that another mechanism of class 1A is to be designed; only a fixed–pinned segment is used, as shown in Figure 10.2c. The values for R and Φ are the same as used in the first example. Suppose that the length of the flexible segment is $L = 203.2$ mm (8 in.) and a force of $F = 1.68$ N (0.38 lb) is desired. The link lengths of the pseudo-rigid-body model are $r_2 = \gamma L = 168.7$ mm (6.640 in.) and $r_3 = Rr_2 = 149.3$ mm (5.878 in.). Substituting the spring constant for the torsional spring ($k = \gamma K_\Theta EI/L$) into equation (10.13) and solving for the required moment of inertia, I, results in

$$I = \frac{FLr_2}{\gamma K_\Theta E\Phi} = 33.86 \text{ mm}^4 \ (8.135 \times 10^{-5} \text{ in.}^4) \quad (10.18)$$

Assuming a rectangular cross section ($I = wh^3/12$) with $w = 12.7$ mm (0.50 in.) yields a thickness of $h = 3.18$ mm (0.125 in.).

Example: Mechanism with Three Flexural Pivots. For a mechanism with three flexural pivots (Figure 10.2n) and a deflection of $\Delta x/(r_2 + r_3) = 0.16$, Table 10.1 suggests that $R = r_3/r_2 = 2.6633$, $K_1 = 1.0$, and $K_2 = 12.67$. If $r_2 = 40.2$ mm (1.583 in.), then $r_3 = Rr_2 = 107.1$ mm (4.216 in.). Recall that $k_i = EI_i/l_i$. If each flexural pivot has the same width (w), length (l), and modulus of elasticity (E), then $K_1 = k_2/k_1 = h_2^3/h_1^3$ and $K_2 = k_3/k_1 = h_3^3/h_1^3$. If $h_1 = 0.76$ mm (0.030 in.), then $h_2 = h_1$, and $h_3 = h_1 K^{1/3}{}_2 = 1.78$ mm (0.070 in.).

The nondimensionalized force term is $\Phi = 3.4016$ (Table 10.1). If the mechanism has a width $w = 12.7$ mm (0.50 in.), modulus of elasticity $E = 1655$ MPa (240,000 lb/in.^{2}), and the flexural pivots each have a length of $l = 5.08$ mm (0.20 in.), then

$$k_1 = \frac{EI}{l} = 151 \ \frac{\text{N-mm}}{\text{rad}} \quad \left(1.35 \ \frac{\text{in.-lb}}{\text{rad}}\right) \quad (10.19)$$

The resulting force is

$$F = \frac{k_1}{r_2}\Phi = 13 \text{ N} \quad (2.9 \text{ lb}) \quad (10.20)$$

10.1.5 Estimation of Flexural Pivot Stress

The small-length flexural pivots must go through a large deformation for the constant-force mechanism to obtain the desired deflection. The Bernoulli–Euler equation states that moment (M) is proportional to the beam curvature ($d\theta/ds$), that is,

$$M = EI\frac{d\theta}{ds} \quad (10.21)$$

where θ is the angular deflection and s is the distance along the beam. Since the length of the flexural pivot is small, the moment arm is short and the moment is nearly constant over its length. Assuming a constant moment, separating variables and integrating (10.21) yields

Compliant Constant-Force Mechanisms

$$\int_0^{\theta_0} d\theta = \int_0^l \frac{M}{EI} ds \qquad (10.22)$$

$$\theta_o = \frac{Ml}{EI} \qquad (10.23)$$

where θ_0 is the angular deflection at the end of the segment. Note that for this special case, no small-deflection assumptions were made. Assuming that the stress is dominated by the moment, the stress is $\sigma = Mh/2I$. Rearranging equation (10.23) to solve for the moment (i.e., $M = \theta_o EI/l$), substituting into the stress equation, and simplifying yields

$$\sigma = \frac{\theta_o E h}{2l} \qquad (10.24)$$

Since θ_o can be approximated using the pseudo-rigid-body model, the stress is easily estimated using equation (10.24).

10.1.6 Examples

Consider the first mechanism of Section 10.1.4 at a slider deflection of $\Delta x = 23.75$ mm (0.935 in.). The angular deflection of the flexural pivot, θ_o, is approximated from equation (10.2) as 0.535 rad (30.7°). Substituting into equation (10.24) estimates a stress of 66.4 MPa (9,640 lb/in.^{2}). A commercial finite element analysis program capable on nonlinear analysis (ANSYS) was also used as a comparison. Using 10 beam elements to model the flexible segment resulted in a stress of 68.2 MPa (9,898 lb/in.^{2}).

A similar comparison was performed for the right flexural pivot of the third mechanism in Section 10.1.4. The stresses estimated using equation (10.24) and finite element analysis are 131 MPa (19,000 lb/in.^{2}), and 123 MPa (17,800 lb/in.^{2}), respectively.

Experimental results have shown that the output force is constant for large input displacements, as predicted. It is suggested that plastics not be used for design of mechanisms of this type without full realization of their limitations, particularly stress relaxation and other nonlinear material characteristics.

Example: Robot End Effector. Individual constant-force mechanisms can be combined in various ways to obtain a desired performance. The end effector shown in Figure 10.3 is an example of a device where several constant-force mechanisms are combined in series and parallel. This configuration was chosen because it allows for considerable motion in a compact space and has acceptable stresses in the flexible members. A constant-force end effector is can be used where the

Figure 10.3. Constant-force robot end effector. (From [179].)

position control is not adequate. This is the case when the work surface is unknown or changing. It can be helpful for such applications as buffing, grinding, polishing, or painting. The device shown in Figure 10.3 was designed for cutting glass.

10.2 PARALLEL MECHANISMS*

A parallel-guiding mechanism is a mechanism in which two opposing links remain parallel throughout the mechanism's motion. Figure 10.4 illustrates the motion of the rigid-link parallel-guiding mechanism. The mechanism is a simple four-bar in which the opposing links have the same length, thus forming a parallelogram.

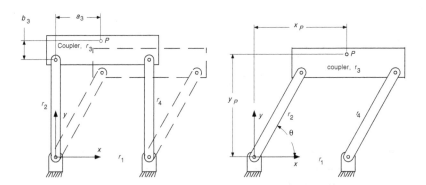

Figure 10.4. Rigid-link parallel-guiding mechanism.

---

*See [10] and [34].

Parallel Mechanisms

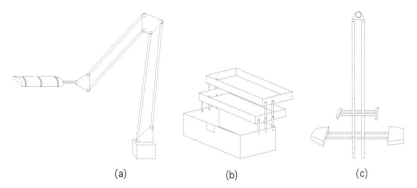

Figure 10.5. Example consumer products that use a parallel-guiding mechanism: (a) desktop lamp, (b) utility/tackle box, and (c) playground swing.

Rigid-link parallel-guiding mechanisms have found use in a variety of practical applications. They are used in high-speed catenaries, positioning of optics, and amusement park rides. They are also common in consumer products such as the desktop lamp mechanisms (Figure 10.5a), floating trays of utility/tackle boxes (Figure 10.5b), children's swing mechanisms (Figure 10.5c), and bicycle derailleur mechanisms.

The rigid-link parallel-guiding mechanism is one of the easiest mechanisms to analyze. The path of an arbitrary point, P, on the coupler link is easily described using the following simple equations:

$$x_P = r_2 \cos \Theta + a_3 \qquad (10.25)$$

$$y_P = r_2 \sin \Theta + b_3 \qquad (10.26)$$

10.2.1 Compliant Parallel-Guiding Mechanisms

Compliant parallel-guiding mechanisms can be designed that retain all the advantages associated with compliant mechanisms, including the elimination of joint friction, backlash, and the need for lubrication, in addition to a reduction in part count, weight, and assembly time. Two common types of parallel-guiding plate-spring mechanisms are shown in Figure 10.6.

10.2.2 Applications

A widely used compliant parallel-guiding mechanism is the parallel-guiding plate-spring mechanism. Figure 10.6 illustrates two types of parallel-guiding plate-spring mechanisms. These mechanisms are used in various fields of application for force–displacement measurement systems, accurate and reproducible motion in optical

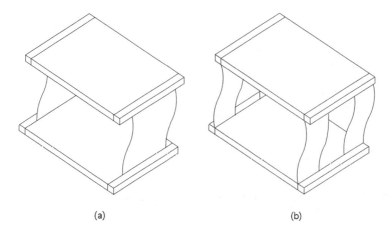

Figure 10.6. Two common types of plate-spring mechanisms.

systems, and guiding parts over small displacements while subject to and without disturbance from dynamic loading forces.

Figure 10.7 illustrates some specific applications that incorporate plate-spring mechanisms. Plate-spring mechanisms provide high-quality, reproducible parallel motion without friction and backlash. Their operation is reliable even under extreme environmental conditions and they are usually made from inexpensive and simple components.

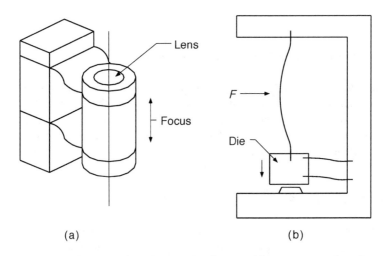

Figure 10.7. (a) Optical lens focusing mechanism used in a compact disc player, and (b) a coining press.

Parallel Mechanisms 349

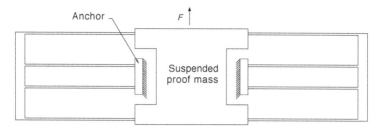

Figure 10.8. Lateral microresonator supported by parallel mechanisms.

Parallel-guiding plate-spring mechanisms have become an important component of devices at the micro scale. Plate springs are commonly used as support structures and flexures for micromechanisms. A microresonator that uses multiple parallel mechanisms is illustrated in Figure 10.8.

The bike brakes shown in Figure 10.9 are one of the commercial versions (Tektro BX-30) of the brakes discussed in earlier examples in the book (see Sections 6.1, and 6.8). The rigid-body version of the mechanism is a four-bar mechanism with a return spring. The compliant mechanism has two fewer pin joints and eliminates the coil return spring. Using this design from Brigham Young University, the company was able to eliminate components that had been a reliability problem for their competition and was able to cut the manufacturing cost by more than half. A compliant bicycle derailleur may have a similar configuration and may also result in a significant cost savings.

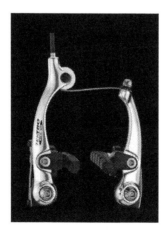

Figure 10.9. Tektro BX-30 compliant parallel-travel bicycle brakes. (Courtesy of Tektro Brake Systems.) The compliant beam replaces two pin joints and the return spring that are in the rigid-body version of the brake.

The orthoplanar spring discussed in several earlier examples in the book (see Sections 3.2.1 and 3.3.11 and Problem 5.10) is an example of a parallel-motion mechanism that has parallel mechanisms combined in series and parallel to achieve the desired motion.

10.2.3 Pseudo-Rigid-Body Model

Using the pseudo-rigid-body model makes the modeling of compliant parallel-guiding mechanisms relatively simple. Figure 10.10 shows the pseudo-rigid-body model for a compliant parallel-guiding mechanism with a horizontal force acting on the coupler. For complaint parallel-guiding mechanisms the pseudo-rigid-body model will always be a parallelogram similar to the rigid-body parallel-guiding mechanism, with the exception of the torsional springs.

Path Analysis. The path of a compliant parallel-guiding mechanism is easily modeled using the pseudo-rigid-body model. By applying standard kinematic position analysis to the pseudo-rigid-body model of the mechanism in Figure 10.10, it is easy to determine the path equations for point P.

$$x_P = \gamma l \cos \Theta + a_3 \tag{10.27}$$

$$y_P = \gamma l \sin \Theta + l(1 - \gamma) + b_3 \tag{10.28}$$

The characteristic radius factor, γ, is determined from the orientation of the input force, F. For a purely horizontal force the value is a constant 0.8517.

Determining the path of motion for other configurations of parallel-guiding mechanisms is only a matter of determining the correct link lengths of the pseudo-rigid-body model. Once these lengths are determined, standard kinematic analysis can be performed on the model to determine the path of the compliant mechanism.

Force–Displacement Analysis. Figure 10.11 illustrates a generalized pseudo-rigid-body model for the parallel-guiding mechanism. A generalized force–displacement relationship is determined by applying the principle of virtual work to this model:

$$\begin{aligned}
&[F_{x2}b_2 + F_{y2}a_2 + F_{x4}b_4 + F_{y4}a_4 + r_2(F_{y3} + F_{y4})] \cos \Theta \\
&-[F_{x2}a_2 + F_{y2}b_2 + F_{x4}a_4 + F_{y4}b_4 + r_2(F_{x3} + F_{x4})] \sin \Theta \\
&+ M_2 + M_4 + T_1 + T_2 + T_3 + T_4 = 0
\end{aligned} \tag{10.29}$$

This equation relates the torque at the joints, the applied forces, and the mechanism's geometry. The torque at each torsional spring, T_1 through T_4, is related to the change in the pseudo-rigid-body angle, Θ, by the following equation:

Parallel Mechanisms

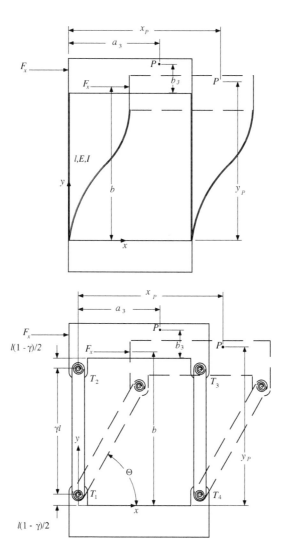

Figure 10.10. Compliant parallel-guiding mechanism and its pseudo-rigid-body model.

$$T_1 = -m_k(\Theta - \Theta_0) \qquad (10.30)$$

where Θ_0 is the pseudo-rigid-body angle of the mechanism in its unforced position. The spring function, m_k, of each torsion spring depends on the type of flexible segment it represents on the compliant mechanism. For the parallel-guiding mechanism a value of $K_\Theta = 2.65$ is satisfactory for an applied force in the range of ±45° from horizontal.

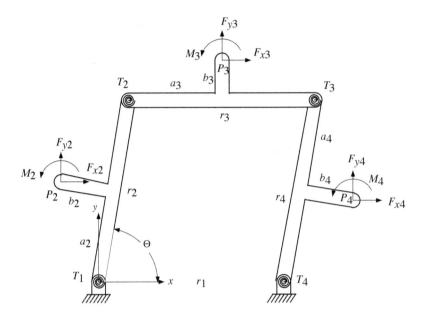

Figure 10.11. Generalized pseudo-rigid-body model for parallel-guiding mechanisms.

When using the generalized pseudo-rigid-body model for determining force–displacement relationships, it is assumed that the applied force acts on a rigid-body segment.

10.2.4 Additional Design Considerations

Ideally, a parallel-guiding mechanism's coupler will remain parallel to the ground throughout the mechanism's entire motion. Assuming this for a compliant parallel-guiding mechanism is a good assumption for a large portion of the mechanism's motion. However, some rotation is caused by a moment acting on the coupler of the mechanism. The pseudo-rigid-body model method used to model the parallel-guiding mechanism does not show any rotation of the coupler. This is because the method being used is a simplified form, which uses a constant value, 0.8518, for the variable, γ (the characteristic radius factor). To model the coupler rotation with the pseudo-rigid-body model accurately it would be necessary to adjust the value of γ according to the orientation of the force acting on the flexible segment. Since the amount of coupler rotation is relatively small for a majority of the mechanism's motion (as has been demonstrated by experimental results), the simplified pseudo-rigid-body model is sufficient for this analysis.

Compliant parallel-guiding mechanisms have few or no pin joints but can still provide motion similar to the rigid-link four-bar mechanism. Because these mecha-

Parallel Mechanisms

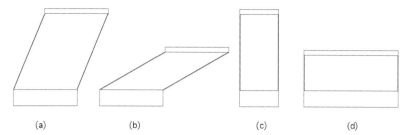

Figure 10.12. Geometric variations of a compliant parallel-guiding mechanism.

nisms obtain their motion though compliance as opposed to pin joints, there is a significant increase in the number of design considerations. The following are some of the most notable considerations and difference when designing a compliant parallel-guiding mechanism.

Unlike the rigid-link parallel-guiding mechanism, the path of motion of a compliant mechanism is determined not only by the geometry of the rigid links, but also by the placement and orientation of the input and output forces. Forcing the mechanism in different locations and directions alters the mechanism's path of motion. For example, with the parallel-guiding mechanism, a downward force applied to the coupler could cause the mechanism to move in a nonparallel fashion by buckling the flexible segments.

The initial geometry of the compliant mechanism is an important factor to consider in the analysis and design of compliant mechanisms. Figure 10.12 illustrates some geometric variations of the mechanism's resting position and the relative lengths of the adjacent segments. Unlike their rigid-link counterparts, these types of variations affect the motion and performance of the mechanism and should be considered in the design.

PROBLEMS

10.1 You are to design a constant-force mechanism with three flexural pivots. The distance between the outer ends of the two end flexural pivots is 10 in. The material is polypropylene with $E = 200,000$ lb/in^2. The material width out of the page is $w = 1$ in. The length of each flexible segment is $l = 0.2$ in. The deflection should be 16% of the length of the pseudo-rigid-body model. See Figure P10.1.
   (a) Specify the lengths of the rigid segments.
   (b) If the flexural pivot on the left has a thickness of $h = 0.03$ in., determine the thicknesses of the other two.
   (c) What would be the advantage of placing two mechanisms together? Would this affect the magnitude of the constant force? If so, how? If not, why not?

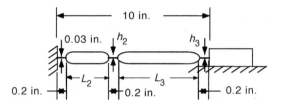

Figure P10.1. Figure for Problem 10.1.

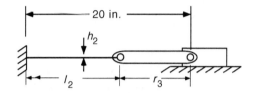

Figure P10.2. Figure for Problem 10.2.

10.2 You are to design a constant-force mechanism with a fixed–pinned segment (class 1A), as shown in Figure P10.2. The material is titanium with $E = 16.5 \times 10^6$ lb/in². The material width out of the page is $w = 1$ in. The deflection will be 16% of the length of the pseudo-rigid-body model.
  (a) Specify the lengths of the segments l_2 and r_3.
  (b) If the flexible segment has a height of $h_2 = 0.03$ in., what is the value of the constant force produced?
  (c) How would the force above change if w where doubled? If h_2 were doubled?

10.3 Consider the compliant mechanism shown in Figure P10.3. The coupler does not rotate in its motion, and the right end of the figure represents a slider.
  (a) Sketch the pseudo-rigid-body model of this mechanism.
  (b) How would you design the mechanism to ensure that the coupler does not rotate in its motion (a parallel motion mechanism)?
  (c) How would you determine the number of degrees of freedom of this mechanism? Calculate the degrees of freedom.

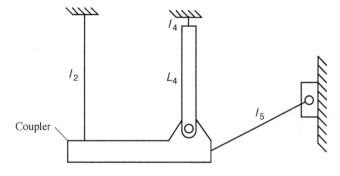

Figure P10.3. Figure for Problem 10.3.

CHAPTER 11

BISTABLE MECHANISMS

Bistable mechanisms tend toward one of their two stable equilibrium positions. Many types of devices benefit from this type of behavior. Common examples of bistable mechanisms are light switches, self-closing gates, cabinet hinges, three-ring binders, and closures. Figure 11.1 is an example of a bistable closure used on many consumer products. The closure is similar to a light switch, three-ring binder, or other bistable devices, in that it will tend to one of two positions if no external forces are acting on it. In this way the closure will remain in the open or closed position without any external forces required to maintain the position. The bistable clasp illustrated in Figure 11.2 is another example of a bistable mechanism.

Compliant mechanisms offer a particularly economical way to achieve bistable behavior. Because flexible segments store energy as they deflect, a compliant mechanism can use the same segments to gain both motion and two stable states, allowing a significant reduction in part count.

This chapter discusses the basic theory of bistable mechanisms and applies it to compliant mechanisms. General equations for various types of mechanisms are presented, and specific configurations required to obtain bistable mechanisms are outlined.

11.1 STABILITY

When a system has no acceleration, it may be said to be in a state of equilibrium. The state of equilibrium is *stable* if a small external disturbance only causes oscillations about the equilibrium state. However, if a small external disturbance causes the system to diverge from its equilibrium state, the equilibrium position is *unsta-*

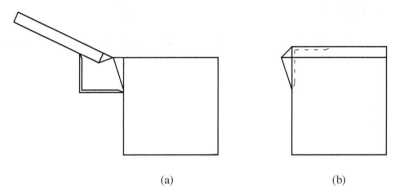

Figure 11.1. Common bistable closure used on many consumer products packaging.

ble. If, on the other hand, the system reacts to the disturbances and stays in the disturbed position, the equilibrium position is *neutral* [71, 72, 181 – 183].

The stability of a system may be illustrated using the "ball-on-the-hill" analogy, as illustrated in Figure 11.3. The ball is shown in position A, which is a stable equilibrium position. If it is shifted from this position by a small amount, it will tend to return to position A or oscillate around it. However, position B is an unstable equilibrium position. Although the ball will stay in position if placed precisely on top of the hill, it will move to a different position if any disturbance occurs. Position C is stable, while position D is neutrally stable, because any disturbance will cause the ball to move to its disturbed position only.

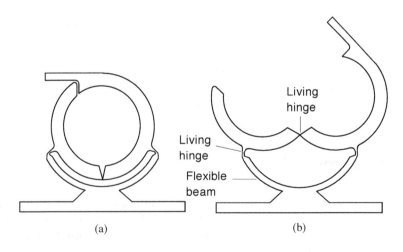

Figure 11.2. Bistable clasp in its (a) closed and (b) open positions.

Compliant Bistable Mechanisms

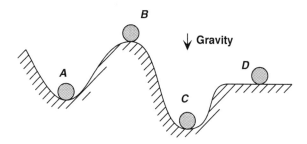

Figure 11.3. Illustration of the "ball-on-the-hill" analogy. Positions *A* and *C* are stable equilibrium positions. Position *B* is an unstable equilibrium position. Position *D* is neutrally stable.

Because this system has two stable equilibrium positions, it is bistable. Because two local minima enclose a local maximum, two stable equilibrium positions will have an unstable position between them. Therefore, a bistable mechanism will have two stable equilibrium positions and at least one unstable equilibrium position. In Figure 11.4 a stop has been placed at *E* to illustrate the creation of a new equilibrium position by the application of an external load. This "new" equilibrium position is also stable.

The energy method, based on the Lagrange–Dirichlet theorem, states that a stable equilibrium position occurs at a position where the potential energy has a local minimum. Therefore, to establish the stability of a mechanism, the potential energy of the mechanism may be plotted over the mechanism's motion and any local minima represent stable positions. The potential energy curve is similar to the hill topography in the ball-on-the-hill analogy [184].

11.2 COMPLIANT BISTABLE MECHANISMS

Because compliant mechanisms inherently store energy in their flexible joints, they are particularly useful as bistable mechanisms. Not only can the mechanism often be made of one piece, but no extra springs are required for energy storage.

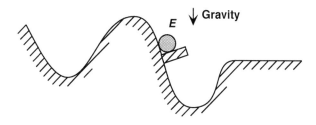

Figure 11.4. In this figure, a stop at position *E* has created a "new" stable equilibrium position.

Using the pseudo-rigid-body model, the potential energy equation of a compliant mechanism can easily be found. For a small-length flexural pivot or a fixed–pinned segment, the potential energy V stored in the segment is

$$V = \frac{1}{2}K\Theta^2 \qquad (11.1)$$

where K is the torsional spring constant, found using the pseudo-rigid-body model, and Θ is the pseudo-rigid-body angle. Using the linear spring model and approximating the spring function using Hooke's law, the potential energy stored in a functionally binary pinned–pinned segment is

$$V = \frac{1}{2}K_s(\Delta x)^2 \qquad (11.2)$$

where Δx is the change in distance between the segment's two pin joints, and K_s is the linear spring constant. Because each compliant segment's energy storage depends only on the deflection of the segment, the total potential energy in the mechanism is the sum of the potential energy stored in each compliant segment.

There are several possible ways to analyze the stability of a mechanism. For cases where the position equations and energy equations are available or readily developed, the following approach is recommended:

- *Position analysis.* Obtain the position equations for the mechanism.
- *Energy equations.* Obtain the equations that express the energy stored in the springs of the mechanism in terms of the displacement variables.
- *First derivative.* Take the first derivative of the energy equations with respect to the generalized coordinate (usually, the input variable).
- *Equilibrium positions.* Solve for all values of the generalized coordinate for which the first derivative of the energy equation is zero. These are the equilibrium positions.
- *Stable positions.* Differentiate the energy equations again to find the second derivative of the energy equations with respect to the generalized coordinate. Evaluate this equation at the equilibrium positions. A positive value of the second derivative at the equilibrium positions means that it is a stable equilibrium position, while a negative value means that it is an unstable position. This could also be done by evaluating the energy equation at locations on either side of the equilibrium position to determine if it was a local minimum (stable position) or a local maximum (unstable position) of potential energy.

For cases where the position or energy equations are not readily expressed or differentiated (such as in cases where the analysis is performed using finite element analysis), the following approach is possible:

Four-Link Mechanisms

- *Input force/moment vs. displacement.* Calculate the input force/moment at various values of the input variable and plot the results.
- *Equilibrium positions.* The equilibrium positions occur at locations where the input force/moment is zero.
- *Stable positions.* Evaluate the input force near the equilibrium positions. If the input force/moment decreases as the generalized coordinate increases, it is a stable equilibrium position. If it increases as the generalized coordinate increases, it is an unstable equilibrium position.

In a bistable mechanism synthesis problem, a designer typically must design a mechanism to be stable at particular locations. The unstable equilibrium position and the maximum force or moment required to move the mechanism from one stable position to another may also be specified. The methods described above may be used for these purposes. However, the first step in the synthesis process is to determine the best mechanism configuration to accomplish the desired task. Some of the possible classes of compliant bistable mechanisms are listed below in terms of their pseudo-rigid-body model.

- Four-link mechanisms
- Slider–crank or slider–rocker mechanisms
- Double-slider mechanisms with a pin joining the sliders
- Double-slider mechanisms with a link joining the sliders
- Snap-through buckled beams (note that this class must be flexible, not rigid)
- Bistable cam mechanisms

Because these classes relate to the pseudo-rigid-body models of the compliant mechanisms, they also apply to rigid-link mechanisms with springs. Each of the classes above is discussed briefly in the following sections.

11.3 FOUR-LINK MECHANISMS

A general four-link mechanism's basic kinematic chain is shown in Figure 11.5a. By fixing any link, the mechanism may be formed, as shown in Figure 11.5b. This mechanism may be further classified according to Grashof's criterion as a Grashof or non-Grashof mechanism, as discussed in Chapter 4. This distinction is important for determining the location of springs that cause bistable behavior. In a Grashof mechanism, the shortest link can rotate through a full revolution with respect to either link connected to it. In a non-Grashof mechanism, no link can rotate through a full revolution with respect to any other links. Recall that Grashof's criterion is stated mathematically as

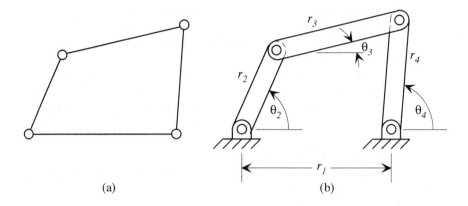

Figure 11.5. General four-link mechanism: (a) basic kinematic chain, and (b) the mechanism.

$$s + l \leq p + q \quad \text{Grashof} \tag{11.3}$$

where s is the length of the shortest link, l is the length of the longest link, and p and q are the lengths of the intermediate links. Crank rockers, double cranks, and double rockers are examples of Grashof mechanisms. If the inequality is not satisfied, the mechanism is non-Grashof (a triple rocker). If the sum of the lengths of the longest and shortest links is equal to the sum of the lengths of the other two links, the mechanism is a special case of a Grashof mechanism known as a change-point mechanism. Mathematically,

$$s + l > p + q \quad \text{non-Grashof} \tag{11.4}$$

$$s + l = p + q \quad \text{change point} \tag{11.5}$$

The position equations for four-bar mechanisms are provided in Section 4.2.

11.3.1 Energy Equations

The model of a fully compliant four-link mechanism is shown in Figure 11.6. The energy equation is

$$V = \frac{1}{2}(K_1\psi_1^2 + K_2\psi_2^2 + K_3\psi_3^2 + K_4\psi_4^2) \tag{11.6}$$

where

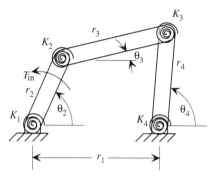

Figure 11.6. Four-link mechanism with a spring at each joint.

$$\begin{aligned}
\psi_1 &= \theta_2 - \theta_{2o} \\
\psi_2 &= \theta_2 - \theta_{2o} - (\theta_3 - \theta_{3o}) \\
\psi_3 &= \theta_4 - \theta_{4o} - (\theta_3 - \theta_{3o}) \\
\psi_4 &= \theta_4 - \theta_{4o}
\end{aligned} \quad (11.7)$$

Choosing θ_2 as the generalized coordinate, the first derivative is

$$\frac{dV}{d\theta_2} = K_1\psi_1 + K_2\psi_2\left(1 - \frac{d\theta_3}{d\theta_2}\right) + K_3\psi_3\left(\frac{d\theta_4}{d\theta_2} - \frac{d\theta_3}{d\theta_2}\right) + K_4\psi_4\frac{d\theta_4}{d\theta_2} \quad (11.8)$$

where $d\theta_i/d\theta_j$ are kinematic coefficients, and

$$\frac{d\theta_3}{d\theta_2} = \frac{r_2 \sin(\theta_4 - \theta_2)}{r_3 \sin(\theta_3 - \theta_4)} \quad (11.9)$$

$$\frac{d\theta_4}{d\theta_2} = \frac{r_2 \sin(\theta_3 - \theta_2)}{r_4 \sin(\theta_3 - \theta_4)} \quad (11.10)$$

The input torque, applied to link 2, and the first derivative of the energy equation are zero at all equilibrium positions. In fact, the torque curve is the first derivative of the energy curve with respect to the crank angle. This may be shown by considering the equation for work put into the system:

$$W = \int_{\theta_0}^{\theta} T_{in}\, d\theta_2 \quad (11.11)$$

By taking the derivative of this equation, it may be seen that

$$\frac{dW}{d\theta_2} = T_{in} \qquad (11.12)$$

Therefore, the applied torque is equal to the first derivative of the energy with respect to crank angle. For this reason, zeros of the torque curve are relative maxima or minima of the energy curve.

11.3.2 Requirements for Bistable Behavior

The following theorems apply for a pseudo-rigid-body four-bar mechanism [185]:

- **Theorem 1**. A compliant mechanism whose pseudo-rigid-body model behaves like a Grashof four-link mechanism with a torsional spring placed at one joint will be bistable if and only if the torsional spring is located opposite the shortest link and the spring's undeflected state does not correspond to a mechanism position in which the shortest link and the other link opposite the spring are collinear.
- **Theorem 2**. A compliant mechanism whose pseudo-rigid-body model behaves like a non-Grashof four-link mechanism with a torsional spring at any one joint will be bistable if and only if the spring's undeflected state does not correspond to a mechanism position in which the two links opposite the spring are collinear.
- **Theorem 3**. A compliant mechanism whose pseudo-rigid-body model behaves like a change-point four-link mechanism with a torsional spring placed at any one joint will be bistable if and only if the spring's undeflected state does not correspond to a mechanism position in which the two links opposite the spring are collinear.

An example of a four-link bistable mechanism with a spring at position 4 is shown in Figure 11.7. For a compliant equivalent, the spring could be replaced by

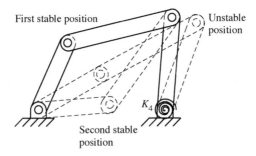

Figure 11.7. Bistable four-link mechanism showing the two stable positions and one unstable position.

Four-Link Mechanisms

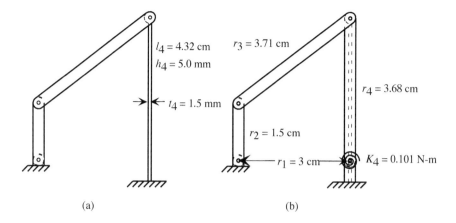

Figure 11.8. Partially compliant bistable mechanism. When the short link on the left is turned, this mechanism acts as a crank–rocker.

either a small-length flexural pivot or a fixed–pinned segment, such as is shown in Figure 11.8.

The theorems above can be used to guarantee bistable behavior if one torsional spring is used. However, other springs may be present in the mechanism, and the mechanism may or may not be bistable. Additional springs may be used to obtain the type of energy profile desired for a bistable mechanism. If only one spring is in the mechanism, it will always have zero potential energy at both stable positions. With more than one spring it is possible to design a mechanism with a stable position that is "cocked." In other words, a stable position may be obtained that requires very little energy to move out of to the unstable position, after which the mechanism releases considerably more energy in returning to the first stable position. For a full analysis of the location of the unstable and stable positions when multiple springs are present, the potential energy equation must be solved for each configuration involving more than one spring.

Example: Compliant Bistable Mechanism. The mechanism illustrated in Figure 11.8a has a pseudo-rigid-body model (Figure 11.8b) that satisfies Grashof's criteria. The mechanism is bistable because the spring is opposite the shortest link (link 2). Using the mechanism shown in Figure 11.8b, the potential energy and crank torque curves may be calculated. The potential energy can be calculated from equation (11.6) with $K_1 = K_2 = K_3 = 0$, and

$$V = \frac{1}{2}K_4(\theta_4 - \theta_{4o})^2 \qquad (11.13)$$

The input torque required on link 2 may be calculated using the general equations for a four-bar mechanism developed in Chapter 6 [equation (6.136)]. Another approach is to use the fact that the input torque will be the derivative of the energy equation [equation (11.6)]. Either approach leads to the following equation for the input torque:

$$T_{in} = \frac{dV}{d\theta_2} = K_4(\theta_4 - \theta_{4o})\frac{r_2 \sin(\theta_3 - \theta_2)}{r_4 \sin(\theta_3 - \theta_4)} \tag{11.14}$$

where the relationship between the angles and link lengths for a four-bar mechanism are given in Section 4.2.1.

Assuming that the flexible beam is polypropylene with a Young's modulus of $E = 1.4$ GPa, and using typical values of $\gamma = 0.85$ and $K_\Theta = 2.65$, the spring constant for the torsional spring can be found from Chapter 5 or Appendix E as

$$K_4 = \gamma K_\Theta \frac{EI}{l_r} = \gamma K_\Theta \frac{Eh_4 t_r^3}{12 l_r} = 0.10 \text{ N-m/rad} \tag{11.15}$$

The potential energy and input torque are plotted as a function of $\Delta\theta_2$ in Figure 11.9, where $\Delta\theta_2$ is defined as $\theta_2 - \theta_{2o}$. These curves show that the mechanism will be stable when $\Delta\theta_2 = -79°$, corresponding to position B in Figure 11.9. The mechanism is shown in this position in Figure 11.10. The mechanism also has an unstable position at $\Delta\theta_2 = -45°$, corresponding to position C.

As the mechanism moves from one stable position to another, the maximum absolute value of the torque may be called the *critical torque* [72]. If a force is used

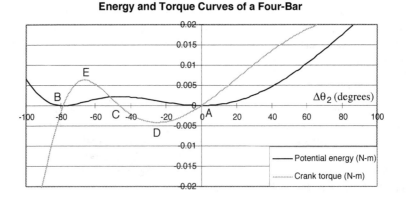

Figure 11.9. Energy and crank torque curves for the mechanism shown in Figure 11.8.

Four-Link Mechanisms

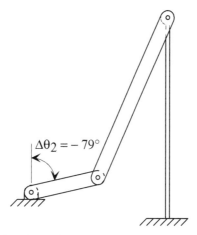

Figure 11.10. Mechanism shown in Figure 11.8 in its second stable position.

instead of a torque, the maximum force is termed the *critical force*. When moving from position A to position B, the critical torque is about 0.004 N-m, as shown at D in Figure 11.9. When moving from position B to position A, the critical torque is about 0.0065 N-m, as shown at E.

Increased Contact Force. At the stable positions, no force or torque is required to keep the mechanism in position. Conversely, the mechanism cannot exert a force on any external body such as electrical contacts for a switch. To allow such a reaction force, the mechanism may be stopped at an intermediate position. This is analogous to the stop at position E shown in Figure 11.4. At this position, the stop provides a reaction force on the crank creating a torque equal to the value predicted by the torque curve of Figure 11.9. In this way, a new stable position has been created in which the mechanism is exerting a force on the external body. This is also demonstrated in the next example.

Example: Bistable Switch. An example of a compliant mechanism with a four-bar mechanism pseudo-rigid-body model is shown in Figure 11.11. The link lengths and the initial angle of link 4 for the pseudo-rigid-body model are listed in Table 11.1, as are the dimensions of the small-length flexural pivot. The out-of-plane thickness is b_4. This is a non-Grashof mechanism, and a spring at any location would cause bistable behavior. A contact force is created by causing the contacts to connect before the second stable equilibrium position is reached. Living hinges are used at the other joints and they can be modeled as pin joints without torsional springs. The mechanism can be used as a fully compliant electrical switch, or for other applications, such as on cabinet door hinges.

366 **Bistable Mechanisms**

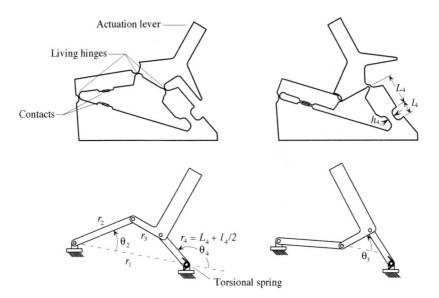

Figure 11.11. Compliant bistable switch and its pseudo-rigid-body model.

Because the spring is in the same location for this example as for the preceding example, the equation for the potential energy is again simplified from equation (11.6) as

$$V = \frac{1}{2}K_4(\theta_4 - \theta_{4o})^2 \qquad (11.16)$$

and

$$\frac{dV}{d\theta_2} = K_4(\theta_4 - \theta_{4o})\frac{r_2 \sin(\theta_3 - \theta_2)}{r_4 \sin(\theta_3 - \theta_4)} \qquad (11.17)$$

The torsional spring represents the small-length flexural pivot, and the torsional spring constant is

TABLE 11.1. Link lengths and initial angles for the pseudo-rigid-body model of the switch example

r_1	r_2	r_3	r_4	θ_{4o}	b_4	h_4	l_4
3.61 cm	2.0 cm	1.0 cm	1.23 cm	142 deg	6.3 mm	2.3 mm	4.2 mm

Four-Link Mechanisms

$$K_4 = \frac{EI}{l_4} = \frac{(1.4 \times 10^9 \text{Pa})(6.3 \times 10^{-3}\text{m})(2.3 \times 10^{-3}\text{m})^3}{12(4.2 \times 10^{-3}\text{m})} = 2\frac{\text{N-m}}{\text{rad}} \quad (11.18)$$

Because there is only one torsional spring, the minimum potential energy occurs where $V = 0$. Equation (11.16) shows that this will occur when $\theta_4 = \theta_{4o}$. Note that equation (11.17) is also zero at this condition, as expected for a stable equilibrium position. The values of the other angles may be found by completing a position analysis of the mechanism using the equations developed in Chapter 4. The resulting angles of the links in the pseudo-rigid-body model are listed in Table 11.2. The unstable equilibrium position also occurs where equation (11.17) is zero. This position is represented by the toggle position when links 2 and 3 are collinear. The angles of the pseudo links are listed in Table 11.2 for this position.

11.3.3 Young Bistable Mechanisms*

The previous two examples included bistable mechanisms with only one torsional spring in their pseudo-rigid-body model, and the springs were placed such that bistable behavior was guaranteed. A micro bistable mechanism can be used to further demonstrate the bistable behavior of devices with more than one spring. A special type of mechanism, the Young mechanism, is used to provide this example.

A Young mechanism is one that:

- Has two revolute joints, and therefore two links, where a link is defined as the continuum between two rigid-body joints
- Has two compliant segments, both part of the same link
- Has a pseudo-rigid-body model that resembles a four-bar mechanism

The first and second conditions, taken together, imply that the two pin joints are connected with one completely rigid link, while the other link consists of two com-

TABLE 11.2. Equilibrium positions (degrees) for the bistable switch example

Equilibrium position	θ_2	θ_3	θ_4
First stable	32	18	142
Second stable	0	50	142
Unstable	18.7	18.7	128.6

---

*See [12] for more details on these mechanisms. The name "Young" was chosen for this mechanism because the original research was conducted at Brigham Young University.

368   Bistable Mechanisms

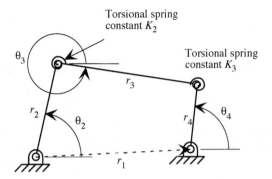

Figure 11.12. Generic model used to design bistable Young mechanisms. Torsional spring constants K_2 and K_3 represent the stiffness of compliant segments in the pseudo-rigid-body model.

pliant segments and one or more rigid segments. A general pseudo-rigid-body model of a Young mechanism is shown in Figure 11.12. In this model, the two revolute joints are connected to ground, while pins A and B represent compliant segments modeled by the pseudo-rigid-body model.

Young mechanisms may be useful in MEMS (microelectromechanical systems) for several reasons. For example, pin joints connected to the substrate (ground) can easily be fabricated with two layers of polysilicon, but true pin joints connecting two moving links require more layers. Also, the two pin joints help the mechanism to achieve larger motion, in general, by reducing the stress in the compliant segments. In addition, the two compliant segments give the mechanism the energy storage elements it needs for bistable behavior. Figure 11.13 illustrates an example of a Young mechanism.

Table 11.3 lists dimensions for the example bistable mechanisms illustrated in Figures 11.13 and 11.14.

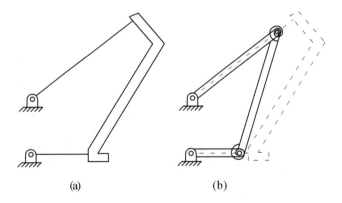

Figure 11.13. (a) Compliant bistable Young mechanism, and (b) its corresponding pseudo-rigid-body model.

Four-Link Mechanisms

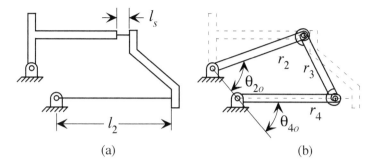

Figure 11.14. (a) Bistable Young mechanism, and (b) its pseudo-rigid-body model.

Design of Bistable Young Mechanisms. The potential energy equation may be found by summing the energy stored in the two torsional springs, or from equation (11.6) as

$$V = \frac{1}{2}(K_2\psi_2^2 + K_3\psi_3^2) \tag{11.19}$$

where V is the potential energy, K_2 and K_3 are the torsional spring constants, and ψ_2 and ψ_3 are the relative deflections of the torsional springs and are given by equation (11.7).

The extrema of equation (11.19) may be found by locating zeros of the first derivative of V. The first derivative of V with respect to θ_2 is

$$\frac{dV}{d\theta_2} = K_2\psi_2\left(1 - \frac{d\theta_3}{d\theta_2}\right) + K_3\psi_3\left(\frac{d\theta_4}{d\theta_2} - \frac{d\theta_3}{d\theta_2}\right) \tag{11.20}$$

where $d\theta_3/d\theta_2$ and $d\theta_4/d\theta_2$ were defined in equations (11.9) and (11.10), respectively.

The maximum nominal stress in the compliant segment during motion is another important quantity to consider. Compliant mechanism theory can be used to find this stress from the maximum angular deflection of each segment, $\psi_{2,\max}$ and

TABLE 11.3. Design parameters for the mechanisms in Figures 11.13 and 11.14

Figures	r_1 (μm)	r_2 (μm)	r_4 (μm)	θ_{2o}	θ_{4o}	I_2 (μm)^4	I_4 (μm)^4	l_s (μm)
11.13 and 11.16a	120	236	109	130°	90°	4.5	4.5	—
11.14 and 11.16b	100	250	250	83°	53°	4.5	4.5	26

$\psi_{3,max}$. For either compliant segment, the maximum nominal stress may be approximated with the classical stress equation

$$\sigma_{o\,max} = \frac{M_{max}c}{I} \qquad (11.21)$$

where M_{max} may be approximated using the pseudo-rigid-body model as the product of K and ψ_{max}. Assuming a rectangular cross section,

$$\sigma_{o\,max} = \frac{6K\psi_{max}}{ht^2} \qquad (11.22)$$

where h is the height of the compliant beam (the dimension out of the plane of motion) and t is its thickness (the dimension within the plane of motion). This nominal stress is the stress calculated without taking stress concentrations into account. It may be used by comparing the nominal stress in the segment to the nominal stress at fracture of previously tested devices with similar stress concentrations.

Example: Bistable Mechanism. The mechanism shown in Figure 11.14 has one small-length flexural pivot and one fixed–pinned segment. The design parameters for this mechanism are listed in Table 11.3. These parameters define the pseudo-rigid-body model shown in Figure 11.14b. The potential energy curve through the mechanism's motion may be generated using equation (11.19). In Figure 11.15 this curve is shown as a function of θ_2. The two relative minima on this curve represent the two stable positions of the mechanism. These minima occur at $\theta_2 = \theta_{2o} = 83°$

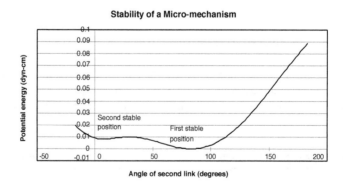

Figure 11.15. Potential energy curve of a class II mechanism (Figure 11.14) as a function of θ_2.

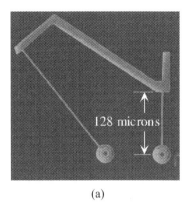
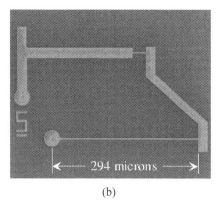

(a) (b)

Figure 11.16. Scanning electron microscope (SEM) photographs of two bistable micromechanisms. One dimension is given to provide an idea of the mechanism's scale. The pseudo-rigid-body model dimensions are provided in Table 11.3.

and $\theta_2 = 7°$. Therefore, the angular deflection of the second link between the two stable positions is approximately 76°. At each point, the first derivative of potential energy, given in equation (11.20), is zero.

Figure 11.16 shows scanning electron microscope (SEM) photographs of example micro bistable mechanisms. Table 11.3 shows the dimensions of these devices. Figure 11.17 shows SEM images of these two mechanisms in their second stable

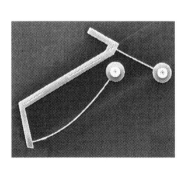
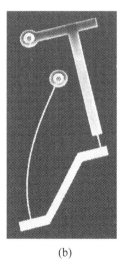

(a) (b)

Figure 11.17. Bistable Young mechanisms (Figures 11.13 and 11.16a) in their second stable equilibrium position: (a) for the mechanism in Figures 11.13 and 11.16a, and (b) for the mechanism in Figures 11.14 and 11.16b.

positions. Note that one of the compliant segments is still deflected in the second stable position, indicating that some energy is stored in that state. Despite this stored energy, the mechanism is at a local minimum of potential energy.

11.4 SLIDER–CRANK OR SLIDER–ROCKER MECHANISMS

The basic kinematic chain of a general slider–crank mechanism is shown in Figure 11.18a. By fixing the long sliding link, the mechanism shown in Figure 11.18b results. This mechanism will have different mobility, depending on the relationship of the link lengths, and Grashof criteria discussed in Chapter 4 may be applied. If the difference between the lengths of links 3 and 2 is greater than the offset, e, the shortest link can revolve 360°. This is called a crank–rocker mechanism. If the link lengths are equal to the offset, it is a change-point mechanism and it has a configuration where all links are collinear. If the difference is less than the offset, it is a slider–rocker mechanism and no links can undergo complete rotation relative to the other links. This can be summarized mathematically as

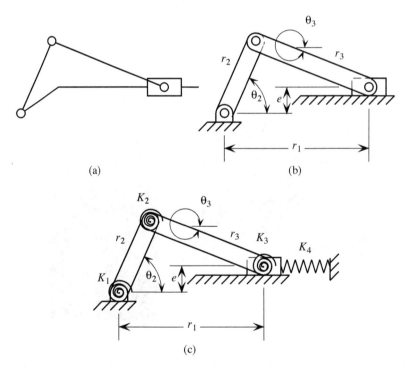

Figure 11.18. Basic kinematic chain of a general slider–crank or slider–rocker mechanism and the mechanism that results when one link is fixed. If $r_3 - r_2 \geq e$, the mechanism is a slider–crank; otherwise, it is a slider–rocker.

Slider–Crank or Slider–Rocker Mechanisms

$$|r_3 - r_2| > e \quad \text{slider–crank} \tag{11.23}$$

$$r_2 - r_3 = e \quad \text{change point} \tag{11.24}$$

$$|r_3 - r_2| < e \quad \text{slider–rocker} \tag{11.25}$$

For convenience, r_2 is chosen arbitrarily as the shortest link. This may be done without loss of generality because the case where $r_2 > r_3$ is merely a kinematic inversion of the case where $r_2 < r_3$. The equations for this mechanism were reviewed in Chapter 4, but they are repeated here for convenience. Choosing θ_2 as the generalized coordinate, the displacement equations are

$$\theta_3 = \mathrm{asin}\,\frac{e - r_2 \sin\theta_2}{r_3} \tag{11.26}$$

$$r_1 = r_2 \cos\theta_2 + r_3 \cos\theta_3 \tag{11.27}$$

11.4.1 Energy Equations

If a compliant segment is added in place of each joint, the mechanism may be modeled as shown in Figure 11.18c. The energy equation is

$$V = \frac{1}{2}(K_1\psi_1^2 + K_2\psi_2^2 + K_3\psi_3^2 + K_4\psi_4^2) \tag{11.28}$$

where

$$\begin{aligned}
\psi_1 &= \theta_2 - \theta_{2o} \\
\psi_2 &= \theta_2 - \theta_{2o} - (\theta_3 - \theta_{3o}) \\
\psi_3 &= \theta_3 - \theta_{3o} \\
\psi_4 &= r_1 - r_{1o}
\end{aligned} \tag{11.29}$$

Using θ_2 as the generalized coordinate, the first derivative of energy is

$$\frac{dV}{d\theta_2} = K_1\psi_1 + K_2\psi_2\frac{d\psi_2}{d\theta_2} + K_3\psi_3\frac{d\psi_3}{d\theta_2} + K_4\psi_4\frac{d\psi_4}{d\theta_2} \tag{11.30}$$

where

$$\frac{d\psi_2}{d\theta_2} = 1 + \frac{r_2 \cos\theta_2}{r_3\sqrt{1 - [(e - r_2 \sin\theta_2)/r_3]^2}} \tag{11.31}$$

$$\frac{d\psi_3}{d\theta_2} = -\frac{r_2 \cos\theta_2}{r_3 \cos\theta_3} \tag{11.32}$$

$$\frac{d\psi_4}{d\theta_2} = (r_2 \cos\theta_2 \tan\theta_3 - r_2 \sin\theta_2) \tag{11.33}$$

11.4.2 Requirements for Bistable Behavior

For slider–crank mechanisms to be bistable, a spring must be opposite the shortest link, and springs attached to the shortest link do not cause bistable behavior. A mechanism may or may not be bistable if other springs are included, depending on their relative stiffnesses, link lengths, and so on. In such cases equations (11.28) and (11.30) should be analyzed to determine the stability of the mechanism.

A spring at any of the four positions of Figure 11.18c will cause bistable behavior for change-point and slider–rocker mechanisms. However, as stated in the theorems, if the undeflected position of the spring at position 4 (or any other spring that causes bistable behavior) is at an extreme position of the mechanism, it will not be bistable.

11.4.3 Examples for Various Spring Positions

As stated earlier, if link 2 is the shortest link, springs at positions 1 and 2 will not cause bistable behavior for slider–crank mechanisms, but will for change-point and slider–rocker mechanisms. A bistable slider–rocker mechanism with a spring at position 1 is shown in Figure 11.19a. In this and the following figures, the second stable equilibrium position and the unstable equilibrium position are shown by dashed lines. One possible compliant mechanism that is based on this model is shown in Figure 11.19b. An example mechanism with the spring at location 2 is shown in Figure 11.20a. Figure 11.21b shows a sample compliant mechanism corresponding to this pseudo-rigid-body model.

Continuing the assumption that link 2 is the shortest link, springs at positions 3 or 4 will cause bistable behavior for all slider mechanism types (unless the undeflected state corresponds to one of the special case situations where it is undeflected at an extreme position of the mechanism). A bistable mechanism with a spring at position 3 is shown in Figure 11.21a. Figure 11.21b illustrates one way that this mechanism could be made compliant. A mechanism with a spring in position 4 is illustrated in Figure 11.22, with one unstable position and the second stable position shown by dashed lines. A compliant equivalent could be achieved by replacing the spring with a FBPP segment.

Slider–Crank or Slider–Rocker Mechanisms

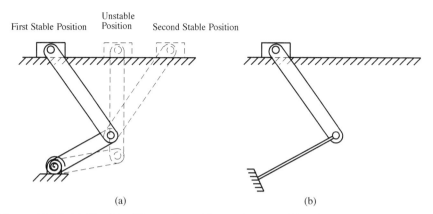

Figure 11.19. (a) Bistable slider–rocker with a spring at position 1. The unstable position and second stable position are also shown. (b) One possible compliant configuration of the mechanism.

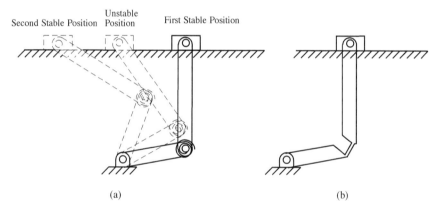

Figure 11.20. Bistable slider–rocker with a spring placed at position 2.

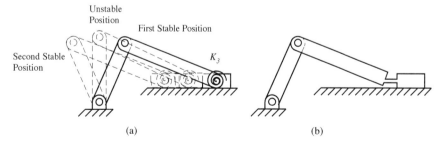

Figure 11.21. (a) Bistable slider–crank with a spring at position 3. The second stable position and one of the unstable positions are shown in dashed lines. (b) Compliant bistable mechanism. Figure 11.21 shows a model of this mechanism.

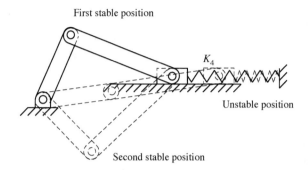

Figure 11.22. Bistable slider–crank with the two stable positions and one unstable position shown. In this case, the spring is placed in position 4.

The compliant bistable clasp mechanism is another example of a bistable slider mechanism. It is shown again in Figure 11.23 with its pseudo-rigid-body model. The symmetry of the mechanism allows each half to be modeled as a slider–rocker mechanism with a spring in position 1, as shown in Figure 11.23c. The living hinges offer very little resistance to motion and are modeled as pin joints, while the stiffer segment is modeled as an initially curved fixed–pinned segment with a torsional spring. The two stable equilibrium positions occur where the spring is undeflected.

Another example is the bistable closure shown in Figure 11.24. Its pseudo-rigid-body model is an inversion of a slider–crank mechanism with a spring at position 4.

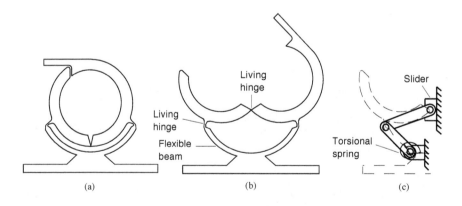

Figure 11.23. Bistable clasp shown in its (a) closed and (b) open positions. The symmetry of the mechanism allows each half of the mechanism to be modeled with a pseudo-rigid-body model that is a slider–rocker mechanism as shown in (c).

Double-Slider Mechanisms

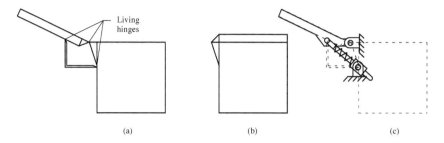

Figure 11.24. Bistable closure shown in its (a) open and (b) closed positions. Its pseudo-rigid-body model, shown in (c), is an inversion of a slider–crank mechanism.

11.5 DOUBLE-SLIDER MECHANISMS

This class consists of mechanisms with four joints, two of which are prismatic (sliding) joints. Two possible kinematic chains exists, one with a pin joint joining the sliders, and the other with a link between the sliders.

11.5.1 Double-Slider Mechanisms with a Pin Joining the Sliders

For the first double-slider mechanism discussed, the two sliders are joined by a pin joint, as shown by the basic kinematic chain in Figure 11.25a. When one link is fixed, the mechanism shown in Figure 11.25b is formed. The position equations for this mechanism may be found by using θ_2 as the generalized coordinate,

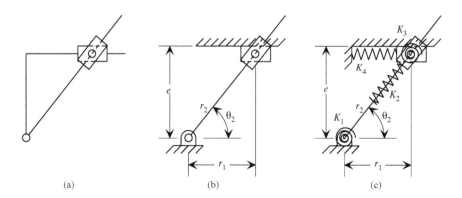

Figure 11.25. (a) Basic kinematic chain of a double-slider mechanism with the sliders jointed by a pin joint. By fixing one link, the mechanism in (b) results. (c) Pseudo-rigid-body model with springs at each sliding and pin joint.

$$r_2 = \frac{e}{\sin\theta_2} \tag{11.34}$$

and

$$r_1 = \frac{e}{\tan\theta_2} \tag{11.35}$$

where r_2 is the distance from the fixed pin joint to the pin joint joining the sliders, r_1 is the horizontal distance between the pin joints, and e and θ_2 are defined in Figure 11.25.

Energy Equations. If a compliant segment is added in place of each joint, the mechanism may be modeled as shown in Figure 11.25c, where a spring has been placed at each joint. The energy equation is found by adding the energy storage terms for each spring:

$$V = \frac{1}{2}(K_1\psi_1^2 + K_2\psi_2^2 + K_3\psi_3^2 + K_4\psi_4^2) \tag{11.36}$$

where the K's are the spring constants as noted in the figure, and the ψ's are the deflections of each spring, given by

$$\begin{aligned} \psi_1 &= \theta_2 - \theta_{2o} \\ \psi_2 &= r_2 - r_{2o} \\ \psi_3 &= \theta_2 - \theta_{2o} \\ \psi_4 &= r_1 - r_{1o} \end{aligned} \tag{11.37}$$

where a "0" subscript indicates the initial, undeflected position. Choosing θ_2 as the generalized coordinate, the derivative of the potential energy with respect to the generalized coordinate is

$$\frac{dV}{d\theta_2} = K_1\psi_1 + K_2\psi_2\frac{d\psi_2}{d\theta_2} + K_3\psi_3 + K_4\psi_4\frac{d\psi_4}{d\theta_2} \tag{11.38}$$

After substituting and rearranging, the equation becomes

$$\frac{dV}{d\theta_2} = (K_1 + K_3)(\theta_2 - \theta_{2o}) \\ - \frac{e^2}{\sin^2\theta_2}\left[K_2\cos\theta_2\left(\frac{1}{\sin\theta_2} - \frac{1}{\sin\theta_{2o}}\right) + K_4\left(\frac{1}{\tan\theta_2} - \frac{1}{\tan\theta_{2o}}\right)\right] \tag{11.39}$$

Double-Slider Mechanisms

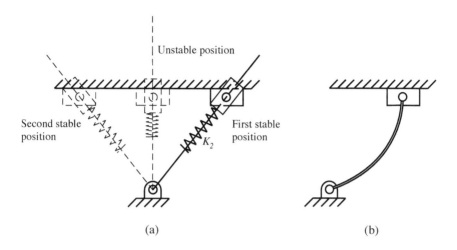

Figure 11.26. (a) Bistable double-slider mechanism with a pin joint joining the sliders. The unstable and second stable positions are shown in dashed lines. (b) Compliant mechanism whose pseudo-rigid-body model is a double-slider with the sliders joined by a pin joint.

These equations may be used to analyze bistable mechanisms. Equations (11.36) and (11.39) may be plotted to investigate the energy-displacement relationship for a mechanism. As before, the valleys and peaks in the potential energy curve [equation (11.36)] represent stable and unstable equilibrium positions. These positions are also represented as zero values of equation (11.39).

Requirements for Bistable Behavior. A spring must be placed between the rotating bar and its slider for the mechanism to be bistable, as illustrated in Figure 11.26a. If this is the only spring in the mechanism, bistable behavior is guaranteed; if other springs (K_1, K_3, or K_4) are present, equation (11.36) or (11.39) can be used to determine whether the mechanism is bistable. A bistable compliant mechanism may be constructed as illustrated in Figure 11.26b, where the spring and slider have been replaced by a functionally binary pinned–pinned segment. This figure represents only one possible compliant configuration.

11.5.2 Double-Slider Mechanisms with a Link Joining the Sliders

The basic kinematic chain for this type of mechanism is shown in Figure 11.27a. By fixing the link between the two prismatic joints, the mechanism in Figure 11.27b is formed. In this figure, x_2 and x_4 are measured from the undeflected state. Choosing θ_3 as the generalized coordinate, the displacement equations are

Bistable Mechanisms

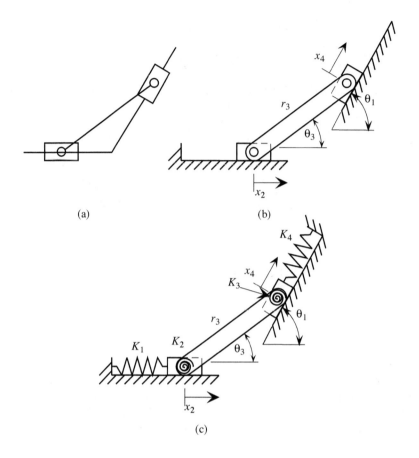

Figure 11.27. Double-slider mechanism with the two sliders joined by a link. The basic kinematic chain is shown in (a), with the mechanism in (b) formed by fixing the link between the slider joints, and (c) is the pseudo-rigid-body model with springs at each joint.

$$x_2 = \frac{r_3[\sin(\theta_1 - \theta_{3o}) + \sin(\theta_3 - \theta_1)]}{\sin \theta_1} \quad (11.40)$$

$$x_4 = \frac{r_3(\sin\theta_3 - \sin\theta_{3o})}{\sin \theta_1} \quad (11.41)$$

Energy Equations. If all joints are replaced with compliant segments, the mechanism may be modeled as shown in Figure 11.27c. The energy equation for this mechanism is

Double-Slider Mechanisms

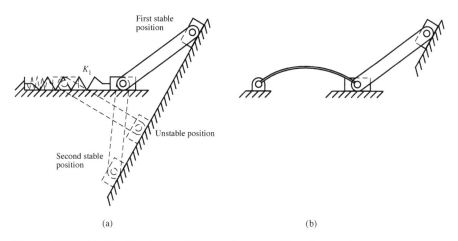

Figure 11.28. (a) Bistable double-slider mechanism with a link joining the sliders. The second stable position and one of the unstable positions are shown. If the mechanism has a spring at position 4 instead of position 1, the motion will be similar. (b) Compliant mechanism whose pseudo-rigid-body model is shown in (a).

$$V = \frac{1}{2}(K_1 \psi_1^2 + K_2 \psi_2^2 + K_3 \psi_3^2 + K_4 \psi_4^2) \tag{11.42}$$

with

$$\begin{aligned}\psi_1 &= x_2 \\ \psi_2 &= \theta_3 - \theta_{3o} \\ \psi_3 &= \theta_3 - \theta_{3o} \\ \psi_4 &= x_4\end{aligned} \tag{11.43}$$

The first derivative is

$$\frac{dV}{d\theta_3} = K_1 x_2 \cos(\theta_3 - \theta_1) + K_2 + K_3 \theta_3 - \theta_{3o} + K_4 x_4 \cos \theta_3 \tag{11.44}$$

11.5.3 Requirements for Bistable Behavior

A double-slider mechanism with a link joining the sliders will be bistable if a spring is placed between either of the sliders and the ground link, as shown in Figure 11.28a. In the figure, one of the two possible springs is shown. A mechanism with a spring at the other position would have similar motion. This figure also shows one of the unstable positions and the second stable position. An equivalent

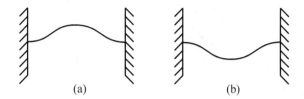

Figure 11.29. Snap-through buckled beam in its two stable positions. While this example shows a fixed–fixed beam, it may also be either pinned at either end or free at one end.

compliant mechanism is shown in Figure 11.28b. If other springs are added to the mechanism, equation (11.42) or (11.44) can be evaluated to ensure that the mechanism is still bistable.

11.6 SNAP-THROUGH BUCKLED BEAMS

Snap-through buckled beams are some of the easiest bistable mechanisms to use; however, their motion is very limited. A snap-through buckled beam is simply a buckled beam, like the one illustrated in Figure 11.29a, which can snap into a second stable position, as shown in Figure 11.29b. The analysis of such beams is based on classical structural mechanics.

11.7 BISTABLE CAM MECHANISMS

If a spring-loaded follower goes through two local minima of potential energy as it travels around the cam, a bistable mechanism results, as illustrated in Figure 11.30. The actual mechanism may be of any class, either rigid or compliant. If the mechanism is compliant, care should be taken so that the energy stored in the compliant segments is not greater than the energy stored in the spring-loaded follower. The principles of cam design are well documented, and bistable cams allow the stable

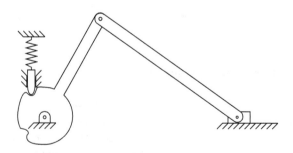

Figure 11.30. Bistable cam mechanism.

Bistable Cam Mechanisms

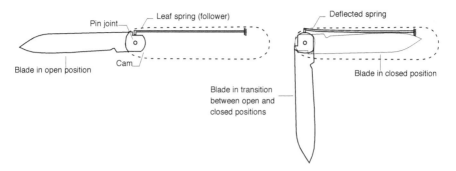

Figure 11.31. Pocket knife with a bistable cam and a compliant follower.

positions to be placed easily anywhere in the mechanism's motion. Multiple stable positions may even be created using this method. However, cam designs do not take advantage of the beneficial aspects of compliant mechanisms, especially the integration of the mechanism's motion and energy storage into one member.

A common example of a bistable cam mechanism with a flexible follower is a pocket knife, as shown in Figure 11.31. The blade is stable in its open and closed positions when the leaf spring is undeflected.

PROBLEMS

11.1 Consider the flapper mechanism in its undeflected position illustrated in Figure P11.1. Is this a bistable mechanism? If so, sketch the positions of links 2 and 3 for the stable equilibrium positions and for an unstable equilibrium position. If not, explain what would have to change to make it a bistable mechanism.

11.2 A bistable mechanism is to be created that has the same general motion as the rigid-link mechanism illustrated in Figure P11.2.

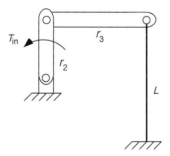

Figure P11.1. Figure for Problem 11.1.

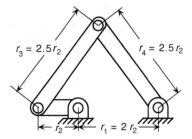

Figure P11.2. Figure for Problem 11.2.

(a) Show what locations a torsional spring could be placed to cause the mechanism to be bistable.
(b) Choose a location for the torsional spring and convert the resulting mechanism to a possible compliant mechanism.

11.3 The bistable mechanism illustrated in Figure P11.3 has two identical initially curved flexible segments that have living hinges at both ends. The device is constrained such that the center shuttle goes through a vertical motion. Because of symmetry, only one of the two pinned–pinned segments need to be analyzed. Each segment is made of polypropylene, has an initial radius of 55.6 mm, a length of 63 mm, a width of 1.22 mm, and an out-of-plane thickness of 3.23 mm. The stiffness of the living hinges is much lower than the other elements.
(a) Sketch the pseudo-rigid-body model.
(b) Calculate and label the link lengths and the torsional spring constant.
(c) Determine the maximum deflection that the segment must go through before it snaps to its second stable equilibrium position.
(d) Plot the required actuation force as a function of the vertical deflection of the center shuttle.

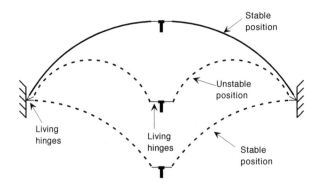

Figure P11.3. Figure for Problem 11.3.

APPENDIX A

REFERENCES

[1] Erdman, A. G., and Sandor, G. N., 1997, *Mechanism Design: Analysis and Synthesis,* Vol. 1, 3rd Ed., Prentice Hall, Upper Saddle River, NJ.
[2] Shigley, J. E., and Uicker, J. J., 1995, *Theory of Machines and Mechanisms,* 2nd Ed., McGraw-Hill, New York.
[3] Howell, L. L., and Midha, A., 1994, "A Method for the Design of Compliant Mechanisms with Small-Length Flexural Pivots," *Journal of Mechanical Design,* Trans. ASME, Vol. 116, No. 1, pp. 280–290.
[4] Her, I., 1986, "Methodology for Compliant Mechanisms Design," Ph.D. dissertation, Purdue University, West Lafayette, IN.
[5] Eijk, J. V., 1985, "On the Design of Plate-Spring Mechanisms," Ph.D. dissertation, Delft University of Technology, Delft, The Netherlands.
[6] Roach, G. M., Lyon, S. M., and Howell, L. L., 1998, "A Compliant, Overrunning Ratchet and Pawl Clutch with Centrifugal Throw-out," *Proceedings of the 1998 ASME Design Engineering Technical Conferences,* DETC98/MECH-5819.
[7] Roach, G. M., and Howell, L. L., 2000, "Compliant Overrunning Clutch with Centrifugal Throw-out," U. S. patent 6,148,979, Nov. 21.
[8] Motsinger, R. N., 1964, "Flexural Devices in Measurement Systems," Chapter 11 in *Measurement Engineering,* by P. K. Stein, Stein Engineering Services, Phoenix, AZ.
[9] Tuttle, S. B., 1967, "Semifixed Flexural Mechanisms," Chapter 8 in *Mechanisms for Engineering Design,* Wiley, New York.
[10] Derderian, J. M., 1996, "The Pseudo-Rigid-Body Model Concept and Its Application to Micro Compliant Mechanisms," M.S. thesis, Brigham Young University, Provo, UT.

[11] Jensen, B. D., Howell, L. L., Gunyan, D. B., and Salmon, L. G., 1997, "The Design and Analysis of Compliant MEMS Using the Pseudo-Rigid-Body Model," *Microelectromechanical Systems (MEMS)*, at the 1997 ASME International Mechanical Engineering Congress and Exposition, DSC-Vol. 62, pp. 119–126.

[12] Jensen, B. D., Howell, L. L., and Salmon, L. G., 1999, "Design of Two-Link, In-Plane, Bistable Compliant Micro-Mechanisms," *Journal of Mechanical Design*, Trans. ASME, Vol. 121, No. 3, pp. 416–423.

[13] Ananthasuresh, G. K., and Kota, S., 1995, "Designing Compliant Mechanisms," *Mechanical Engineering*, Vol. 117, No. 11, pp. 93–96.

[14] Ananthasuresh, G. K., Kota, S., and Gianchandani, Y., 1994, "A Methodical Approach to the Design of Compliant Micromechanisms," *Solid-State Sensor and Actuator Workshop*, Hilton Head Island, SC, pp. 189–192.

[15] Larsen, V. D., Sigmund, O., and Bouwstra, S., 1996, "Design and Fabrication of Compliant Micromechanisms and Structures with Negative Poisson's Ratio," *Proceedings of the 9th Annual International Workshop on Micro Electro Mechanical Systems*, San Diego, CA, Feb. 11–15, pp. 365–371.

[16] Baker, M. S., Lyon, S. M., and Howell, L. L., 2000, "A Linear Displacement Bistable Micromechanism," *Proceedings of the 26th Biennial Mechanisms and Robotics Conference*, 2000 ASME Design Engineering Technical Conferences, DETC2000/MECH-14117.

[17] Clements, D., 2000, "Implementing Compliant Mechanisms in Micro-Electro-Mechanical Systems (MEMS)," M.S. thesis, Brigham Young University, Provo, UT.

[18] Parkinson, M. B., Jensen, B. D., and Roach, G. M., 2000, "Optimization-Based Design of a Fully-Compliant Bistable Micromechanism," *Proceedings of the 26th Biennial Mechanisms and Robotics Conference*, 2000 ASME Design Engineering Technical Conferences, DETC2000/MECH-14119.

[19] Salmon, L. G., Gunyan, D. B., Derderian, J. M., Opdahl, P. G., and Howell, L. L., 1996, "Use of the Pseudo-Rigid Body Model to Simplify the Description of Compliant Micro-Mechanisms," *1996 IEEE Solid-State and Actuator Workshop*, Hilton Head Island, SC, pp. 136–139.

[20] McEwen, E., Miller, R. L., and Bergman, C. A., 1991, "Early Bow Design and Construction," *Scientific American*, Vol. 264, June, pp. 76–82.

[21] De Camp, L. S., 1974, *Ancient Engineers*, Ballantine, New York.

[22] Smith, C. G., and Rees, G., 1978, *The Inventions of Leonardo Da Vinci*, Phaidon Press, Oxford.

[23] Weinstein, W. D., 1965, "Flexural Pivot Bearings," *Machine Design*, June 10, pp. 150–157.

[24] Paros, J. M., and Weisbord, L., 1965, "How to Design Flexural Hinges," *Machine Design*, Nov. 25, pp. 151–156.

[25] Chow, W. W., 1981, "Hinges and Straps," Chapter 12 in *Plastics Products Design Handbook*, Part B, E. Miller, ed., Marcel Dekker, New York.

References

[26] Ananthasuresh, G. K., and Howell, L. L., 1996, "Case Studies and a Note on the Degree-of-Freedom in Compliant Mechanisms," *Proceedings of the 1996 ASME Design Engineering Technical Conferences*, 96-DETC/MECH-1217.

[27] Berglund, M., Magleby, S. P., and Howell, L. L., 2000, "Design Rules for Selecting and Designing Compliant Mechanisms for Rigid-Body Replacement Synthesis," *Proceedings of the 26th Design Automation Conference*, 2000 ASME Design Engineering Technical Conferences, DETC2000/DAC-14225.

[28] Boronkay, T. G., and Mei, C., 1970, "Analysis and Design of Multiple Input Flexible Link Mechanisms," *Journal of Mechanisms*, Vol. 5, No. 1, pp. 29–49.

[29] Burns, R. H., 1964, "The Kinetostatic Synthesis of Flexible Link Mechanisms," Ph.D. dissertation, Yale University, New Haven, CT.

[30] Burns, R. H., and Crossley, F. R. E., 1966, "Structural Permutations of Flexible Link Mechanisms," ASME Paper 66-Mech-5.

[31] Burns, R. H., and Crossley, F. R. E., 1968, "Kinetostatic Synthesis of Flexible Link Mechanisms," ASME Paper 68-Mech-36.

[32] Byers, F. K., and Midha, A., 1991, "Design of a Compliant Gripper Mechanism," *Proceedings of the 2nd National Applied Mechanisms and Robotics Conference*, Cincinnati, OH, Paper XC-1.

[33] Crane, N. B., Howell, L. L., and Weight, B. L., 2000, "Design and Testing of a Compliant Floating-Opposing-Arm (FOA) Centrifugal Clutch," *Proceedings of 8th International Power Transmission and Gearing Conference*, 2000 ASME Design Engineering Technical Conferences, DETC2000/PTG-14451.

[34] Derderian, J. M., Howell, L. L., Murphy, M. D., Lyon, S. M., and Pack, S. D., 1996, "Compliant Parallel-Guiding Mechanisms," *Proceedings of the 1996 ASME Design Engineering Technical Conferences*, 96-DETC/MECH-1208.

[35] Edwards, B. J., 1996, "Functionally Binary Pinned–Pinned Segments," M.S. thesis, Brigham Young University, Provo, UT.

[36] Edwards, B. J., 1999, "A Pseudo-Rigid-Body Model for Functionally Binary Pinned–Pinned Segments Used in Compliant Mechanisms," *Proceedings of the 25th Design Automation Conference*, 1999 ASME Design Engineering Technical Conferences, DETC99/DAC-8619.

[37] Frecker, M. I., Kikuchi, N., and Kota, S., 1996, "Optimal Synthesis of Compliant Mechanisms to Meet Structural and Kinematic Requirements: Preliminary Results," *Proceedings of the 1996 ASME Design Engineering Technical Conferences*, 96-DETC/DAC-1497.

[38] Her, I., and Midha, A., 1987, "A Compliance Number Concept for Compliant Mechanisms, and Type Synthesis," *Journal of Mechanisms, Transmissions, and Automation in Design*, Trans. ASME, Vol. 109, No. 3, pp. 348–355.

[39] Her, I., Midha, A., and Salamon, B. A., 1992, "A Methodology for Compliant Mechanisms Design, Part II: Shooting Method and Application," *Proceedings of the 18th ASME Design Automation Conference*, D. A. Hoeltzel, ed., DE-Vol. 44-2, pp. 39–45.

[40] Hetrick, J. A. and Kota, S., 2000, "Topological and Geometric Synthesis of Compliant Mechanisms," *Proceedings of the 26th Biennial Mechanisms and Robotics Conference*, 2000 ASME Design Engineering Technical Conferences, DETC2000/MECH-14140.

[41] Hill, T. C., 1987, "Applications in the Analysis and Design of Compliant Mechanisms," M.S. thesis, Purdue University, West Lafayette, IN.

[42] Hill, T. C., and Midha, A., 1990, "A Graphical User-Driven Newton–Raphson Technique for Use in the Analysis and Design of Compliant Mechanisms," *Journal of Mechanical Design*, Trans. ASME, Vol. 112, No. 1, pp. 123–130.

[43] Howell, L. L., 1991, "The Design and Analysis of Large-Deflection Members in Compliant Mechanisms," M.S. thesis, Purdue University, West Lafayette, IN.

[44] Howell, L. L., 1993, "A Generalized Loop-Closure Theory for the Analysis and Synthesis of Compliant Mechanisms," Ph.D. dissertation, Purdue University, West Lafayette, IN.

[45] Howell, L. L., and Midha, A., 1993, "Compliant Mechanisms," Section 9.10 in *Modern Kinematics: Developments in the Last Forty Years*, (A. G. Erdman, ed.), Wiley, New York, pp. 422–428.

[46] Howell, L. L., Rao, S. S., and Midha, A., 1994, "The Reliability-Based Optimal Design of a Bistable Compliant Mechanism," *Journal of Mechanical Design*, Trans. ASME, Vol. 116, No. 4, pp. 1115–1121.

[47] Howell, L. L., and Midha, A., 1994, "A Loop-Closure Theory for the Analysis and Synthesis of Compliant Mechanisms," *Journal of Mechanical Design*, Trans. ASME, Vol. 118, No. 1, pp. 121–125.

[48] Howell, L. L., and Midha, A., 1994, "The Development of Force–Deflection Relationships for Compliant Mechanisms," *Machine Elements and Machine Dynamics: Proceedings of the 1994 ASME Mechanisms Conference*, DE-Vol. 71, pp. 501–508.

[49] Howell, L. L., Midha, A., and Murphy, M. D., 1994, "Dimensional Synthesis of Compliant Constant-Force Slider Mechanisms," *Machine Elements and Machine Dynamics: Proceedings of the 1994 ASME Mechanisms Conference*, DE-Vol. 71, pp. 509–515.

[50] Howell, L. L., and Midha, A., 1995, "Parametric Deflection Approximations for End-Loaded, Large-Deflection Beams in Compliant Mechanisms," *Journal of Mechanical Design*, Trans. ASME, Vol. 117, No. 1, pp. 156–165.

[51] Howell, L. L., and Midha, A., 1995, "Determination of the Degrees of Freedom of Compliant Mechanisms Using the Pseudo-Rigid-Body Model Concept," *Proceedings of the 9th World Congress on the Theory of Machines and Mechanisms*, Milano, Italy, Vol. 2, pp. 1537–1541.

[52] Howell, L. L., and Midha, A., 1996, "Parametric Deflection Approximations for Initially Curved, Large-Deflection Beams in Compliant Mechanisms," *Proceedings of the 1996 ASME Design Engineering Technical Conferences*, 96-DETC/MECH-1215.

[53] Howell, L. L., Midha, A., and Norton, T. W., 1996, "Evaluation of Equivalent Spring Stiffness for Use in a Pseudo-Rigid-Body Model of Large-Deflection Compliant Mechanisms," *Journal of Mechanical Design*, Trans. ASME, Vol. 118, No. 1, pp. 126–131.

[54] Howell, L. L., and Leonard, J. N., 1997, "Optimal Loading Conditions for Non-linear Deflections," *International Journal of Non-linear Mechanics*, Vol. 32, No. 3, pp. 505–514.

[55] Lyon, S. M., Evans, M. S., Erickson, P. A., and Howell, L. L., 1999, "Prediction of the First Modal Frequency of Compliant Mechanisms Using the Pseudo-Rigid-Body Model," *Journal of Mechanical Design*, Trans. ASME, Vol. 121, No. 2, pp. 309–313.

[56] Mettlach, G. A., and Midha, A., 1996, "Using Burmester Theory in the Design of Compliant Mechanisms," *Proceedings of the 1996 ASME Design Engineering Technical Conferences*, 96-DETC/MECH-1181.

[57] Midha, A., Her, I., and Salamon, B. A., 1992, "A Methodology for Compliant Mechanisms Design, Part I: Introduction and Large-Deflection Analysis," in *Advances in Design Automation*, D. A. Hoeltzel, ed., 18th ASME Design Automation Conference, DE-Vol. 44-2, pp. 29–38.

[58] Midha, A., Norton, T. W., and Howell, L. L., 1994, "On the Nomenclature, Classification, and Abstractions of Compliant Mechanisms," *Journal of Mechanical Design*, Trans. ASME, Vol. 116, pp. 270–279.

[59] Midha, A., and Howell, L. L., 2000, "Limit Positions of Compliant Mechanisms Using the Pseudo-Rigid-Body Model," *Mechanism and Machine Theory*, Vol. 35, No. 1, pp. 99–115.

[60] Millar, A. J., Howell, L. L., and Leonard, J. N., 1996 "Design and Evaluation of Compliant Constant-Force Mechanisms," *Proceedings of the 1996 ASME Design Engineering Technical Conferences*, 96-DETC/MECH-1209.

[61] Mortensen, C. R., Weight, B. L., Howell, L. L., and Magleby, S. P., 2000, "Compliant Mechanism Prototyping," *Proceedings of the 26th Biennial Mechanisms and Robotics Conference*, 2000 ASME Design Engineering Technical Conferences, DETC2000/MECH-14204.

[62] Murphy, M. D., 1993, "A Generalized Theory for the Type Synthesis and Design of Compliant Mechanisms," Ph.D. dissertation, Purdue University, West Lafayette, IN.

[63] Murphy, M. D., Midha, A., and Howell, L. L., 1996, "The Topological Synthesis of Compliant Mechanisms," *Mechanism and Machine Theory*, Vol. 31, No. 2, pp. 185–199.

[64] Murphy, M. D., Midha, A., and Howell, L. L., 1994, "On the Mobility of Compliant Mechanisms," *Machine Elements and Machine Dynamics: Proceedings of the 1994 ASME Mechanisms Conference*, DE-Vol. 71, pp. 475–479.

[65] Murphy, M. D., Midha, A., and Howell, L. L., 1994, "The Topological Analysis of Compliant Mechanisms," *Machine Elements and Machine Dynamics: Proceedings of the 1994 ASME Mechanisms Conference*, DE-Vol. 71, pp. 481–489.

[66] Murphy, M. D., Midha, A., and Howell, L. L., 1994, "Type Synthesis of Compliant Mechanisms Employing a Simplified Approach to Segment Type," *Mechanism Synthesis and Analysis: Proceedings of the 1994 ASME Mechanisms Conference*, DE-Vol. 70, pp. 51–60.

[67] Murphy, M. D., Midha, A., and Howell, L. L., 1994, "Methodology for the Design of Compliant Mechanisms Employing Type Synthesis Techniques with Example," *Mechanism Synthesis and Analysis: Proceedings of the 1994 ASME Mechanisms Conference*, DE-Vol. 70, pp. 61–66.

[68] Nahvi, H., 1991, "Static and Dynamic Analysis of Compliant Mechanisms Containing Highly Flexible Members," Ph.D. dissertation, Purdue University, West Lafayette, IN.

[69] Nielson, A. J. and Howell, L. L., 1998, "Compliant Pantographs via the Pseudo-Rigid-Body Model," *Proceedings of the 1998 ASME Design Engineering Technical Conferences*, DETC98/MECH-5930.

[70] Norton, T. W., 1991, "On the Nomenclature and Classification, and Mobility of Compliant Mechanisms," M.S. thesis, Purdue University, West Lafayette, IN.

[71] Opdahl, P. G., 1996, "Modelling and Analysis of Compliant Bi-stable Mechanisms Using the Pseudo-Rigid-Body Model," M.S. thesis, Brigham Young University, Provo, UT.

[72] Opdahl, P. G., Jensen, B. D., and Howell, L. L., 1998, "An Investigation into Compliant Bistable Mechanisms," *Proceedings of the 1998 ASME Design Engineering Technical Conferences*, DETC98/MECH-5914.

[73] Parkinson, M. B., Howell, L. L., and Cox, J. J., 1997, "A Parametric Approach to the Optimization-Based Design of Compliant Mechanisms," *Proceedings of the 23rd Design Automation Conference*, DETC97/DAC-3763.

[74] Salamon, B. A., 1989, "Mechanical Advantage Aspects in Compliant Mechanisms Design," M.S. thesis, Purdue University, West Lafayette, IN.

[75] Salamon, B. A., and Midha, A., 1992, "An Introduction to Mechanical Advantage in Compliant Mechanisms," in *Advances in Design Automation*, D. A. Hoeltzel, ed., 18th ASME Design Automation Conference, DE-Vol 44-2, pp. 47-51.

[76] Shoup, T. E., and McLarnan, C. W., 1971, "A Survey of Flexible Link Mechanisms Having Lower Pairs," *Journal of Mechanisms*, Vol. 6, No. 3, pp. 97–105.

[77] Shoup, T. E., and McLarnan, C. W., 1971, "On the Use of the Undulating Elastica for the Analysis of Flexible Link Devices," *Journal of Engineering for Industry*, Trans. ASME, pp. 263–267.

[78] Shoup, T. E., 1972, "On the Use of the Nodal Elastica for the Analysis of Flexible Link Devices," *Journal of Engineering for Industry*, Trans. ASME, Vol. 94, No. 3, pp. 871–875.

[79] Sevak, N. M., and McLarnan, C. W., 1974, "Optimal Synthesis of Flexible Link Mechanisms with Large Static Deflections," ASME Paper 74-DET-83.

[80] Sigmund, O., 1996, "On the Design of Compliant Mechanisms Using Topology Optimization," Report 535, Danish Center for Applied Mathematics and Mechanics, Technical University of Denmark, Copenhagen.

[81] Vogel, S., 1995, "Better Bent Than Broken," *Discover*, May, pp. 62–67.

[82] Winter, S. J., and Shoup, T. E., 1972, "The Displacement Analysis of Path Generating Flexible-Link Mechanisms," *Mechanism and Machine Theory*, Vol. 7, No. 4, pp. 443–451.

[83] Ananthasuresh, G. K., 1994, "A New Design Paradigm for Micro-Electro-Mechanical Systems and Investigations on the Compliant Mechanism Synthesis," Ph.D. dissertation, University of Michigan, Ann Arbor, MI.

[84] Ananthasuresh, G. K., Kota, S., and Gianchandani, Y., 1994, "A Methodical Approach to the Design of Compliant Micromechanisms," *Solid-State Sensor and Actuator Workshop*, Hilton Head Island, SC, pp. 189–192.

[85] Kota, S., Ananthasuresh, G. K., Crary, S. B., and Wise, K. D., 1994, "Design and Fabrication of Microelectromechanical Systems," ASME *Journal of Mechanical Design*, Vol. 116, No. 4, pp. 1081–1088.

[86] Madou, M., 1997, *Fundamentals of Microfabrication*, CRC Press, Boca Raton, FL.

[87] Clements, D., 2000, "Implementing Compliant Mechanisms in Micro-Electro-Mechanical Systems (MEMS)," M.S. thesis, Brigham Young University, Provo, UT.

[88] Nielsen, L. E., and Landel, R. F., 1994, *Mechanical Properties of Polymers and Composites*, 2nd Ed., Marcel Decker, New York.

[89] Young, W. C., 1989, *Roark's Formulas for Stress and Strain*, 6th Ed., McGraw-Hill, New York.

[90] Bisshopp, K. E., and Drucker, D. C., 1945, "Large Deflection of Cantilever Beams," *Quarterly of Applied Mathematics*, Vol. 3, No. 3, pp. 272–275.

[91] Coulter, B. A., and Miller, R. E., 1988, "Numerical Analysis of a Generalized Plane Elastica with Non-linear Material Behavior," *International Journal for Numerical Methods in Engineering*, Vol. 26, pp. 617–630.

[92] Frisch-Fay, R., 1962, *Flexible Bars*, Butterworth, Washington, DC.

[93] Gorski, W., 1976, "A Review of Literature and a Bibliography on Finite Elastic Deflections of Bars," *Transactions of the Institution of Engineers, Australia, Civil Engineering Transactions*, Vol. 18, No. 2, pp. 74–85.

[94] Harrison, H. B., 1973, "Post-buckling Analysis of Non-Uniform Elastic Columns," *International Journal for Numerical Methods in Engineering*, Vol. 7, pp. 195–210.

[95] Mattiasson, K., 1981, "Numerical Results from Large Deflection Beam and Frame Problems Analyzed by Means of Elliptic Integrals," *International Journal for Numerical Methods in Engineering*, Vol. 17, pp. 145–153.

[96] Miller, R. E., 1980, "Numerical Analysis of a Generalized Plane Elastica," *International Journal for Numerical Methods in Engineering*, Vol. 15, pp. 325–332.

[97] Byrd, P. F., and Friedman, M. D., 1954, *Handbook of Elliptic Integrals for Engineers and Physicists*, Springer-Verlag, Berlin.

[98] King, L. V., 1924, *On the Direct Numerical Calculation of Elliptic Functions and Integrals*, Cambridge University Press, London.

[99] Hancock, H., 1958, *Elliptic Integrals*, Dover, New York.

[100] Greenhill, A. G., 1959, *The Applications of Elliptic Functions*, Dover, New York.

[101] Parise, J. J, 1999, "Ortho-planar Mechanisms," M.S. thesis, Brigham Young University, Provo, UT.

[102] Pilkey, W. D., 1997, *Peterson's Stress Concentration Factors*, Wiley, New York.

[103] Norton, R. L., 2000, *Machine Design*, 2nd Ed., Prentice Hall, Upper Saddle River, NJ.

[104] Shigley, J. E., and Mischke, C. R., 2001, *Mechanical Engineering Design*, 6th Ed., McGraw-Hill, New York.

[105] Neuber, H., 1946, *Theory of Notch Stresses*, J.W. Edwards, Ann Arbor, MI.

[106] Kuhn, P., and Hardrath, H. F., 1952, *An Engineering Method for Estimating Notch-Size Effect in Fatigue Tests on Steel*, Technical Note 2805, National Advisory Committee on Aeronautics, Washington, DC.

[107] Shigley, J. E., and Mitchell, L. D., 1983, *Mechanical Engineering Design*, 4th Ed., McGraw-Hill, New York.

[108] Dowling, N. E., 1993, *Mechanical Behavior of Materials*, Prentice Hall, Upper Saddle River, NJ.

[109] Juvinall, R. C., 1983, *Fundamentals of Machine Component Design*, Wiley, New York.

[110] Juvinall, R. C., 1967, *Stress, Strain, and Strength,* McGraw-Hill, New York.

[111] Forrest, P. G., 1962, *Fatigue of Metals*, Pergamon Press, Elmsford, NY.

[112] *Modern Plastics Encyclopedia*, McGraw-Hill, New York.

[113] Marin, J., 1962, *Mechanical Behavior of Engineering Materials*, Prentice Hall, Upper Saddle River, NJ.

[114] Shigley, J. E., and Mischke, C. R., 1996, *Standard Handbook of Machine Design*, 2nd Ed., McGraw-Hill, New York.

[115] Haugen, E. B., and Wirsching, P. H., 1975, "Probabilistic Design," *Machine Design*, Vol. 47, pp. 10–14.

[116] Fontana, M. G., 1986, *Corrosion Engineering*, 3rd Ed., McGraw-Hill, New York.

[117] Hertzberg, R. W., and Manson, J. A., 1980, *Fatigue of Engineering Plastics*, Academic Press, San Diego, CA.

[118] Trantina, G., and Nimmer, R., 1994, *Structural Analysis of Thermoplastic Components*, McGraw-Hill, New York.

[119] Nielsen, L. E., and Landel, R. F., 1994, *Mechanical Properties of Polymers and Composites*, 2nd Ed., Marcel Dekker, New York.

References

[120] Miller, E., ed., 1981, *Plastics Products Design Handbook*, Marcel Dekker, New York.

[121] Paul, B., 1979, *Kinematics and Dynamics of Planar Machinery*, Prentice Hall, Upper Saddle River, NJ

[122] Norton, R. L., 1999, *Design of Machinery*, 2nd Ed., McGraw-Hill, New York.

[123] Sandor, G. N., and Erdman, A. G., 1984, *Mechanism Design: Analysis and Synthesis*, Vol. 2, Prentice Hall, ,Upper Saddle River, NJ.

[124] Hall, A. S., Jr., 1986, *Kinematics and Linkage Design*, Waveland Press, New York.

[125] Norton, T. W., Midha, A., and Howell, L. L., 1994, "Graphical Synthesis for Limit Positions of a Four-Bar Mechanism Using the Triangle Inequality Concept," *Journal of Mechanical Design*, Trans. ASME, Vol. 116, No. 4, pp. 1132–1140.

[126] Olson, D. G., Erdman, A. G., and Riley, D. R., 1985, "A Systematic Procedure for Type Synthesis of Mechanisms with Literature Review," *Mechanism and Machine Theory*, Vol. 20, No. 4, pp. 285–295.

[127] Lyon, S. M., Howell, L. L., and Roach, G. M., 2000, "Modeling Fixed–Fixed Flexible Segments via the Pseudo-Rigid-Body Model," *Proceedings of the ASME Dynamics and Control Division*, at the 2000 ASME International Mechanical Engineering Congress and Exposition, Orlando, FL, Nov. 5-10, DSC-Vol. 69-2, pp. 883-890.

[128] Saxena, A. and Kramer, S. N., 1998, "A Simple and Accurate Method for Determining Large Deflections in Compliant Mechanisms Subjected to End Forces and Moments," *Journal of Mechanical Design*, Trans. ASME, Vol. 120, No. 3, pp. 392–400, erratum, Vol. 121, No. 2, p. 194.

[129] Amoco Chemical Company, 1992, "PP Hinges and Hinge Strengths," Amoco Polypropylene technical report.

[130] Goldfarb, M., and Speich, J. E., 1999, "A Well-Behaved Revolute Flexure Joint for Compliant Mechanism Design," *Journal of Mechanical Design*, Trans. ASME, Vol. 121, No. 3, pp. 424–429.

[131] Yang, T. Y., 1973, "Matrix Displacement Solution to Elastica Problems of Beams and Frames," *International Journal of Solids and Structures*, Vol. 9, No. 7, pp. 829–842.

[132] Bathe, K.-J., and Bolourch, S., 1979, "Large Displacement Analysis of Three-Dimensional Beam Structures," *International Journal for Numerical Methods in Engineering*, Vol. 14, pp. 961–986.

[133] Coulter, B. A., and Miller, R. E., 1988, "Numerical Analysis of a Generalized Plane Elastica with Non-linear Material Behavior," *International Journal for Numerical Methods in Engineering*, Vol. 26, pp. 617–630.

[134] Yang, T. Y., 1986, *Finite Element Structural Analysis*, Prentice Hall, Upper Saddle River, NJ.

[135] Bathe, K. J., 1996, *Finite Element Procedures*, Prentice Hall, Upper Saddle River, NJ.

[136] Forsythe, G. E. and Wasow, W. R., 1960, *Finite Difference Methods for Partial Differential Equations*, Wiley, New York.

[137] Wright, J. R. and Baron, M. L., 1973, "A Survey of Finite-Difference Methods for Partial Differential Equation," in *Numerical and Computer Methods in Structural Mechanics*, S. J. Fenve, N. Perrone, A. R. Robinson, and W. C. Schnobrich, eds., Academic Press, San Diego, CA, pp. 265–289.

[138] Kythe, P. K., 1995, *An Introduction to Boundary Element Methods*, CRC Press, Boca Raton, FL.

[139] Barr, V., 1992, "Alexandre Gustave Eiffel: A Towering Engineering Genius," *Mechanical Engineering*, Feb., pp. 58–65.

[140] Lakes, R., 1993, "Materials with Structural Hierarchy," *Nature*, Vol. 361, Feb. 11, pp. 511–515.

[141] Luenberger, D. G., 1984, *Introduction to Linear and Non-linear Programming*, Addison-Wesley, Reading, MA.

[142] Gelfand, I. M., and Fomin, S. V., 1963, *Calculus of Variations*, translated and edited by R. A. Silverman, Prentice Hall, Upper Saddle River, NJ.

[143] Haftka, R. T., and Gürdal, Z., 1990, *Elements of Structural Optimization*, Kluwer Academic, Dordrecht, The Netherlands.

[144] Rao, S. S., 1996, *Engineering Optimization: Theory and Practice*, Wiley-Interscience, New York.

[145] Shield, R. T., and Prager, W., 1970, "Optimal Structural Design for Given Deflection," *Journal of Applied Mathematics and Physics*, Vol. 21, pp. 513–523.

[146] Saxena, A., and Ananthasuresh, G. K., 1998, "Topology Synthesis of Compliant Mechanisms Using the Optimality Criteria Method," *Proceedings of the AIAA/USAF/NASA/ISSMO 1998 Symposium on Multi-disciplinary Analysis and Optimization*, St. Louis, MO, Sept. 2–4, Vol. 3, pp. 1990–1910.

[147] Sigmund, O., 1997, "On the Design of Compliant Mechanisms Using Topology Optimization," *Mechanics of Structures and Machines*, Vol. 25, No. 4, pp. 495–526.

[148] Hetrick, J., and Kota, S., 1999, "An Energy Formulation for Parametric Size and Shape Optimization of Compliant Mechanisms," *Journal of Mechanical Design*, Vol. 21, No. 2, pp. 229–234.

[149] Frecker, M. I., 1997, *Optimal Design of Compliant Mechanisms*, Ph.D. dissertation, University of Michigan, Ann Arbor, MI.

[150] Frecker, M. I., Ananthasuresh, G. K., Nishiwaki, S., Kikuchi, N., and Kota, S., 1997, "Topological Synthesis of Compliant Mechanisms Using Multi-criteria Optimization," *Journal of Mechanical Design*, Vol. 119, pp. 238–245.

[151] Nishiwaki, S., Frecker, M. I., Min, S. J., and Kikuchi, N., 1998, "Topology Optimization of Compliant Mechanisms Using the Homogenization Method," *International Journal for Numerical Methods in Engineering*, Vol. 42, No. 3, pp. 535–559.

References

[152] Barnett, R. L., 1961, "Minimum Weight Design of Beams for Deflection," *Journal of Engineering Mechanics Division, Proceedings of the American Society of Civil Engineers*, Feb., Vol. EM1, pp. 75–109.

[153] Haftka, R. T., and Grandhi, R. V., 1986, "Structural Optimization: A Survey," *Computer Methods in Applied Mechanics and Engineering*, Vol. 57, pp. 91–106.

[154] Parkinson, M. B., Howell, L. L., and Cox, J. J., 1997, "A Parametric Approach to the Optimization-Based Design of Compliant Mechanisms," *Proceedings of the 23rd Design Automation Conference*, DETC97/DAC-3763.

[155] Michell, A. G. M., 1904, "The Limits of Economy of Material in Frame Structures," *Philosophical Magazine*, Ser. 6, No. 8, pp. 589–597.

[156] Kirsch, U., 1989, "Optimal Topologies of Structures," *Applied Mechanics Reviews*, Vol. 42, No. 8, pp. 223–239.

[157] Bendsøe, M. P., 1995, *Optimization of Structural Topology, Shape, and Material*, Springer-Verlag, New York.

[158] Rozvany, G. I. N., 1989, *Structural Design via Optimality Criteria*, Kluwer Academic, Boston.

[159] Strasys, 1999, *Fused Deposition Modeling System*, Stratasys, Inc., Eden Prairie, MN.

[160] Mlejnek, H. P., and Schirrmacher, R., 1993, "An Engineer's Approach to Optimal Material Distribution and Shape Finding," *Computer Methods in Applied Mechanics and Engineering*, Vol. 106, pp. 1–26.

[161] Bendsøe, M. P., and Kikuchi, N., 1988, "Generating Optimal Topologies in Structural Design Using a Homogenization Method," *Computer Methods in Applied Mechanics and Engineering*, Vol. 71, pp. 197–224.

[162] Suzuki, K., and Kikuchi, N., 1990, "Shape and Topology Optimization for Generalized Layout Problems Using the Homogenization Method," *Computer Methods in Applied Mechanics and Engineering*, Vol. 93, pp. 291–318.

[163] Ananthasuresh, G. K., Kota, S., and Kikuchi, N., 1994, "Strategies for Systematic Synthesis of Compliant MEMS," *Proceedings of the 1994 ASME Winter Annual Meeting, Symposium on MEMS: Dynamics Systems and Control*, Chicago, DSC-Vol. 55-2, Nov., pp. 677–686.

[164] Vanderplaats, G. N., 1984, *Numerical Optimization Techniques for Engineering Design*, McGraw-Hill, New York.

[165] Kelly, J. E., 1960, "The Cutting Plane Method for Solving Convex Programs," *Journal of SIAM*, Vol. 8, pp. 702–712.

[166] Biggs, M. C., 1972, "Constrained Minimization Using Recursive Equality Quadratic Programming," in *Numerical Methods for Nonlinear Optimization*, F. A. Lootsma, ed., Academic Press, London, pp. 411–428.

[167] Fleury, C., and Braibant, V., 1986, "Structural Optimization: A New Dual Method Using Mixed Variables Information," *International Journal for Numerical Methods in Engineering*, Vol. 23, pp. 409–428.

[168] Svanberg, K., 1987, "The Method of Moving Asymptotes: A New Method for Structural Optimization," *International Journal for Numerical Methods in Engineering*, Vol. 24, pp. 359–373.

[169] Chickermane, H., and Gea, H.C., 1996, "Structural Optimization Using a New Local Approximation Method," *International Journal for Numerical Methods in Engineering*, Vol. 39, pp. 829–846.

[170] Xie, Y. M., and Steven, G. P., 1997, *Evolutionary Structural Optimization*, Springer-Verlag, New York.

[171] Haug, E. J., Choi, K. K., and Kpmkov, V., 1980, *Design Sensitivity Analysis in Structural Systems*, Academic Press, San Diego, CA.

[172] Saxena, A., and Ananthasuresh, G. K., 2000, "On an Optimal Property of Compliant Topologies," *Structural and Multidisciplinary Optimization*, Vol. 19, pp. 36-49.

[173] Nathan, R. H., 1985, "A Constant Force Generation Mechanism," *Journal of Mechanisms, Transmissions, and Automation in Design*, Trans. ASME, Vol. 107, Dec., pp. 508–512.

[174] Jenuwine, J. G., and Midha, A., 1989, "Design of an Exact Constant Force Generating Mechanism," *Proceedings of the First National Applied Mechanisms and Robotics Conference*, Cincinnati, OH, Vol. II, pp. 10B-4-1 to 10B-4-5.

[175] Jenuwine, J. G., and Midha, A., 1994, "Synthesis of Single-Input and Multiple-Output Port Mechanisms with Springs for Specified Energy Absorption," *Journal of Mechanical Design*, Trans. ASME, Vol. 116, No. 3, Sept., pp. 937–943.

[176] Wahl, A., 1963, *Mechanical Springs*, 2nd ed. McGraw-Hill, New York.

[177] Midha, A., Murphy, M. D., and Howell, L. L., 1995, "Compliant Constant-Force Mechanism and Devices Formed Therein," US patent 5,649,454.

[178] Rao, S. S., 1984, *Optimization: Theory and Applications*, Wiley Eastern, New Delhi, India.

[179] Evans, M. S., and Howell, L. L., 1999, "Constant-Force End-Effector Mechanism," *Proceedings of the IASTED International Conference, Robotics and Applications*, pp. 250–256.

[180] Timoshenko, S. P., 1958, *Strength of Materials,* Part II: *Advanced Theory and Problems,* Van Nostrand Reinhold, New York.

[181] Timoshenko, S. P., 1961, *Theory of Elastic Stability,* 2nd ed., McGraw-Hill, New York.

[182] Timoshenko, S. P., and Young, D. H., 1951, *Engineering Mechanics,* 3rd ed., McGraw-Hill, New York.

[183] Simitses, G. J., 1976, *An Introduction to the Elastic Stability of Structures,* Prentice Hall, Upper Saddle River, NJ.

[184] Leipholz, H., 1970, *Stability Theory,* Academic Press, San Diego, CA.

[185] Jensen, B. D., and Howell, L. L., 2000, "Identification of Compliant Pseudo-Rigid-Body Mechanism Configurations Resulting in Bistable Behavior," *Proceedings of the ASME 2000 Design Engineering Technical Conferences*, Baltimore, Sept., DETC2000/MECH-14147.

References

[186] *Metals Handbook*, American Society for Metals, Materials Park, OH.

[187] *SAE Handbook*, Society of Automotive Engineers, Warrendale PA.

[188] Military Handbook, *Metallic Materials and Elements for Aerospace Vehicle Structures*, MIL-HDBK-5,Vol. 2, U.S. Department of Defense, Washington, DC.

[189] *Properties of Some Metals and Alloys*, International Nickel Co., New York.

[190] Freudenstein, F., and Dobrjanskyj, L., 1964, "On a Theory for the Type Synthesis of Mechanisms," *Proceedings of the 11th International Conference of Applied Mechanics*, pp. 420–428.

[191] Uicker, J. J., and Raicu, A., 1975, "A Method for the Identification of and Recognition of Equivalence of Kinematic Chains," *Mechanism and Machine Theory*, Vol. 10, No. 10, pp. 375–383.

[192] Yan, H. S., and Hwang, W. M., 1983, "A Method for the Identification of Planar Linkages," *Journal of Mechanisms, Transmissions, and Automation in Design*, Trans. ASME, Vol. 105, No. 4, pp. 658–662.

[193] Yan, H. S., 1992, "A Methodology for Creative Mechanism Design," *Mechanism and Machine Theory*, Vol. 27, No. 3, pp. 235–242.

[194] Erdman, A. G., Nelson, E., Peterson, J., and Bowen, J., 1980, "Type and Dimensional Synthesis of Casement Window Mechanisms," ASME Paper 80-DET-78.

[195] Kota, S., 1993, "Type Synthesis and Creative Design," Chapter 3 in *Modern Kinematics: Developments in the Last Forty Years*, A. G. Erdman, ed., Wiley, New York.

APPENDIX B

PROPERTIES OF SECTIONS

A = area   The centroid is at the intersection of the x and y axes.

$$I_x = \int x^2 dA; \quad I_y = \int y^2 dA \quad J_z = \int r^2 dA = \int (x^2 + y^2) dA = I_x + I_y$$

B.1 RECTANGLE

$A = bh$

$I_x = \dfrac{bh^3}{12}$

$I_y = \dfrac{b^3 h}{12}$

$J_z = I_x + I_y = \dfrac{bh}{12}(h^2 + b^2)$

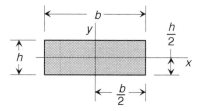

B.2 CIRCLE

$A = \dfrac{\pi D^2}{4} = \pi R^2$

$I_x = I_y = \dfrac{\pi D^4}{64} = \dfrac{\pi R^4}{4}$

$J_z = I_x + I_y = \dfrac{\pi D^4}{32} = \dfrac{\pi R^4}{2}$

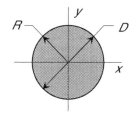

B.3 HOLLOW CIRCLE

$$A = \frac{\pi}{4}(D^2 - d^2)$$

$$I_x = I_y = \frac{\pi}{64}(D^4 - d^4)$$

$$J_z = I_x + I_y = \frac{\pi}{32}(D^4 - d^4)$$

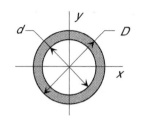

B.4 SOLID SEMICIRCLE

$$A = \frac{\pi D^2}{8}$$

$$I_x = \frac{R^4}{72\pi}(9\pi^2 - 64)$$

$$I_y = \frac{\pi}{8}R^4$$

$$J_z = I_x + I_y = \frac{R^4}{36\pi}(9\pi^2 - 32)$$

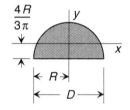

B.5 RIGHT TRIANGLE

$$A = \frac{bh}{2}$$

$$I_x = \frac{bh^3}{36}$$

$$I_y = \frac{b^3 h}{36}$$

$$J_z = I_x + I_y = \frac{bh}{36}(h^2 + b^2)$$

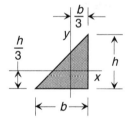

B.6 I BEAM WITH EQUAL FLANGES

$$A = 2bt + wd$$

$$I_x = \frac{bh^3}{12} - \frac{(b-w)(h-2t)^3}{12}$$

$$I_y = \frac{b^3 t}{6} + \frac{w^3(h-2t)}{12}$$

$$J_z = I_x + I_y$$

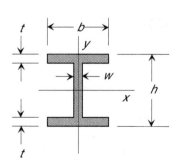

APPENDIX C

MATERIAL PROPERTIES

This appendix contains approximate material property values for use with the problems in this book. Consult the manufacturer of the materials for more accurate and up-to-date information.

TABLE C.1. Physical properties

Material	Young's modulus [Mpsi (GPa)]		Poisson's ratio	Density (lb/in^3)	Density (Mg/m^3)
Aluminum alloys	10.4	(71.7)	0.34	0.10	2.8
Beryllium copper	18.5	(127.6)	0.29	0.30	8.3
Brass, bronze	16.0	(110.3)	0.33	0.31	8.6
Copper	17.5	(120.7)	0.35	0.32	8.9
Magnesium alloys	6.5	(44.8)	0.33	0.07	1.8
Steel, carbon	30.0	(206.8)	0.28	0.28	7.8
Steel, alloys	30.0	(206.8)	0.28	0.28	7.8
Steel, stainless	27.5	(189.6)	0.28	0.28	7.8
Titanium alloys	16.5	(113.8)	0.34	0.16	4.4
Zinc alloys	12.0	(82.7)	0.33	0.24	6.6

Source: Data from [103], [104], [186], and [187].

TABLE C.2. Mechanical properties for some engineering plastics

Property	Polypropylene	HDPE[a]	HMWPE[b]	Nylon 6/6	Acetal[c]
Melting temperature (°C)	160–175	130–137	125–135	255–265	172–184
Tensile strength at break (psi)	4500–6000	3200–4500	2500–4300	13,700	9700–10,000
Elongation at break, (%)	100–600	10–1200	170–800	15–80	10–75
Tensile yield strength	4500–5400	3800–4800	2800–3900	8000–12,000	9500–12,000
Compressive strength	5500–8000	2700–3600	—	12,500–15,000	15,600–18,000 @10%
Flexural strength	6000–8000	—	—	17,900	13,600–16,000
Tensile modulus (kpsi)	165–225	155–158	136	230–550	400–520
Compressive modulus (kpsi)	150–300	—	—	—	670
Flexural modulus (kpsi) At 73° At 200° At 250°	170–250 50 35	145–225	125–175	410–470	380–490 120–135 75–90
Coefficient of thermal expansion (10^{-6} in./in./°F)	81–100	59–110	70–110	80	50–112
Deflection temp. under flexural load (°F) At 264 psi	120–140	—	—	158–212	253–277
At 66 psi	225–250	175–196	154–158	425–474	324–342
Specific gravity	0.9–0.91	0.952–0.965	0.947–0.955	1.13–1.15	1.42

Source: Data from [112].

[a.] High-density polyethylene.
[b.] High-molecular-weight polyethylene.
[c.] Delrin (a registered tradename of E.I. Dupont Co.) is a common acetal.

TABLE C.3. Mechanical properties for some aluminum alloys

Alloy	Tensile yield strength [kpsi (MPa)]		Ultimate tensile strength [kpsi (MPa)]		Elongation over 2 in. (%)	Brinell hardness
1100 O	5	(35)	13	(90)	35	23
2024 O	11	(76)	27	(186)	22	47
2024 T3	50	(345)	70	(482)	16	120
2024 T4	47	(324)	68	(469)	19	120
6061 O	8	(55)	18	(124)	25	30
6061 T6	40	(276)	45	(310)	12	95
7075 O	15	(103)	33	(228)	16	60
7075 T6	73	(503)	83	(572)	11	150

Source: Data from [103], [104], [186], and [187].

TABLE C.4. Mechanical properties for some titanium alloys

	Tensile yield strength [kpsi (MPa)]		Ultimate tensile strength [kpsi (MPa)]		Elongation of 2 in. (%)
Ti-5Al-2.5Sn annealed	113	(779)	120	(827)	10
Ti-8Al-1Mo-1V annealed	135	(931)	145	(1000)	8
Ti-6Al-4V annealed	120	(827)	130	(896)	10
Ti-6Al-4V solution treated and aged	145	(1000)	160	(1103)	8
Ti-13V-11Cr-3Al solution treated	120	(827)	125	(862)	8–10
Ti-13V-11Cr-3Al solution treated and aged	160	(1103)	170	(1172)	4

Source: Data from [188].

TABLE C.5. Mechanical properties for some carbon steels

Steel designation	Tensile yield strength [kpsi (MPa)]		Ultimate tensile strength [kpsi (MPa)]		Elongation over 2 in. (%)	Brinell hardness
1010 HR	26	(180)	47	(320)	28	95
1010 CD	44	(300)	53	(370)	20	105
1020 HR	30	(210)	55	(380)	25	111
1020 CD	51	(350)	68	(470)	15	131
1030 HR	38	(259)	68	(470)	20	137
1030 CD	64	(440)	76	(520)	12	149
1030 Normalized	50	(345)	75	(517)	32	149
1030 Q&T 1200	64	(441)	85	(586)	32	207
1030 Q&T 800	84	(579)	106	(731)	23	302
1030 Q&T 400	94	(648)	123	(848)	17	495
1040 HR	42	(290)	76	(520)	18	149
1040 CD	71	(490)	85	(590)	12	170
1040 Normalized	54	(372)	86	(590)	28	170
1040 Q&T 1200	63	(434)	92	(634)	29	192
1040 Q&T 800	80	(552)	110	(758)	21	241
1040 Q&T 400	86	(593)	113	(779)	19	262
1050 HR	50	(345)	90	(620)	15	179
1050 CD	84	(580)	100	(690)	10	197
1050 Normalized	62	(427)	108	(745)	20	217
1050 Q&T 1200	78	(538)	104	(717)	28	235
1050 Q&T 800	115	(793)	158	(1090)	13	444
1050 Q&T 400	117	(807)	163	(1120)	9	514
1060 HR	54	(370)	98	(680)	12	200
1060 Normalized	61	(421)	112	(772)	18	229

TABLE C.5. (continued)

Steel designation	Tensile yield strength [kpsi (MPa)]		Ultimate tensile strength [kpsi (MPa)]		Elongation over 2 in. (%)	Brinell hardness
1060 Q&T 1200	76	(524)	116	(800)	23	229
1060 Q&T 800	111	(765)	156	(1080)	14	311
1095 HR	66	(460)	120	(830)	10	248
1095 Normalized	72	(500)	147	(1014)	9	293
1095 Q&T 1200	80	(552)	130	(896)	21	269
1095 Q&T 800	112	(772)	176	(1210)	12	363
1095 Q&T 600	118	(814)	183	(1260)	10	375
4130 Q&T 1200	102	(703)	118	(814)	22	245
4130 Q&T 800	173	(1190)	186	(1280)	13	380
4130 Q&T 400	212	(1462)	236	(1627)	10	467

Source: Data from [103], [104], [186], and [187].

TABLE C.6. Mechanical properties for some stainless steels

Stainless steel designation	Tensile yield strength [kpsi (MPa)]		Ultimate tensile strength [kpsi (MPa)]		Elongation over 2 in. (%)
Type 301 annealed	40	(276)	110	(758)	60
Type 301 CR	165	(1138)	200	(1379)	8
Type 410 annealed	45	(310)	70	(483)	25
Type 410 heat treated	140	(965)	180	(1241)	15
AM-350 (AMS 5548)	145	(1000)	180	(1241)	8
17-4 PH (AISI 630) hardened	185	(1276)	200	(1379)	14
17-7 PH (AISI 631) hardened	220	(1517)	235	(1620)	6

Source: Data from [188] and [189].

APPENDIX D

LINEAR ELASTIC BEAM DEFLECTIONS

D.1 CANTILEVER BEAM WITH A FORCE AT THE FREE END

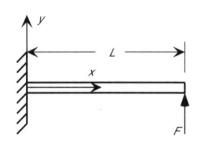

$$y = \frac{Fx^2}{6EI}(3L - x), \quad y_{max} = \frac{FL^3}{3EI} \text{ at } x = L$$

$$\theta = \frac{Fx}{2EI}(2L - x), \quad \theta_{max} = \frac{FL^2}{2EI}$$

$$M_{max} = FL \text{ at } x = 0$$

D.2 CANTILEVER BEAM WITH A FORCE ALONG THE LENGTH

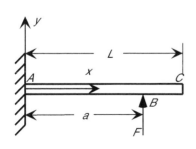

$$y_{AB} = \frac{Fx^2}{6EI}(3a - x), \quad y_{BC} = \frac{Fa^2}{6EI}(3x - a),$$

$$y_{max} = \frac{Fa^2}{6EI}(3L - a) \text{ at } x = L$$

$$\theta_{AB} = \frac{Fx}{2EI}(2a - x), \quad \theta_{max} = \theta_{BC} = \frac{Fa^2}{2EI}$$

$$M_{max} = Fa \text{ at } x = 0$$

D.3 CANTILEVER BEAM WITH A UNIFORMLY DISTRIBUTED LOAD

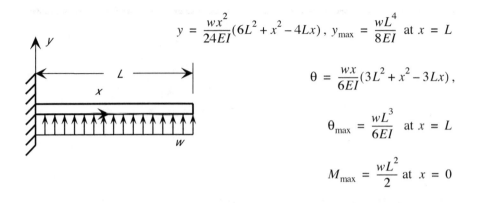

$$y = \frac{wx^2}{24EI}(6L^2 + x^2 - 4Lx), \quad y_{max} = \frac{wL^4}{8EI} \text{ at } x = L$$

$$\theta = \frac{wx}{6EI}(3L^2 + x^2 - 3Lx),$$

$$\theta_{max} = \frac{wL^3}{6EI} \text{ at } x = L$$

$$M_{max} = \frac{wL^2}{2} \text{ at } x = 0$$

D.4 CANTILEVER BEAM WITH A MOMENT AT THE FREE END

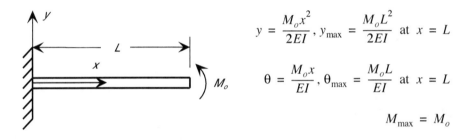

$$y = \frac{M_o x^2}{2EI}, \quad y_{max} = \frac{M_o L^2}{2EI} \text{ at } x = L$$

$$\theta = \frac{M_o x}{EI}, \quad \theta_{max} = \frac{M_o L}{EI} \text{ at } x = L$$

$$M_{max} = M_o$$

D.5 SIMPLY SUPPORTED BEAM WITH A FORCE AT THE CENTER

$$y_{AB} = \frac{Fx}{48EI}(3L^2 - 4x^2), \quad y_{max} = \frac{FL^3}{48EI} \text{ at } x = \frac{L}{2}$$

$$\theta_{AB} = \frac{F}{16EI}(L^2 - 4x^2),$$

$$\theta_{max} = \frac{FL^2}{16EI} \text{ at } x = 0$$

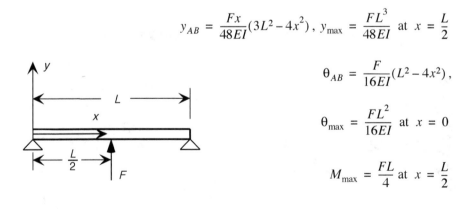

$$M_{max} = \frac{FL}{4} \text{ at } x = \frac{L}{2}$$

Linear Elastic Beam Deflections

D.6 SIMPLY SUPPORTED BEAM WITH A FORCE ALONG THE LENGTH

$$a < b, \quad y_{AB} = \frac{Fbx}{6EIL}(L^2 - x^2 - b^2), \quad y_{BC} = \frac{Fa(L-x)}{6EIL}(2Lx - x^2 - a^2)$$

$$\theta_{AB} = \frac{Fb}{6EIL}(L^2 - 3x^2 - b^2),$$

$$\theta_{BC} = \frac{Fa}{6EIL}(3x^2 + a^2 + 2L^2 - 6Lx)$$

$$M_{AB} = \frac{Fbx}{L}, \quad M_{BC} = \frac{Fa}{L}(L-x)$$

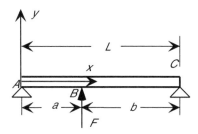

D.7 SIMPLY SUPPORTED BEAM WITH A UNIFORMLY DISTRIBUTED LOAD

$$y = \frac{wx}{24EI}(L^3 - 2Lx^2 + x^3), \quad y_{max} = \frac{5wL^4}{384EI} \text{ at } x = \frac{L}{2}$$

$$\theta = \frac{w}{24EI}(L^3 - 6Lx^2 + 4x^3),$$

$$\theta_{max} = \frac{wL^3}{24EI} \text{ at } x = 0$$

$$M_{max} = \frac{wL^2}{8} \text{ at } x = \frac{L}{2}$$

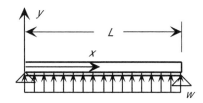

D.8 BEAM WITH ONE END FIXED AND THE OTHER END SIMPLY SUPPORTED

$$y_{AB} = \frac{Fbx^2}{12EIL^3}[3L(L^2 - b^2) + x(b^2 - 3L^2)], \quad y_{BC} = y_{AB} + \frac{F(x-a)^3}{6EI}$$

$$\theta_{AB} = \frac{Fbx}{12EIL^3}[3x(b^2 - 3L^2) + 6L(L^2 - b^2)],$$

$$\theta_{BC} = \theta_{AD} + \frac{F(x-a)^2}{2EI}$$

$$M_{AB} = \frac{Fb}{2L^3}[L^3 - b^2L + x(b^2 - 3l^2)], \quad M_{BC} = \frac{Fa^2}{2L^3}(3Lx - 3L^2 - ax + aL)$$

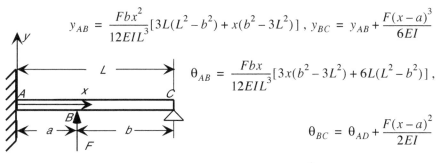

D.9 BEAM WITH FIXED ENDS AND A CENTER LOAD

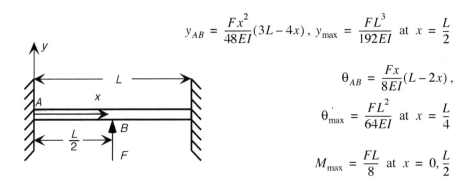

$$y_{AB} = \frac{Fx^2}{48EI}(3L - 4x), \quad y_{max} = \frac{FL^3}{192EI} \text{ at } x = \frac{L}{2}$$

$$\theta_{AB} = \frac{Fx}{8EI}(L - 2x),$$

$$\theta_{max} = \frac{FL^2}{64EI} \text{ at } x = \frac{L}{4}$$

$$M_{max} = \frac{FL}{8} \text{ at } x = 0, \frac{L}{2}$$

D.10 BEAM WITH FIXED ENDS AND A UNIFORMLY DISTRIBUTED LOAD

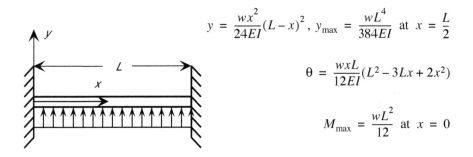

$$y = \frac{wx^2}{24EI}(L - x)^2, \quad y_{max} = \frac{wL^4}{384EI} \text{ at } x = \frac{L}{2}$$

$$\theta = \frac{wxL}{12EI}(L^2 - 3Lx + 2x^2)$$

$$M_{max} = \frac{wL^2}{12} \text{ at } x = 0$$

D.11 BEAM WITH ONE END FIXED AND THE OTHER END GUIDED

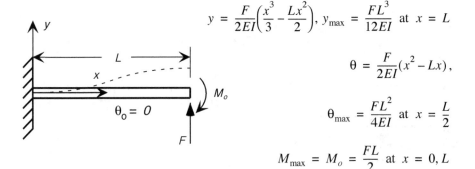

$$y = \frac{F}{2EI}\left(\frac{x^3}{3} - \frac{Lx^2}{2}\right), \quad y_{max} = \frac{FL^3}{12EI} \text{ at } x = L$$

$$\theta = \frac{F}{2EI}(x^2 - Lx),$$

$$\theta_{max} = \frac{FL^2}{4EI} \text{ at } x = \frac{L}{2}$$

$$M_{max} = M_o = \frac{FL}{2} \text{ at } x = 0, L$$

APPENDIX E

PSEUDO-RIGID-BODY MODELS

The pseudo-rigid-body model is used to predict the deflection of large-deflection beams. It is assumed that the flexible part of the beams have a constant cross section, are rigid in shear, have homogeneous material properties, and operate in the elastic range.

E.1 SMALL-LENGTH FLEXURAL PIVOT

Description: a flexible segment that is small in length compared to the rigid segments to which it is attached [i.e., $l \ll L$ and $(EI)_l \ll (EI)_L$]. The characteristic pivot is located at the center of the flexible beam.

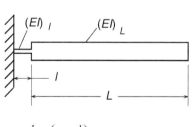
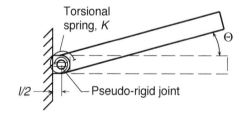

$$a = \frac{l}{2} + \left(L + \frac{l}{2}\right)\cos\Theta$$

$$b = \left(L + \frac{l}{2}\right)\sin\Theta$$

$$\theta_o = \frac{Ml}{EI}$$

$$K = \frac{EI}{l}$$

$$\sigma_{max} = \begin{cases} \dfrac{M_o c}{I} & \text{(loaded with an end moment, } M_o\text{)} \\ \dfrac{Pac}{I} & \text{(loaded with a vertical force at the free end, } P\text{)} \\ \pm\dfrac{P(a+nb)c}{I} - \dfrac{nP}{A} & \text{(for vertical force, } P\text{, and horizontal force, } nP\text{)} \end{cases}$$

where the maximum stress occurs at the fixed end and c is the distance from the neutral axis to the outer surface of the beam (i.e., half the beam height for rectangular beams, the radius of circular cross-section beams, etc.)

E.2 VERTICAL FORCE AT THE FREE END OF A CANTILEVER BEAM

Description: a special case of the model of Section E.3 that applies to a cantilever beam with a vertical force at the free end ($n = 0$).

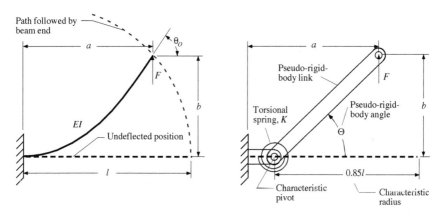

$a = l[1 - 0.85(1 - \cos \Theta)]$ $\qquad b = 0.85l \sin \Theta$

$\Theta < 64.3°$ for accurate position prediction

$\theta_o = 1.24\Theta$

$K = 2.25\dfrac{EI}{l}$

$\Theta < 58.5°$ $\qquad$ for accurate force prediction

$\sigma_{max} = \dfrac{Pac}{I}$ $\qquad$ at the fixed end

where c is the distance from the neutral axis to the outer surface of the beam (i.e., half the beam height for rectangular beams, the radius of circular cross-section beams, etc.)

E.3 CANTILEVER BEAM WITH A FORCE AT THE FREE END

Description: a beam for which the angle of the force is described by the ratio of the horizontal to vertical components, n. In a compliant mechanism, this represents a flexible beam with a pin joint at one end.

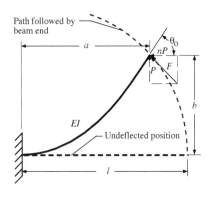
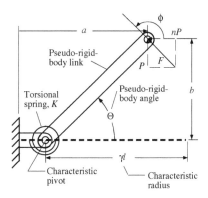

$a = l[1 - \gamma(1 - \cos\Theta)]$ $\qquad b = \gamma l \sin\Theta$

$\Theta < \Theta_{max}(\gamma)$ $\qquad$ for accurate position prediction

$\theta_o = c_\theta \Theta$

$K = \gamma K_\Theta \dfrac{EI}{l}$

$\Theta_{max} < \Theta_{max}(K_\Theta)$ $\qquad$ for accurate force prediction

$\phi = \operatorname{atan}\dfrac{1}{-n}$

$\gamma = \begin{cases} 0.841655 - 0.0067807n + 0.000438n^2 & (0.5 < n < 10.0) \\ 0.852144 - 0.0182867n & (-1.8316 < n < 0.5) \\ 0.912364 + 0.0145928n & (-5 < n < -1.8316) \end{cases}$

$K_\Theta = \begin{cases} 3.024112 + 0.121290n + 0.003169n^2 & (-5 < n \le -2.5) \\ 1.967647 - 2.616021n - 3.738166n^2 - 2.649437n^3 - 0.891906n^4 \\ \qquad -0.113063n^5 & (-2.5 < n \le -1) \\ 2.654855 - 0.509896 \times 10^{-1}n + 0.126749 \times 10^{-1}n^2 \\ \qquad - 0.142039 \times 10^{-2}n^3 + 0.584525 \times 10^{-4}n^4 & (-1 < n \le 10) \end{cases}$

Or, for a quick approximation: $K_\Theta \approx \pi\gamma$. Numerical values for γ and K_Θ are also listed in Table E.1.

$$\sigma_{max} = \pm \frac{P(a+nb)c}{I} - \frac{nP}{A} \quad \text{at the fixed end}$$

where c is the distance from the neutral axis to the outer surface of the beam (i.e., half the beam height for rectangular beams, the radius of circular cross-section beams, etc.)

TABLE E.1. Numerical data for γ, c_θ, and K_Θ for various angles of force

n	ϕ	γ	$\Theta_{max}(\gamma)$	c_θ	K_Θ	$\Theta_{max}(K_\Theta)$
0.0	90.0	0.8517	64.3	1.2385	2.67617	58.5
0.5	116.6	0.8430	81.8	1.2430	2.63744	64.1
1.0	135.0	0.8360	94.8	1.2467	2.61259	67.5
1.5	146.3	0.8311	103.8	1.2492	2.59289	65.8
2.0	153.4	0.8276	108.9	1.2511	2.59707	69.0
3.0	161.6	0.8232	115.4	1.2534	2.56737	64.6
4.0	166.0	0.8207	119.1	1.2548	2.56506	66.4
5.0	168.7	0.8192	121.4	1.2557	2.56251	67.5
7.5	172.4	0.8168	124.5	1.2570	2.55984	69.0
10.0	174.3	0.8156	126.1	1.2578	2.56597	69.7
-0.5	63.4	0.8612	47.7	1.2348	2.69320	44.4
-1.0	45.0	0.8707	36.3	1.2323	2.72816	31.5
-1.5	33.7	0.8796	28.7	1.2322	2.78081	23.6
-2.0	26.6	0.8813	23.2	1.2293	2.80162	18.6
-3.0	18.4	0.8669	16.0	1.2119	2.68893	12.9
-4.0	14.0	0.8522	11.9	1.1971	2.58991	9.8
-5.0	11.3	0.8391	9.7	1.1788	2.49874	7.9

E.4 FIXED–GUIDED BEAM

Description: a beam that is fixed at one end; the other end goes through a deflection such that the angular deflection at the end remains constant, and the beam shape is antisymmetric about the center. This type of beam occurs in parallel motion mechanisms. The moment, M_o, is a reaction force required to maintain the constant beam end angle.

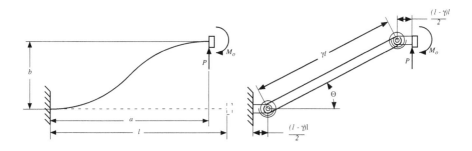

$a = l[1 - \gamma(1 - \cos\Theta)]$ $b = \gamma l \sin\Theta$

$\Theta < \Theta_{max}(\gamma)$ for accurate position prediction

$\theta_o = 0$

$K = 2\gamma K_\Theta \dfrac{EI}{l}$

$\Theta_{max} < \Theta_{max}(K_\Theta)$ for accurate force prediction

See Section E.3 for values of γ, K_Θ, $\Theta_{max}(\gamma)$, and $\Theta_{max}(K_\Theta)$

$\sigma_{max} = \dfrac{Pac}{2I}$ at both ends of the beam

where c is the distance from the neutral axis to the outer surface of the beam (i.e., half the beam height for rectangular beams, the radius of circular cross-section beams, etc.)

E.5 CANTILEVER BEAM WITH AN APPLIED MOMENT AT THE FREE END

Description: a flexible cantilever beam that is loaded with a moment at the free end.

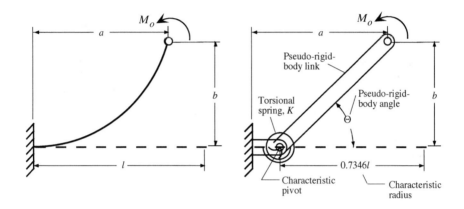

$a = l[1 - 0.7346(1 - \cos \Theta)]$ $b = 0.7346 l \sin \Theta$

$\theta_o = 1.5164 \Theta$

$K = 1.5164 \dfrac{EI}{l}$

$\sigma_{max} = \dfrac{M_o c}{I}$

where c is the distance from the neutral axis to the outer surface of the beam (i.e., half the beam height for rectangular beams, the radius of circular cross-section beams, etc.)

E.6 INITIALLY CURVED CANTILEVER BEAM

Description: a cantilever beam with an undeflected shape that has a constant radius of curvature, and a force at the free end.

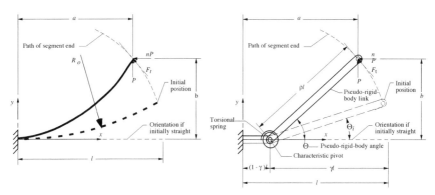

$$\kappa_o = \frac{l}{R_i}$$

$$\Theta_i = \operatorname{atan} \frac{b_i}{a_i - l(1 - \gamma)}$$

$$\rho = \left[\left(\frac{a_i}{l} - (1 - \gamma) \right)^2 + \left(\frac{b_i}{l} \right)^2 \right]^{1/2}$$

$$\frac{a_i}{l} = \frac{1}{\kappa_o} \sin \kappa_o \qquad \frac{b_i}{l} = \frac{1}{\kappa_o}(1 - \cos \kappa_o)$$

$$\frac{a}{l} = 1 - \gamma + \rho \cos \Theta \qquad \frac{b}{l} = \rho \sin \Theta$$

$$K = \rho K_\Theta \frac{EI}{l}$$

$$\sigma_{max} = \pm \frac{P(a + nb)c}{I} - \frac{nP}{A} \qquad \text{at the fixed end}$$

where c is the distance from the neutral axis to the outer surface of the beam (i.e., half the beam height for rectangular beams, the radius of circular cross-section beams, etc.)

Table E.2 lists values for γ, ρ, and K_Θ for various values of κ_o.

Pseudo-Rigid-Body Models

TABLE E.2. Values for γ, ρ, and K_Θ for various values of κ_o

κ_o	γ	ρ	K_Θ
0.00	0.85	0.850	2.65
0.10	0.84	0.840	2.64
0.25	0.83	0.829	2.56
0.50	0.81	0.807	2.52
1.00	0.81	0.797	2.60
1.50	0.80	0.775	2.80
2.00	0.79	0.749	2.99

E.7 PINNED–PINNED SEGMENTS

Description: flexible segments that have forces at the ends only and no applied moments. These segments can be modeled as a spring pinned at both ends. The spring constant depends on the specific geometry and material properties used. Section E.7.1 provides a model for a common type of pinned–pinned segment.

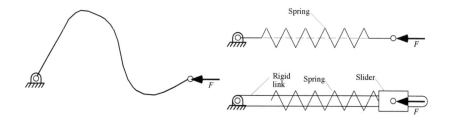

E.7.1 Initially Curved Pinned–Pinned Segments

Description: an initially curved beam with an undeflected shape that has a constant radius of curvature, and both ends are pinned.

Pseudo-Rigid-Body Models

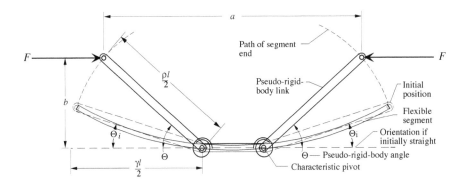

Initial coordinates:

$$\frac{a_i}{l} = \frac{2}{\kappa_o}\sin\frac{\kappa_o}{2} \qquad \frac{b_i}{l} = \frac{1}{\kappa_o}\left(1 - \cos\frac{\kappa_o}{2}\right)$$

$$\Theta_i = \operatorname{atan}\frac{2b_i}{a_i - l(1-\gamma)}$$

$$a = l(1 - \gamma + \rho\cos\Theta) \qquad b = \frac{l}{2}\rho\sin\Theta$$

$$K = 2\rho K_\Theta \frac{EI}{l}$$

$$\rho = \left[\left(\frac{a_i}{l} - (1-\gamma)\right)^2 + \left(\frac{2b_i}{l}\right)^2\right]^{1/2}$$

$$\gamma = \begin{cases} 0.8005 - 0.0173\kappa_0 & 0.500 \le \kappa_0 \le 0.595 \\ 0.8063 - 0.0265\kappa_0 & 0.595 \le \kappa_0 \le 1.500 \end{cases}$$

$$K_\Theta = 2.568 - 0.028\kappa_0 + 0.137\kappa_0^2 \quad \text{for} \quad 0.5 \le \kappa_0 \le 1.5$$

Table E.3 lists numerical values for γ, K_Θ, and $\Delta\Theta_{max}$ for each for various values of κ_o.

$$\sigma_{max} = \pm\frac{Fbc}{I} - \frac{F}{A} \qquad \text{at midlength of segment}$$

where c is the distance from the neutral axis to the outer surface of the beam (i.e., half the beam height for rectangular beams, the radius of circular cross-section beams, etc.)

TABLE E.3. Pseudo-rigid-body link characteristics for initially curved pinned–pinned segment

κ_o	γ	ρ	$\Delta\Theta_{max}(\gamma)$	K_Θ	$\Delta\Theta_{max}(K_\Theta)$
0.50	0.793	0.791	1.677	2.59	0.99
0.75	0.787	0.783	1.456	2.62	0.86
1.00	0.783	0.775	1.327	2.68	0.79
1.25	0.779	0.768	1.203	2.75	0.71
1.50	0.775	0.760	1.070	2.83	0.63

E.8 COMBINED FORCE–MOMENT END LOADING

Description: an initially straight flexible segment with a force and moment at the end, such as occurs when both ends are fixed to rigid segments that can move relative to each other. This approximation is much less accurate than the other pseudo-rigid-body models discussed above, but it is presented here as a starting point for problems with flexible segments that have this type of loading condition.

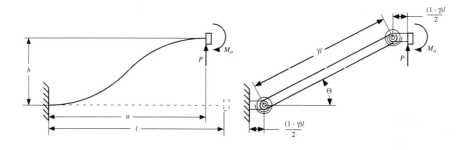

$$a = l[1 - \gamma(1 - \cos\Theta)] \qquad b = \gamma l \sin\Theta$$

$$K = 2\gamma K_\Theta \frac{EI}{l}$$

See Table E.3 for values of γ and K_Θ.

APPENDIX F

EVALUATION OF ELLIPTIC INTEGRALS

Several methods exist which calculate the numerical values of elliptic integrals. The series expansion is common but very inefficient for large values of the amplitude of the elliptic integral. King [98] describes several alternative techniques, one of which is Landen's scale of increasing amplitudes. This method requires the use of Gauss's scale of arithmeticogeometrical means, which proceeds as follows:

$$a_1 = \frac{1}{2}(a_0 + b_0) \qquad b_1 = \sqrt{a_0 b_0} \qquad c_1 = \frac{1}{2}(a_0 - b_0)$$
$$a_2 = \frac{1}{2}(a_1 + b_1) \qquad b_2 = \sqrt{a_1 b_1} \qquad c_2 = \frac{1}{2}(a_1 - b_1)$$
$$\ldots\ldots\ldots \qquad \ldots\ldots\ldots \qquad \ldots\ldots\ldots \qquad \text{(F.1)}$$
$$a_n = \frac{1}{2}(a_{n-1} + b_{n-1}) \qquad b_n = \sqrt{a_{n-1} b_{n-1}} \qquad c_n = \frac{1}{2}(a_{n-1} - b_{n-1})$$
$$\ldots\ldots\ldots \qquad \ldots\ldots\ldots \qquad \ldots\ldots\ldots$$

The a's and b's tend to the same limit, while c tends to zero:

$$\lim_{n \to \infty} a_n = \lim_{n \to \infty} b_n = \alpha \qquad \text{(F.2)}$$

$$\lim_{n \to \infty} c_n = 0 \qquad \text{(F.3)}$$

Landen's scale of increasing amplitudes requires

$$a_o = 1 \tag{F.4}$$

$$b_o = q' = \sqrt{1-q^2} \tag{F.5}$$

$$c_o = q \tag{F.6}$$

where q is the modulus of the elliptic integral. The algorithm is based on the recurrence formula

$$\tan(\beta_{n+1} - \beta_n) = \frac{b_n}{a_n} \tan \beta_n \tag{F.7}$$

where β_0 is the amplitude of the elliptic integral. As n increases, a_n and b_n go to the same limit, while $\beta_n/2^n$ goes to another finite limit. The algorithm continues until c_n is sufficiently close to zero. The values of the complete and incomplete integral of the first kind may be calculated from

$$F\left(\frac{\pi}{2}, q\right) = \frac{\pi}{2a_n} \tag{F.8}$$

$$F(\beta, q) = \frac{1}{a_n} \frac{\beta_n}{2^n} \tag{F.9}$$

The numerical values for the complete and incomplete elliptic integrals of the second kind are

$$E\left(\frac{\pi}{2}, q\right) = F\left(\frac{\pi}{2}, q\right)\left[1 - \frac{1}{2}(c_0^2 + 2c_1^2 + 4c_2^2 + \cdots + 2^n c_n^2 + \cdots)\right] \tag{F.10}$$

$$E(\beta, q) = F(\beta, q) + (c_1 \sin \beta_1 + c_2 \sin \beta_2 + \cdots + c_n \sin \beta_n + \cdots)$$
$$- \frac{1}{2}(c_0^2 + 2c_1^2 + 4c_2^2 + \cdots + 2^n c_n^2 + \cdots)F(\beta, q) \tag{F.11}$$

A problem arises in that there are an infinite number of solutions for β_{n+1} in the recurrence formula (F.7):

$$\tan(\beta_{n+1} - \beta_n) = \tan[(\zeta + j\pi) - \beta_n] \quad j = 0, 1, \ldots, \infty \tag{F.12}$$

Mattiasson [95] mentions that $\beta_{n+1} > \beta_n$, but this condition is not enough to ensure that the proper β_{n+1} has been selected. The problem is readily solved by realizing that the quantity $\beta_n/(a_n 2^n)$ increases as n increases. Should this value ever decrease, an incorrect β was calculated. This can be corrected by using the following formula:

Evaluation of Elliptic Integrals

$$m = \text{rounding of } \frac{2\beta_n - \beta_{n+1}}{\pi} \text{ to nearest whole number} \quad \text{(F.13)}$$

$$(\beta_{n+1})_{\text{corrected}} = \beta_{n+1} + m\pi \quad \text{(F.14)}$$

A stopping criteria of $c_n \leq 10^{-10}$ is usually adequate. Values for $E(\beta, q)$ and $F(\beta, q)$ out of the range of $0 \leq \beta \leq \pi/2$ may be found by the following relationships [97]:

$$E(-\beta, q) = -E(\beta, q) \quad \text{(F.15)}$$

$$F(-\beta, q) = -F(\beta, q) \quad \text{(F.16)}$$

$$E(m\pi \pm \beta, q) = 2mE\left(\frac{\pi}{2}, q\right) \pm E(\beta, q) \quad \text{(F.17)}$$

$$F(m\pi \pm \beta, q) = 2mF\left(\frac{\pi}{2}, q\right) \pm F(\beta, q) \quad \text{(F.18)}$$

When $E(\beta, q)$ and $F(\beta, q)$ are out of the range $0 \leq q \leq 1$, the following formulas are useful:

$$E(\beta, q) = q_1[q^2 E(\beta_1, q_1) + q^2 F(\beta_1, q_1)] \quad \text{(F.19)}$$

$$F(\beta, q) = q_1 F(\beta_1, q_1) \quad \text{(F.20)}$$

where

$$q_1 = \frac{1}{q} \quad \text{(F.21)}$$

$$\sin\beta_1 = q \sin\beta \quad \text{(F.22)}$$

Additional information on elliptic integrals may be found in [99] and [100].

APPENDIX G

TYPE SYNTHESIS OF COMPLIANT MECHANISMS

Morgan D. Murphy
Delphi Automotive Systems

As the challenges of compliant mechanism analysis are addressed, and more resilient materials are developed, the field of compliant mechanism design is expected to grow rapidly. To expedite the application of compliant mechanisms, a systematic design process for compliant mechanisms has been developed that employs type synthesis techniques. The first step of the design process is the formulation of a mathematical model to represent the structure of a mechanism. In rigid-body kinematics, graph theory provides a mathematically rigorous representation of a mechanism structure through the use of matrices. The matrix representation for compliant mechanisms builds on the foundation established in rigid-body kinematics by adding information regarding segment type and the connectivity between segments to the matrices that represent a mechanism's topology.

G.1 MATRIX REPRESENTATION FOR RIGID-LINK MECHANISMS

To provide a basis for a discussion of the mathematical model for compliant mechanisms, a description of rigid-body models is provided. Freudenstein and Dobrjanskyj [190] referred to an assembly of interconnected links when closed, connected, and constrained as a basic kinematic chain (BKC). A mechanism is formed when one link of the basic kinematic chain is fixed (ground link). Based on graph theory, Freudenstein and Dobrjanskyj [190] developed a mathematical representation for basic kinematic chains. In a kinematic graph, vertices (circles) represent links and joints are represented by edges (lines). One advantage of a graph representation is that the kinematic chain can be described in matrix form. The number of links determines the size (order) of the matrix. For example, an 8×8 matrix will repre-

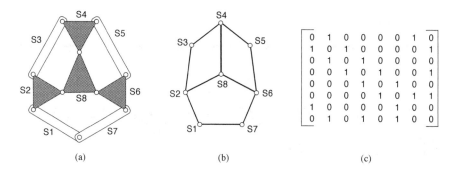

Figure G.1. (a) Basic kinematic chain, (b) its kinematic graph, and (c) its vertex–vertex adjacency matrix.

sent an eight-link mechanism and a 4×4 matrix will represent a four-link mechanism. If a kinematic chain contains n vertices (links), the vertex–vertex adjacency matrix is an nth-order matrix whose element $a(i, j)$ is equal to 1 if the ith vertex (link) is connected to the jth vertex (link); otherwise, it is equal to zero. It should be noted that the diagonal elements are all equal to zero, $a(i, i) = 0$. Figure G.1 shows an eight-link, 10-pair (joint) kinematic chain (Figure G.1a), its kinematic graph (Figure G.1b), and the corresponding vertex–vertex adjacency matrix (Figure G.1c).

As noted earlier, a mechanism is formed when one of the links of a basic kinematic chain is fixed. It is standard notation that each mechanism has only one ground link. To complete the matrix representation for a mechanism, an indication of which link has been fixed (grounded), to form the mechanism, is required. To provide this indication, the vertex–vertex adjacency matrix is modified by setting the diagonal element that represents the ground link equal to -1, or $a(i, i) = -1$. This notation is useful in the development of a matrix representation for compliant mechanisms described in subsequent sections of this appendix. Figure G.2 shows two rigid-link mechanisms and their corresponding rigid-link mechanism matrices. As can be seen from the examples in Figure G.2, the matrix notation provides a compact representation of a mechanism's structure that is fairly straightforward. Other than the element representing the ground link, the diagonal element diagonal elements are zero. Because this is a simple modification of the vertex–vertex adjacency matrix, if two links are connected, the corresponding rigid-link matrix elements are equal to 1, or $a(i, j) = a(j, i) = 1$; if they are not connected, then $a(i, j) = a(j, i) = 0$.

G.2 COMPLIANT MECHANISM MATRICES

As described in Chapter 1, the elements that make up a compliant mechanism are designated as segments. In the simplest terms, these segments can be compliant or

Compliant Mechanism Matrices

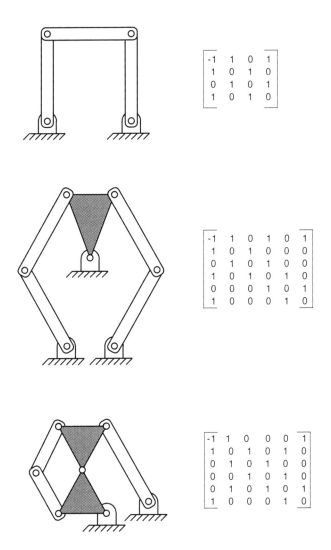

Figure G.2. Rigid-link mechanisms and their corresponding rigid-link mechanism matrices.

rigid (including ground) segments. Moreover, these elements can be attached to each other using flexural pivots and fixed connections, in addition to kinematic pairs (turning pairs and slider joints). By modifying the mathematical formulation used in rigid-body kinematics to include compliance information and dividing the elements of a mechanism at segments, a compliant mechanism matrix [CM] is developed. A new segment is defined at a material or geometric discontinuity, a shift in loading pattern (or application location of a point load), or a change in function. It should be noted that a link might be made up of several segments.

TABLE G.1. Segment-type indication

Segment type	Diagonal element value, c(i,i)
Ground segment	-1
Rigid segment	0
Compliant segment	1

G.2.1 Segment-Type Designation

The first step in the development of a compliant mechanism matrix notation is to provide an indication of segment type (rigid, ground, or compliant). In keeping with the notation established for rigid-body kinematics, the diagonal element corresponding to a rigid segment will be equal to zero, $c(i, i) = 0$, unless it is the ground segment, in which case it is equal to -1, $c(i, i) = -1$. By its nature, the ground segment does not deform and must be rigid. Therefore, a convention of compliant element notation is that only a rigid segment can be denoted as the ground segment. As a result, every compliant mechanism will have at minimum one rigid segment (the ground segment). Compliant segments are denoted by a 1, $c(i, i) = 1$, at their corresponding matrix element. Table G.1 provides a summary of the segment types and their corresponding matrix element value.

G.2.2 Connection-Type Designation

There are three basic types of connections that can be made between the segments in a compliant mechanism: (1) kinematic pairs (or joints), (2) flexural pivots, and (3) clamped or fixed connections. It is important to note, however, that a clamped connection between two rigid segments is not allowed because it simply creates a longer rigid segment. To convey information regarding the type of connection between segments, the nondiagonal elements of the compliant mechanism matrix are used. If segment i is not connected to segment j, its element $c(i, j) = 0$. If segment i is connected to segment j with a kinematic pair (or joint), $c(i, j) = 1$. If a flexural pivot connects segment i to segment j, $c(i, j) = 2$. Finally, if segment i is clamped to segment j, $c(i, j) = 3$. Table G.2 provides a summary of the connection types and their corresponding matrix element value. One consequence of this notation is that two segments can only be joined at one location. If two segments are connected at more than one location, one of the original segments (a compliant segment) can be divided into two separate segments, joined at a fixed (clamped) connection. This will allow the topology of the mechanism to be described, while

TABLE G.2. Connection-type indication

Connection type	Matrix element value, c(i,j)
None	0
Kinematic pair	1
Flexural pivot	2
Clamped connection	3

maintaining the integrity of the mathematical formulation. If rigid segments are connected to each other at more than one location, there is no relative motion and they can be considered as one segment.

G.2.3 Examples

Figure G.3 depicts several mechanisms and their corresponding compliant mechanism matrices [CM]. It is interesting to note that a rigid-link mechanism (Figure G.3b) has the same values for its compliant mechanism matrix as it would for its rigid-link mechanism matrix.

G.3 DETERMINATION OF ISOMORPHIC MECHANISMS

Two mechanisms are isomorphic if there is a one-to-one correspondence in their topology without regard to labeling. In other words, if two mechanisms have the same structure but are labeled differently, they are considered isomorphic. Many calculations may be required to determine that two mechanisms are not isomorphic. For example, the operation of permuting the rows and columns of the compliant mechanism matrix will require $n!$ permutations, where n is the number of segments. Therefore, a computationally efficient means to determine if mechanisms are isomorphic has been developed for compliant mechanisms. As before, the technique presented here is based on developments in rigid-body kinematics. It should be noted that if there are only few mechanisms under consideration, simply looking at the mechanisms' topologies may be enough to determine if they are isomorphic. The need for the systematic process comes from the large number of variations in topology that the addition of compliance creates. For example, there are 135 nonisomorphic four-segment compliant chains that must be determined from the 1215 possible combinations of segments and connections. This is compared to the single topology for a rigid-body four-bar chain.

Type Synthesis of Compliant Mechanisms

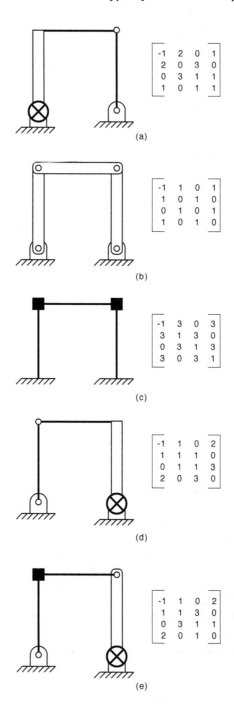

Figure G.3. Compliant mechanisms and their corresponding compliant mechanism matrices.

G.3.1 Rigid-Body Isomorphic Detection Techniques

The application of algebraic graph theory can be used to facilitate the process of determining if two mechanisms are isomorphic. Uicker and Raicu [191] introduced the idea of the characteristic polynomial for the identification of isomorphic kinematic chains. The characteristic polynomial (CP) is calculated as

$$\text{CP} = |xI - D| \tag{G.1}$$

where x is a variable, D is the distance (or vertex–vertex adjacency) matrix, and I is an identity matrix of order n. Basically, we are evaluating the determinant of a "new" matrix created by subtracting the matrix representing the mechanism topology from the identity matrix with a variable (x) at each diagonal element. If two kinematic chains are isomorphic, they will have the same characteristic polynomial [191].

Using the concept of characteristic polynomials, Yan and Hwang [192] developed a systematic method to determine if two kinematic chains are isomorphic. As a first step, the possibility for isomorphisms is investigated by examining the number and type of links. By performing simple evaluations first, the total number of calculations can be reduced. After this initial check, the characteristic polynomials of several different matrices that represent the topology of a kinematic chain are investigated sequentially.

G.3.2 Isomorphism Detection for Compliant Mechanisms

A systematic approach to determine if compliant mechanisms are isomorphic has been developed that employs the concepts developed for rigid-body mechanisms. The approach employs simple evaluations to group the mechanisms. Next, the characteristic polynomials of the compliant mechanism matrices are evaluated.

The first step in the evaluation to determine if mechanisms are isomorphic is to evaluate the number of segments in the mechanisms and group the mechanisms by the number of segments. Quite simply, two mechanisms are not isomorphic if they contain a different number of segments. The next step is to evaluate the number of each type of segment. Since the diagonal elements of the compliant mechanism matrix [CM] provide information regarding the compliance content, the trace (sum of the diagonal element values) of the matrix can be used to evaluate the number of compliant segments. The number of compliant segments can be determined by

$$\text{number of compliant segments} = \text{trace}[\text{CM}] + 1 \tag{G.2}$$

If the total number of segments is known, the trace of [CM] can be used to determine the number of compliant and rigid segments. Table G.3 provides an eval-

TABLE G.3. Trace[CM] for the mechanisms in Figure G.3

Figure	Trace[CM]
G.3a	1
G.3b	-1
G.3c	2
G.3d	1
G.3e	1

uation of the trace of the compliant mechanism matrices for the mechanisms depicted in Figure G.3.

After the mechanisms are grouped by the total number of segments and by the number of compliant segments (trace[CM] + 1), the next evaluation is to determine the number of each connection type between segments. This evaluation is made by constructing a vector where the first element represents the number of kinematic pairs, the second element the number of flexural pivots, and the third element the number of fixed connections. The resulting vector is called the connection-type vector (CTV). If two compliant mechanisms are isomorphic, they will have the same connection-type vector. The connection-type vectors for the mechanisms in Figure G.3 are provided in Table G.4.

The final evaluation presented here is the use of the characteristic polynomial of the compliant mechanism matrices for the mechanisms under investigation. The characteristic polynomial (CP) is calculated as

$$CP = |xI - CM| \qquad (G.3)$$

where x is a variable, CM is the compliant mechanism matrix, and I is an identity matrix of order n. If two kinematic chains are isomorphic, they will have the same characteristic polynomial. Table G.5 provides the characteristic polynomials for the mechanisms presented in Figure G.3. From the sequential evaluations presented

TABLE G.4. Connection-type vectors

Figure	Connection-type vectors
G.3a	(2, 1, 1)
G.3b	(4, 0, 0)
G.3c	(0, 0, 4)
G.3d	(2, 1, 1)
G.3e	(2, 1, 1)

TABLE G.5. Connection-type vectors for the mechanisms in Figure G.3

Figure	Characteristic polynomial
G.3a	$x^4 - x^3 - x^2 + x - 1$
G.3b	$x^4 + x^3$
G.3c	$x^4 - 2x^3 + 2x^2 + 1$
G.3d	$x^4 - x^3 - x^2 + x - 1$
G.3e	$x^4 - x^3 - x^2 + x - 25$

here it can be seen that the mechanisms in Figures G.3d and a are isomorphic, whereas all the other mechanisms are unique.

If two mechanisms are isomorphic, they will have the same evaluations for the number of segments, trace[CM], connection-type vector, and characteristic polynomial. Although these evaluations will be sufficient to determine whether mechanisms are isomorphic, in most cases that will be encountered, Murphy [62] identified some cases in which nonisomorphic mechanisms were not separated using only the techniques presented. Therefore, additional evaluations have been developed that can be used to separate these mechanisms. Those techniques are beyond the scope of this book, and the reader is referred to Murphy [62] for further description. Overall, the simple evaluations presented here provide an effective way to distinguish nonisomorphic mechanisms for most cases.

By modifying the terminology employed in rigid-body kinematics, a matrix representation for compliant mechanisms has been developed. The compliant mechanism [CM] matrix provides information regarding segment type and the connectivity between segments in a compact manner. Moreover, standard linear algebra matrix evaluations can be performed to determine if two mechanisms are isomorphic. As a further benefit, the mathematically rigorous formulation lends itself to automated manipulation that will be employed in the type synthesis process to generate "new" mechanism designs. This process is discussed next.

G.4 TYPE SYNTHESIS

The development of new products is largely based on the intuition and experience of the designer. Often, creative leaps in design and technology are attributed to an innate ability of a fortunate few. However, further analysis shows that creativity can be learned and that systematic approaches can help facilitate the process. Many kinematicians have used type synthesis procedures to develop unique mechanisms or check the uniqueness of a new design. For example, Yan [193] developed three unique six-bar, rear-wheel suspension designs for off-road motorcycles, with a design methodology that employs type synthesis techniques. Erdman et al. [194] enumerated six-link, single-degree-of-freedom mechanisms in the pursuit of novel casement window mechanisms. Kota [195] reported that James Watt used systematic enumeration to scrutinize patents on parts of the steam engine. These are only a few examples that can be used to illustrate the benefits of systematic mechanism design processes.

The inclusion of compliant elements provides the mechanism designer a wide range of solutions to design problems. In fact, this dramatic increase in the number of possible solutions can be overwhelming and make the selection of the best design for a particular application difficult. Therefore, a systematic design process for compliant mechanisms has been developed to guide the designer. This process begins with an original mechanism topology and its description in the matrix formulation described previously. Next, the design requirements (goals of the process) are established. After the requirements have been determined, new mechanism topologies are enumerated by modifying the components that make up the original mechanism. This enumeration procedure is referred to as topological synthesis. Throughout the topological synthesis process, the design requirements are applied in an attempt to limit the number of mechanisms that must be analyzed. Finally, the designs that meet the requirements are forwarded for mechanism synthesis and analysis.

It is important to note that many type synthesis approaches include a topological analysis process, which is used to enumerate all the possible inputs and outputs to a mechanism. Because the procedure described here begins with an initial design, the inputs and outputs to the systems are typically established as part of the original design requirements. Therefore, topological analysis is not discussed in this book. For further description of topological analysis techniques for compliant mechanisms, the reader is referred to Murphy [62].

G.5 DETERMINATION OF DESIGN REQUIREMENTS

Once an original design has been selected and described in the mathematical model proposed in previous sections, the design requirements need to be established. As with any design process, this step is often the most important and least understood. The requirements determine the extent of the enumeration process in the topological synthesis phase of the design process. In most cases, the requirements depend on the application under consideration. However, there are several issues that should be considered in most applications.

After the compliant mechanism matrix [CM] is formed, the original mechanism is investigated to determine the topological requirements. These requirements relate to the numbers of each type of segment and connections that are desired. For example, if a goal of the enumeration process is to eliminate kinematic pairs (joints), a topological requirement may be that kinematic pairs are not allowed. For many applications it may be desirable to specify the ground segment. In general, the following questions may help the designer establish design goals:

- Is a particular segment required to be the ground segment?
- Are any connections required to be kinematic pairs, flexural pivots, or clamped connections?
- Are kinematic pairs allowed?
- Does the design require that symmetry be maintained or established?
- Are segments required to be rigid (or compliant)?
- Are flexural pivots allowed as connections to compliant segments?

Responding to these questions helps to establish the topological requirements. As designers gain more experience with compliant mechanisms, the rules for determining topological requirements will be refined. However, when first employing the systematic design techniques, several iterations may be required to either widen or limit the scope of the design process.

G.6 TOPOLOGICAL SYNTHESIS OF COMPLIANT MECHANISMS

Once the requirements have been established, the topological synthesis of compliant mechanisms is accomplished through a four-step process. The first step is to enumerate the possible combinations of segment type (rigid or compliant) without regard to the ground segment or the type of connections between segments. After the possible segment combinations have been enumerated, the design requirements are investigated and isomorphic chains are removed from further consideration. The second step of the topological synthesis process is to enumerate all the possible combinations of connections between segments without regard to the segment types being connected. Although isomorphisms are not investigated after this phase of the design process, resulting compliant chains are investigated for conformance to requirements. For example, if a design requirement is that each mechanism must contain one kinematic pair, the enumerated chains are investigated to ensure that at least one kinematic pair is present. The third step of the topological synthesis process is to combine the results of the segment and connection-type enumeration processes. The subsequent kinematic chains are grouped by the original compliant chain from the segment enumeration process. This grouping will help limit the extent of later isomorphism investigations. The connections between segments are now examined to remove any fixed connections between rigid segments and the chains are investigated to remove any isomorphisms. The fourth, and final, step of the topological synthesis process is to sequentially fix, or ground, each rigid segment to form mechanisms. If more than one mechanism is formed from a particular compliant chain, the mechanisms formed from that chain need to be investigated to ensure that they are unique (nonisomorphic). As with all the steps of the topological synthesis process, the applicable design requirements are enforced. The resulting mechanisms are forwarded for further investigation, which may include a topological analysis or ranking to determine which mechanisms will be selected for a particular application. The sections that follow provide more detailed descriptions of each step in the topological synthesis process. The results for a four-bar compliant chain without specific design requirements will be provided to illustrate each step. A four-bar mechanism was selected because it is considered the most basic mechanism topology, from which many other topologies are derived.

G.6.1 Segment-Type Enumeration

As noted earlier, the first step of the topological synthesis process is the enumeration of possible segment combinations without regard to the selection of the ground segment or the type of connections between segments. It is important to remember that at least one rigid segment will be required for each kinematic chain so that a fixed (ground) segment can be specified to form a mechanism. Because the ground segment has not been specified, the resulting matrices are referred to as compliant element, [CE], matrices. For this process the number of possible kinematic chains, t, can be calculated as

$$t = \sum_{i=1}^{n} 2^{(n-i)} \tag{G.4}$$

where n is the number of segments. For a four-bar chain, there are 15 possible chains derived from this process.

After the possible kinematic chains have been determined, they are analyzed to remove any isomorphisms. This is accomplished by first determining the number of compliant segments present, n_c, in each chain. Since the ground segment has not been specified, this is simply the trace of the compliant element [CE] matrix, or

$$n_c = \text{trace}[\text{CE}] \tag{G.5}$$

If two chains have the same value for n_c, the characteristic polynomials for their compliant element matrices (CECP: compliant element characteristic polynomials) are calculated as follows:

$$\text{CECP} = |xI - \text{CE}| \tag{G.6}$$

where x is a variable, I is the identity matrix, and CE is the compliant element matrix. If two compliant chains are isomorphic, they will have the same n_c and CECP. Figure G.4 shows the compliant element matrices [CE], the number of compliant segments, n_c, and the characteristic polynomials of the compliant element

Topological Synthesis of Compliant Mechanisms

$$\begin{bmatrix} 0 & 1 & 0 & 1 \\ 1 & 0 & 1 & 0 \\ 0 & 1 & 0 & 1 \\ 1 & 0 & 1 & 0 \end{bmatrix} \quad \begin{array}{l} n_c = 0 \\ \text{CECP} = x^4 + 0x^3 - 4x^2 + 0x + 0 \end{array}$$

(a)

$$\begin{bmatrix} 0 & 1 & 0 & 1 \\ 1 & 1 & 1 & 0 \\ 0 & 1 & 0 & 1 \\ 1 & 0 & 1 & 0 \end{bmatrix} \quad \begin{array}{l} n_c = 1 \\ \text{CECP} = x^4 - 1x^3 - 4x^2 + 2x + 0 \end{array}$$

(b)

$$\begin{bmatrix} 0 & 1 & 0 & 1 \\ 1 & 1 & 1 & 0 \\ 0 & 1 & 0 & 1 \\ 1 & 0 & 1 & 1 \end{bmatrix} \quad \begin{array}{l} n_c = 2 \\ \text{CECP} = x^4 - 2x^3 - 3x^2 + 4x + 0 \end{array}$$

(c)

$$\begin{bmatrix} 0 & 1 & 0 & 1 \\ 1 & 1 & 1 & 0 \\ 0 & 1 & 1 & 1 \\ 1 & 0 & 1 & 0 \end{bmatrix} \quad \begin{array}{l} n_c = 2 \\ \text{CECP} = x^4 - 2x^3 - 3x^2 + 4x - 1 \end{array}$$

(d)

$$\begin{bmatrix} 0 & 1 & 0 & 1 \\ 1 & 1 & 1 & 0 \\ 0 & 1 & 1 & 1 \\ 1 & 0 & 1 & 1 \end{bmatrix} \quad \begin{array}{l} n_c = 3 \\ \text{CECP} = x^4 - 3x^3 - 1x^2 + 5x - 2 \end{array}$$

(e)

Figure G.4. Compliant element matrices (CE), the number of compliant segments (n_c), and the characteristic polynomials (CECP) for the five nonisomorphic compliant chains from a four-bar chain.

matrices for the five nonisomorphic compliant chains derived from a four-bar chain using the segment enumeration process.

G.6.2 Connection-Type Enumeration

After all the nonisomorphic compliant chains have been enumerated with respect to segment compliance, the connections between segments are considered. As noted earlier, in addition to kinematic pairs, the inclusion of compliance requires consideration of flexural pivots and fixed connections. This procedure leads to 3^p combinations, where p is the total number of connections between segments. It is important to remember that fixed (or clamped) connections are not allowed between rigid segments.

To reduce the total number of computations, the connection-type enumeration will initially be performed on the rigid-body kinematic chain with the same basic topology as the original mechanism. The vertex-vertex adjacency matrix for this chain is formulated by making all the segments rigid $c(i, i) = 0$, and considering all the connections as kinematic pairs, $c(i, j) = 1$ if connected, otherwise $c(i, j) = 0$. The procedure simply involves letting each connection take on the values for kinematic pairs, $c(i, j) = 1$, flexural pivots, $c(i, j) = 2$, and clamped connections, $c(i, j) = 3$ in a sequential manner to form all of the 3^p combinations. This enumeration can easily be automated by employing embedded loops, where each connection is allowed to take on values for each of the three connection types. The compliant element matrices for the 81 chains derived from a four-bar chain are provided in Figure G.5.

G.6.3 Combined Segment and Connection-Type Results

The procedure for combining the results of the segment and connection-type enumeration processes simply involves replacing the diagonal elements of the compliant element matrices from the connection-type enumeration process, with each of the resulting diagonal elements of the compliant element matrices from the segment enumeration process. For a four-bar chain this will initially result in 405 "new" compliant element matrices (five sets from the segment enumeration combined with 81 sets from the connection-type enumeration). For example, by combining the diagonal elements of the matrix in Figure G.4b with the nondiagonal elements of the matrix in Figure G.5o, the new compliant chain, Figure G.6, is enumerated.

Once the results from the separate enumeration processes have been combined, the fixed connections between rigid segments need to be removed from further consideration. In matrix form, if $c(i, i) = 0$ and $c(j, j) = 0$, then $c(i, j) \neq 3$. This convention greatly reduces the number of compliant chains that are considered further and can be automated easily.

After the compliant chains with clamped connections between rigid segments are removed, the resulting compliant chains are investigated to remove any isomorphic combinations. The procedure established in previous sections is applied to the chains that are grouped as a result of the segment enumeration. Figure G.7 shows the compliant element matrices for the 135 nonisomorphic compliant chains resulting from the application of this procedure to a four-bar chain. This example serves to illustrate the importance of design requirements to the enumeration process. In

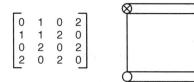

Figure G.6. Compliant chain formed by combining the diagonal elements from Figure G.4b and the nondiagonal elements from Figure G.5o.

Topological Synthesis of Compliant Mechanisms

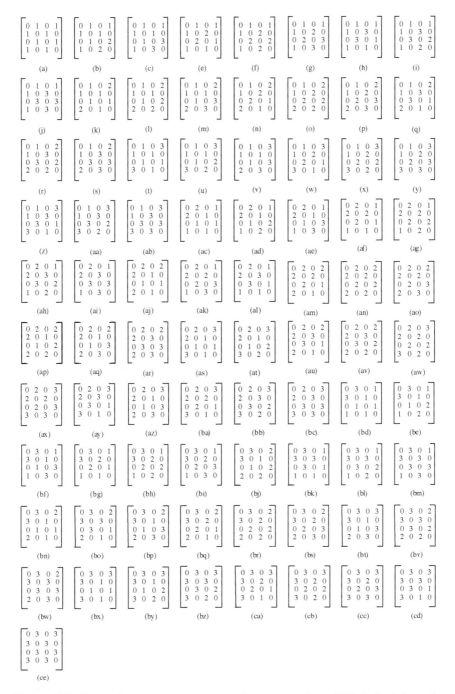

Figure G.5. Connection-type vertex–vertex adjacency matrices (CT) derived from the connection enumeration process for a four-bar chain.

(a:1) $\begin{bmatrix} 0&1&0&1\\1&0&1&0\\0&1&0&1\\1&0&1&0 \end{bmatrix}$
(a:2) $\begin{bmatrix} 0&1&0&1\\1&0&1&0\\0&1&0&2\\1&0&2&0 \end{bmatrix}$
(a:3) $\begin{bmatrix} 0&1&0&1\\1&0&2&0\\0&2&0&2\\1&0&2&0 \end{bmatrix}$
(a:4) $\begin{bmatrix} 0&1&0&2\\1&0&2&0\\0&2&0&1\\2&0&1&0 \end{bmatrix}$
(a:5) $\begin{bmatrix} 0&1&0&2\\1&0&2&0\\0&2&0&2\\2&0&2&0 \end{bmatrix}$
(a:6) $\begin{bmatrix} 0&2&0&2\\2&0&2&0\\0&2&0&2\\2&0&2&0 \end{bmatrix}$

(b:1) $\begin{bmatrix} 0&1&0&1\\1&0&1&0\\0&1&0&1\\1&0&1&1 \end{bmatrix}$
(b:2) $\begin{bmatrix} 0&1&0&1\\1&0&1&0\\0&1&0&2\\1&0&2&1 \end{bmatrix}$
(b:3) $\begin{bmatrix} 0&1&0&1\\1&0&1&0\\0&1&0&3\\1&0&3&1 \end{bmatrix}$
(b:4) $\begin{bmatrix} 0&1&0&1\\1&0&2&0\\0&2&0&1\\1&0&1&1 \end{bmatrix}$
(b:5) $\begin{bmatrix} 0&1&0&1\\1&0&2&0\\0&2&0&2\\1&0&2&1 \end{bmatrix}$
(b:6) $\begin{bmatrix} 0&1&0&1\\1&0&2&0\\0&2&0&3\\1&0&3&1 \end{bmatrix}$
(b:7) $\begin{bmatrix} 0&1&0&2\\1&0&1&0\\0&1&0&2\\2&0&2&1 \end{bmatrix}$
(b:8) $\begin{bmatrix} 0&1&0&2\\1&0&1&0\\0&1&0&3\\2&0&3&1 \end{bmatrix}$
(b:9) $\begin{bmatrix} 0&1&0&2\\1&0&2&0\\0&2&0&1\\2&0&1&1 \end{bmatrix}$

(b:10) $\begin{bmatrix} 0&1&0&2\\1&0&2&0\\0&2&0&2\\2&0&2&1 \end{bmatrix}$
(b:11) $\begin{bmatrix} 0&1&0&2\\1&0&2&0\\0&2&0&3\\2&0&3&1 \end{bmatrix}$
(b:12) $\begin{bmatrix} 0&1&0&3\\1&0&1&0\\0&1&0&3\\3&0&3&1 \end{bmatrix}$
(b:13) $\begin{bmatrix} 0&1&0&3\\1&0&2&0\\0&2&0&1\\3&0&1&1 \end{bmatrix}$
(b:14) $\begin{bmatrix} 0&1&0&3\\1&0&2&0\\0&2&0&2\\3&0&2&1 \end{bmatrix}$
(b:15) $\begin{bmatrix} 0&1&0&3\\1&0&2&0\\0&2&0&3\\3&0&3&1 \end{bmatrix}$
(b:16) $\begin{bmatrix} 0&2&0&1\\2&0&2&0\\0&2&0&1\\1&0&1&1 \end{bmatrix}$
(b:17) $\begin{bmatrix} 0&2&0&1\\2&0&2&0\\0&2&0&2\\1&0&2&1 \end{bmatrix}$
(b:18) $\begin{bmatrix} 0&2&0&1\\2&0&2&0\\0&2&0&3\\1&0&3&1 \end{bmatrix}$

(b:19) $\begin{bmatrix} 0&2&0&2\\2&0&2&0\\0&2&0&2\\2&0&2&1 \end{bmatrix}$
(b:20) $\begin{bmatrix} 0&2&0&2\\2&0&2&0\\0&2&0&3\\2&0&3&1 \end{bmatrix}$
(b:21) $\begin{bmatrix} 0&2&0&3\\2&0&2&0\\0&2&0&3\\3&0&3&1 \end{bmatrix}$

(c:1) $\begin{bmatrix} 0&1&0&1\\1&0&1&0\\0&1&1&1\\1&0&1&1 \end{bmatrix}$
(c:2) $\begin{bmatrix} 0&1&0&1\\1&0&1&0\\0&1&1&2\\1&0&2&1 \end{bmatrix}$
(c:3) $\begin{bmatrix} 0&1&0&1\\1&0&1&0\\0&1&1&3\\1&0&3&1 \end{bmatrix}$
(c:4) $\begin{bmatrix} 0&1&0&1\\1&0&2&0\\0&2&1&1\\1&0&1&1 \end{bmatrix}$
(c:5) $\begin{bmatrix} 0&1&0&1\\1&0&2&0\\0&2&1&2\\1&0&2&1 \end{bmatrix}$
(c:6) $\begin{bmatrix} 0&1&0&1\\1&0&2&0\\0&2&1&3\\1&0&3&1 \end{bmatrix}$
(c:7) $\begin{bmatrix} 0&1&0&1\\1&0&3&0\\0&3&1&1\\1&0&1&1 \end{bmatrix}$
(c:8) $\begin{bmatrix} 0&1&0&1\\1&0&3&0\\0&3&1&2\\1&0&2&1 \end{bmatrix}$
(c:9) $\begin{bmatrix} 0&1&0&1\\1&0&3&0\\0&3&1&3\\1&0&3&1 \end{bmatrix}$

(c:10) $\begin{bmatrix} 0&1&0&2\\1&0&2&0\\0&2&1&1\\2&0&1&1 \end{bmatrix}$
(c:11) $\begin{bmatrix} 0&1&0&2\\1&0&2&0\\0&2&1&2\\2&0&2&1 \end{bmatrix}$
(c:12) $\begin{bmatrix} 0&1&0&2\\1&0&2&0\\0&2&1&3\\2&0&3&1 \end{bmatrix}$
(c:13) $\begin{bmatrix} 0&1&0&2\\1&0&3&0\\0&3&1&1\\2&0&1&1 \end{bmatrix}$
(c:14) $\begin{bmatrix} 0&1&0&2\\1&0&3&0\\0&3&1&2\\2&0&2&1 \end{bmatrix}$
(c:15) $\begin{bmatrix} 0&1&0&2\\1&0&3&0\\0&3&1&3\\2&0&3&1 \end{bmatrix}$
(c:16) $\begin{bmatrix} 0&1&0&3\\1&0&3&0\\0&3&1&1\\3&0&1&1 \end{bmatrix}$
(c:17) $\begin{bmatrix} 0&1&0&3\\1&0&3&0\\0&3&1&2\\3&0&2&1 \end{bmatrix}$
(c:18) $\begin{bmatrix} 0&1&0&3\\1&0&3&0\\0&3&1&3\\3&0&3&1 \end{bmatrix}$

(c:19) $\begin{bmatrix} 0&2&0&1\\2&0&1&0\\0&1&1&1\\1&0&1&1 \end{bmatrix}$
(c:20) $\begin{bmatrix} 0&2&0&1\\2&0&1&0\\0&1&1&2\\1&0&2&1 \end{bmatrix}$
(c:21) $\begin{bmatrix} 0&2&0&1\\2&0&1&0\\0&1&1&3\\1&0&3&1 \end{bmatrix}$
(c:22) $\begin{bmatrix} 0&2&0&1\\2&0&2&0\\0&2&1&1\\1&0&1&1 \end{bmatrix}$
(c:23) $\begin{bmatrix} 0&2&0&1\\2&0&2&0\\0&2&1&2\\1&0&2&1 \end{bmatrix}$
(c:24) $\begin{bmatrix} 0&2&0&1\\2&0&2&0\\0&2&1&3\\1&0&3&1 \end{bmatrix}$
(c:25) $\begin{bmatrix} 0&2&0&1\\2&0&3&0\\0&3&1&1\\1&0&1&1 \end{bmatrix}$
(c:26) $\begin{bmatrix} 0&2&0&1\\2&0&3&0\\0&3&1&2\\1&0&2&1 \end{bmatrix}$
(c:27) $\begin{bmatrix} 0&2&0&1\\2&0&3&0\\0&3&1&3\\1&0&3&1 \end{bmatrix}$

(c:28) $\begin{bmatrix} 0&2&0&2\\2&0&2&0\\0&2&1&1\\2&0&1&1 \end{bmatrix}$
(c:29) $\begin{bmatrix} 0&2&0&2\\2&0&2&0\\0&2&1&2\\2&0&2&1 \end{bmatrix}$
(c:30) $\begin{bmatrix} 0&2&0&2\\2&0&2&0\\0&2&1&3\\2&0&3&1 \end{bmatrix}$
(c:31) $\begin{bmatrix} 0&2&0&2\\2&0&3&0\\0&3&1&1\\2&0&1&1 \end{bmatrix}$
(c:32) $\begin{bmatrix} 0&2&0&2\\2&0&3&0\\0&3&1&2\\2&0&2&1 \end{bmatrix}$
(c:33) $\begin{bmatrix} 0&2&0&2\\2&0&3&0\\0&3&1&3\\2&0&3&1 \end{bmatrix}$
(c:34) $\begin{bmatrix} 0&2&0&3\\2&0&3&0\\0&3&1&1\\3&0&1&1 \end{bmatrix}$
(c:35) $\begin{bmatrix} 0&2&0&3\\2&0&3&0\\0&3&1&2\\3&0&2&1 \end{bmatrix}$
(c:36) $\begin{bmatrix} 0&2&0&3\\2&0&3&0\\0&3&1&3\\3&0&3&1 \end{bmatrix}$

(d:1) $\begin{bmatrix} 0&1&0&1\\1&1&1&0\\0&1&0&1\\1&0&1&1 \end{bmatrix}$
(d:2) $\begin{bmatrix} 0&1&0&1\\1&1&1&0\\0&1&0&2\\1&0&2&1 \end{bmatrix}$
(d:3) $\begin{bmatrix} 0&1&0&1\\1&1&1&0\\0&1&0&3\\1&0&3&1 \end{bmatrix}$
(d:4) $\begin{bmatrix} 0&1&0&1\\1&1&2&0\\0&2&0&2\\1&0&2&1 \end{bmatrix}$
(d:5) $\begin{bmatrix} 0&1&0&1\\1&1&2&0\\0&2&0&3\\1&0&3&1 \end{bmatrix}$
(d:6) $\begin{bmatrix} 0&1&0&1\\1&1&3&0\\0&3&0&3\\1&0&3&1 \end{bmatrix}$
(d:7) $\begin{bmatrix} 0&1&0&2\\1&1&1&0\\0&1&0&2\\2&0&2&1 \end{bmatrix}$
(d:8) $\begin{bmatrix} 0&1&0&2\\1&1&1&0\\0&1&0&3\\2&0&3&1 \end{bmatrix}$
(d:9) $\begin{bmatrix} 0&1&0&2\\1&1&2&0\\0&2&0&1\\2&0&1&1 \end{bmatrix}$

(d:10) $\begin{bmatrix} 0&1&0&2\\1&1&2&0\\0&2&0&2\\2&0&2&1 \end{bmatrix}$
(d:11) $\begin{bmatrix} 0&1&0&2\\1&1&2&0\\0&2&0&3\\2&0&3&1 \end{bmatrix}$
(d:12) $\begin{bmatrix} 0&1&0&2\\1&1&3&0\\0&3&0&1\\2&0&1&1 \end{bmatrix}$
(d:13) $\begin{bmatrix} 0&1&0&2\\1&1&3&0\\0&3&0&2\\2&0&2&1 \end{bmatrix}$
(d:14) $\begin{bmatrix} 0&1&0&2\\1&1&3&0\\0&3&0&3\\2&0&3&1 \end{bmatrix}$
(d:15) $\begin{bmatrix} 0&1&0&3\\1&1&1&0\\0&1&0&3\\3&0&3&1 \end{bmatrix}$
(d:16) $\begin{bmatrix} 0&1&0&3\\1&1&2&0\\0&2&0&2\\3&0&2&1 \end{bmatrix}$
(d:17) $\begin{bmatrix} 0&1&0&3\\1&1&2&0\\0&2&0&3\\3&0&3&1 \end{bmatrix}$
(d:18) $\begin{bmatrix} 0&1&0&3\\1&1&3&0\\0&3&0&1\\3&0&1&1 \end{bmatrix}$

(d:19) $\begin{bmatrix} 0&1&0&3\\1&1&3&0\\0&3&0&2\\3&0&2&1 \end{bmatrix}$
(d:20) $\begin{bmatrix} 0&1&0&3\\1&1&3&0\\0&3&0&3\\3&0&3&1 \end{bmatrix}$
(d:21) $\begin{bmatrix} 0&2&0&2\\2&1&2&0\\0&2&0&3\\2&0&3&1 \end{bmatrix}$
(d:22) $\begin{bmatrix} 0&2&0&2\\2&1&2&0\\0&2&0&3\\2&0&3&1 \end{bmatrix}$
(d:23) $\begin{bmatrix} 0&2&0&2\\2&1&3&0\\0&3&0&3\\2&0&3&1 \end{bmatrix}$
(d:24) $\begin{bmatrix} 0&2&0&3\\2&1&2&0\\0&2&0&3\\3&0&3&1 \end{bmatrix}$
(d:25) $\begin{bmatrix} 0&2&0&3\\2&1&3&0\\0&3&0&3\\3&0&2&1 \end{bmatrix}$
(d:26) $\begin{bmatrix} 0&2&0&3\\2&1&3&0\\0&3&0&3\\3&0&3&1 \end{bmatrix}$
(d:27) $\begin{bmatrix} 0&3&0&3\\3&1&3&0\\0&3&0&3\\3&0&3&1 \end{bmatrix}$

Figure G.7. Compliant element matrices (CE's) for the nonisomorphic compliant chains that can be derived from a rigid-body four-link kinematic chain.

Topological Synthesis of Compliant Mechanisms

$$\begin{matrix}
\begin{bmatrix}0&1&0&1\\1&1&1&0\\0&1&1&1\\1&0&1&1\end{bmatrix} & \begin{bmatrix}0&1&0&1\\1&1&1&0\\0&1&1&2\\1&0&2&1\end{bmatrix} & \begin{bmatrix}0&1&0&1\\1&1&1&0\\0&1&1&3\\1&0&3&1\end{bmatrix} & \begin{bmatrix}0&1&0&1\\1&1&2&0\\0&2&1&2\\1&0&2&1\end{bmatrix} & \begin{bmatrix}0&1&0&1\\1&1&2&0\\0&2&1&3\\1&0&3&1\end{bmatrix} & \begin{bmatrix}0&1&0&1\\1&1&3&0\\0&3&1&3\\1&0&3&1\end{bmatrix} & \begin{bmatrix}0&1&0&2\\1&1&1&0\\0&1&1&1\\2&0&1&1\end{bmatrix} & \begin{bmatrix}0&1&0&2\\1&1&1&0\\0&1&1&2\\2&0&2&1\end{bmatrix} & \begin{bmatrix}0&1&0&2\\1&1&1&0\\0&1&1&3\\2&0&3&1\end{bmatrix}\\
(e:1) & (e:2) & (e:3) & (e:4) & (e:5) & (e:6) & (e:7) & (e:8) & (e:9)
\end{matrix}$$

$$\begin{matrix}
\begin{bmatrix}0&1&0&2\\1&1&2&0\\0&2&1&1\\2&0&1&1\end{bmatrix} & \begin{bmatrix}0&1&0&2\\1&1&2&0\\0&2&1&2\\2&0&2&1\end{bmatrix} & \begin{bmatrix}0&1&0&2\\1&1&2&0\\0&2&1&3\\2&0&3&1\end{bmatrix} & \begin{bmatrix}0&1&0&2\\1&1&3&0\\0&3&1&1\\2&0&1&1\end{bmatrix} & \begin{bmatrix}0&1&0&2\\1&1&3&0\\0&3&1&2\\2&0&2&1\end{bmatrix} & \begin{bmatrix}0&1&0&2\\1&1&3&0\\0&3&1&3\\2&0&3&1\end{bmatrix} & \begin{bmatrix}0&1&0&3\\1&1&1&0\\0&1&1&1\\3&0&1&1\end{bmatrix} & \begin{bmatrix}0&1&0&3\\1&1&1&0\\0&1&1&2\\3&0&2&1\end{bmatrix} & \begin{bmatrix}0&1&0&3\\1&1&1&0\\0&1&1&3\\3&0&3&1\end{bmatrix}\\
(e:10) & (e:11) & (e:12) & (e:13) & (e:14) & (e:15) & (e:16) & (e:17) & (e:18)
\end{matrix}$$

$$\begin{matrix}
\begin{bmatrix}0&1&0&3\\1&1&2&0\\0&2&1&1\\3&0&1&1\end{bmatrix} & \begin{bmatrix}0&1&0&3\\1&1&2&0\\0&2&1&2\\3&0&2&1\end{bmatrix} & \begin{bmatrix}0&1&0&3\\1&1&2&0\\0&2&1&3\\3&0&3&1\end{bmatrix} & \begin{bmatrix}0&1&0&3\\1&1&3&0\\0&3&1&1\\3&0&1&1\end{bmatrix} & \begin{bmatrix}0&1&0&3\\1&1&3&0\\0&3&1&2\\3&0&2&1\end{bmatrix} & \begin{bmatrix}0&1&0&3\\1&1&3&0\\0&3&1&3\\3&0&3&1\end{bmatrix} & \begin{bmatrix}0&2&0&2\\2&1&1&0\\0&1&1&1\\2&0&1&1\end{bmatrix} & \begin{bmatrix}0&2&0&2\\2&1&1&0\\0&1&1&2\\2&0&2&1\end{bmatrix} & \begin{bmatrix}0&2&0&2\\2&1&1&0\\0&1&1&3\\2&0&3&1\end{bmatrix}\\
(e:19) & (e:20) & (e:21) & (e:22) & (e:23) & (e:24) & (e:25) & (e:26) & (e:27)
\end{matrix}$$

$$\begin{matrix}
\begin{bmatrix}0&2&0&2\\2&1&2&0\\0&2&1&2\\2&0&2&1\end{bmatrix} & \begin{bmatrix}0&2&0&2\\2&1&2&0\\0&2&1&3\\2&0&3&1\end{bmatrix} & \begin{bmatrix}0&2&0&2\\2&1&3&0\\0&3&1&3\\2&0&3&1\end{bmatrix} & \begin{bmatrix}0&2&0&3\\2&1&1&0\\0&1&1&1\\3&0&1&1\end{bmatrix} & \begin{bmatrix}0&2&0&3\\2&1&1&0\\0&1&1&2\\3&0&2&1\end{bmatrix} & \begin{bmatrix}0&2&0&3\\2&1&1&0\\0&1&1&3\\3&0&3&1\end{bmatrix} & \begin{bmatrix}0&2&0&3\\2&1&2&0\\0&2&1&1\\3&0&1&1\end{bmatrix} & \begin{bmatrix}0&2&0&3\\2&1&2&0\\0&2&1&2\\3&0&2&1\end{bmatrix} & \begin{bmatrix}0&2&0&3\\2&1&2&0\\0&2&1&3\\3&0&3&1\end{bmatrix}\\
(e:28) & (e:29) & (e:30) & (e:31) & (e:32) & (e:33) & (e:34) & (e:35) & (e:36)
\end{matrix}$$

$$\begin{matrix}
\begin{bmatrix}0&2&0&3\\2&1&3&0\\0&3&1&1\\3&0&1&1\end{bmatrix} & \begin{bmatrix}0&2&0&3\\2&1&3&0\\0&3&1&2\\3&0&2&1\end{bmatrix} & \begin{bmatrix}0&2&0&3\\2&1&3&0\\0&3&1&3\\3&0&3&1\end{bmatrix} & \begin{bmatrix}0&2&0&3\\3&1&1&0\\0&1&1&1\\3&0&1&1\end{bmatrix} & \begin{bmatrix}0&3&0&3\\3&1&1&0\\0&1&1&2\\3&0&2&1\end{bmatrix} & \begin{bmatrix}0&3&0&3\\3&1&1&0\\0&1&1&3\\3&0&3&1\end{bmatrix} & \begin{bmatrix}0&3&0&3\\3&1&2&0\\0&2&1&2\\3&0&2&1\end{bmatrix} & \begin{bmatrix}0&3&0&3\\3&1&2&0\\0&2&1&3\\3&0&3&1\end{bmatrix} & \begin{bmatrix}0&3&0&3\\3&1&3&0\\0&3&1&3\\3&0&3&1\end{bmatrix}\\
(e:37) & (e:38) & (e:39) & (e:40) & (e:41) & (e:42) & (e:43) & (e:44) & (e:45)
\end{matrix}$$

Figure G.7 (continued).

rigid-body kinematics, there is only one unique rigid-body four-bar chain, whereas the designer is faced with 135 unique combinations for compliant four-bar chains. As with the other steps of the topological synthesis process, the applicable design requirements are enforced before continuing to the final step.

G.6.4 Formation of Compliant Mechanisms

The final step of the topological synthesis process is to form all the unique compliant mechanisms that can be derived from the compliant chains from the previous step. This is accomplished by sequentially designating one of the rigid segments as the ground segment for each compliant chain. Next, the mechanisms created from each compliant chain are examined to ensure that they are nonisomorphic. Finally, the mechanisms are examined to ensure that they meet the design requirements specified.

To illustrate the enumeration process, consider the compliant chain represented by the matrix in Figure G.7a:6. Because this compliant chain has four rigid segments to be fixed, four possible mechanisms can be specified. Their respective compliant mechanism matrices are presented in Figure G.8. Upon further examination, it is discovered that all of these mechanisms are isomorphic. Therefore, only one unique mechanism can be derived from the compliant chain represented in Figure G.7a:6. The compliant chains represented by the matrices in Figure G.7e:1 through Figure G.7e:45, each contain only one rigid segment. Therefore, only one mechanism can be derived from each compliant chain. Because the compliant chains are nonisomorphic, the resulting compliant mechanisms are also nonisomorphic. The

$$\begin{bmatrix} -1 & 2 & 0 & 2 \\ 2 & 0 & 2 & 0 \\ 0 & 2 & 0 & 2 \\ 2 & 0 & 2 & 0 \end{bmatrix} \begin{bmatrix} 0 & 2 & 0 & 2 \\ 1 & -1 & 2 & 0 \\ 0 & 2 & 0 & 2 \\ 2 & 0 & 2 & 0 \end{bmatrix} \begin{bmatrix} 0 & 2 & 0 & 2 \\ 2 & 0 & 2 & 0 \\ 0 & 2 & -1 & 2 \\ 2 & 0 & 2 & 0 \end{bmatrix} \begin{bmatrix} 0 & 2 & 0 & 2 \\ 2 & 0 & 2 & 0 \\ 0 & 2 & 0 & 2 \\ 2 & 0 & 2 & -1 \end{bmatrix}$$

Figure G.8. Four compliant mechanisms that can be formed from the compliant chain represented in Figure G.7a:6.

compliant chain in Figure G.7c:4 contains two rigid segments. Fixing each of these segments sequentially results in two nonisomorphic mechanisms. To provide a summary for this process, the compliant mechanism matrices for the 209 unique compliant four-bar mechanisms are provided in Figure G.9.

The discussion presented illustrates the large number of unique solutions that can be enumerated using topological synthesis for compliant mechanisms. As discussed earlier, design requirements are important to limit the extent of the enumeration process. To illustrate this, two examples are provided below.

G.7 EXAMPLES

Examples play an important role in the learning process because they serve as guides when new problems are encountered. The two examples provided, build on the previous presentation by employing four-bar compliant mechanisms as their solution set. The first example is the topological synthesis of the 28 unique compliant near-constant-force mechanisms. The dimensional synthesis and force–deflection relationships for these mechanisms are discussed in Section 10.1. The second topological synthesis example is the enumeration of the unique compliant parallel-guiding mechanisms discussed in Section 10.2. The goal of these examples is to illustrate the importance design requirements play in reducing the extent of the enumeration process.

Example: Constant-Force Mechanisms. The use of springs as energy absorbers in mechanisms has a wide range of industrial applications. Many of these applications would benefit significantly if the output force remained nearly constant regardless of the amount of deflection at the input. Using a straightforward synthesis technique, Jenuwine and Midha [174] developed the exact constant force mechanism shown in Figure G.10. This mechanism employs two orthogonal linear springs acting on a rigid-body topology to obtain an exact-constant-force output. Inspired by this mechanism, a design variation employing flexural pivots in place of the linear springs was developed. Employing traditional design techniques, and the analysis methodology presented by Jenuwine and Midha [174], a very near-constant-force mechanism was developed. This mechanism and its compliant mechanism matrix [CM] are shown in Figure G.11. A discussion of the force–deflection relationships for this mechanism is presented in Section 10.1.

Examples

$$(a{:}1.1)\begin{bmatrix}-1&1&0&1\\1&0&1&0\\0&1&0&1\\1&0&1&0\end{bmatrix}\quad(a{:}2.1)\begin{bmatrix}-1&1&0&1\\1&0&1&0\\0&1&0&2\\1&0&2&0\end{bmatrix}\quad(a{:}2.2)\begin{bmatrix}0&1&0&1\\1&0&1&0\\0&1&0&2\\1&0&2&-1\end{bmatrix}\quad(a{:}3.1)\begin{bmatrix}-1&1&0&1\\1&0&2&0\\0&2&0&2\\1&0&2&0\end{bmatrix}\quad(a{:}3.2)\begin{bmatrix}0&1&0&1\\1&-1&2&0\\0&2&0&2\\1&0&2&0\end{bmatrix}\quad(a{:}3.3)\begin{bmatrix}0&1&0&1\\1&0&2&0\\0&2&-1&2\\1&0&2&0\end{bmatrix}\quad(a{:}4.1)\begin{bmatrix}-1&1&0&2\\1&0&2&0\\0&2&0&1\\2&0&1&0\end{bmatrix}\quad(a{:}5.1)\begin{bmatrix}-1&1&0&2\\1&0&2&0\\0&2&0&2\\2&0&2&0\end{bmatrix}\quad(a{:}5.2)\begin{bmatrix}0&1&0&2\\1&0&2&0\\0&2&0&2\\2&0&2&-1\end{bmatrix}$$

$$(a{:}6.1)\begin{bmatrix}-1&2&0&2\\2&0&2&0\\0&2&0&2\\2&0&2&0\end{bmatrix}$$

$$(b{:}1.1)\begin{bmatrix}-1&1&0&1\\1&0&1&0\\0&1&0&1\\1&0&1&1\end{bmatrix}\quad(b{:}1.2)\begin{bmatrix}0&1&0&1\\1&-1&1&0\\0&1&0&1\\1&0&1&1\end{bmatrix}\quad(b{:}2.1)\begin{bmatrix}-1&1&0&1\\1&-1&1&0\\0&1&0&2\\1&0&2&1\end{bmatrix}\quad(b{:}2.2)\begin{bmatrix}0&1&0&1\\1&-1&1&0\\0&1&0&2\\1&0&2&1\end{bmatrix}\quad(b{:}2.3)\begin{bmatrix}0&1&0&1\\1&0&1&0\\0&1&-1&2\\1&0&2&1\end{bmatrix}\quad(b{:}3.1)\begin{bmatrix}-1&1&0&1\\1&0&1&0\\0&1&0&3\\1&0&3&1\end{bmatrix}\quad(b{:}3.2)\begin{bmatrix}0&1&0&1\\1&-1&1&0\\0&1&0&3\\1&0&3&1\end{bmatrix}\quad(b{:}3.3)\begin{bmatrix}0&1&0&1\\1&0&1&0\\0&1&-1&3\\1&0&3&1\end{bmatrix}\quad(b{:}4.1)\begin{bmatrix}-1&1&0&1\\1&0&2&0\\0&2&0&1\\1&0&1&1\end{bmatrix}$$

$$(b{:}4.2)\begin{bmatrix}0&1&0&1\\1&-1&2&0\\0&2&0&1\\1&0&1&1\end{bmatrix}\quad(b{:}5.1)\begin{bmatrix}-1&1&0&1\\1&0&2&0\\0&2&0&2\\1&0&2&1\end{bmatrix}\quad(b{:}5.2)\begin{bmatrix}0&1&0&1\\1&-1&2&0\\0&2&0&2\\1&0&2&1\end{bmatrix}\quad(b{:}5.3)\begin{bmatrix}0&1&0&1\\1&0&2&0\\0&2&-1&2\\1&0&2&1\end{bmatrix}\quad(b{:}6.1)\begin{bmatrix}-1&1&0&1\\1&0&2&0\\0&2&0&3\\1&0&3&1\end{bmatrix}\quad(b{:}6.2)\begin{bmatrix}0&1&0&1\\1&-1&2&0\\0&2&0&3\\1&0&3&1\end{bmatrix}\quad(b{:}6.3)\begin{bmatrix}0&1&0&1\\1&0&2&0\\0&2&-1&3\\1&0&3&1\end{bmatrix}\quad(b{:}7.1)\begin{bmatrix}-1&1&0&2\\1&0&1&0\\0&1&0&2\\2&0&2&1\end{bmatrix}\quad(b{:}7.2)\begin{bmatrix}0&1&0&2\\1&-1&1&0\\0&1&0&2\\2&0&2&1\end{bmatrix}$$

$$(b{:}8.1)\begin{bmatrix}-1&1&0&2\\1&0&1&0\\0&1&0&3\\2&0&3&1\end{bmatrix}\quad(b{:}8.2)\begin{bmatrix}0&1&0&2\\1&-1&1&0\\0&1&0&3\\2&0&3&1\end{bmatrix}\quad(b{:}8.3)\begin{bmatrix}0&1&0&2\\1&0&1&0\\0&1&-1&3\\2&0&3&1\end{bmatrix}\quad(b{:}9.1)\begin{bmatrix}-1&1&0&2\\1&0&2&0\\0&2&0&1\\2&0&1&1\end{bmatrix}\quad(b{:}9.2)\begin{bmatrix}0&1&0&2\\1&-1&2&0\\0&2&0&1\\2&0&1&1\end{bmatrix}\quad(b{:}10.1)\begin{bmatrix}-1&1&0&2\\1&0&2&0\\0&2&0&2\\2&0&1&1\end{bmatrix}\quad(b{:}10.2)\begin{bmatrix}-1&1&0&2\\1&-1&2&0\\0&2&0&2\\2&0&2&1\end{bmatrix}\quad(b{:}10.3)\begin{bmatrix}0&1&0&2\\1&0&2&0\\0&2&-1&2\\2&0&2&1\end{bmatrix}\quad(b{:}11.1)\begin{bmatrix}-1&1&0&2\\1&0&2&0\\0&2&0&3\\2&0&3&1\end{bmatrix}$$

$$(b{:}11.2)\begin{bmatrix}0&1&0&2\\1&-1&2&0\\0&2&0&3\\2&0&3&1\end{bmatrix}\quad(b{:}11.3)\begin{bmatrix}0&1&0&2\\1&0&2&0\\0&2&-1&3\\2&0&3&1\end{bmatrix}\quad(b{:}12.1)\begin{bmatrix}-1&1&0&3\\1&0&1&0\\0&1&0&3\\3&0&3&1\end{bmatrix}\quad(b{:}12.2)\begin{bmatrix}0&1&0&3\\1&-1&1&0\\0&1&0&3\\3&0&3&1\end{bmatrix}\quad(b{:}13.1)\begin{bmatrix}-1&1&0&3\\1&0&2&0\\0&2&0&1\\3&0&1&1\end{bmatrix}\quad(b{:}13.2)\begin{bmatrix}0&1&0&3\\1&-1&2&0\\0&2&0&1\\3&0&3&1\end{bmatrix}\quad(b{:}13.3)\begin{bmatrix}0&1&0&3\\1&0&2&0\\0&2&-1&1\\3&0&3&1\end{bmatrix}\quad(b{:}14.1)\begin{bmatrix}-1&1&0&3\\1&0&2&0\\0&2&0&2\\3&0&2&1\end{bmatrix}\quad(b{:}14.2)\begin{bmatrix}0&1&0&3\\1&-1&2&0\\0&2&0&2\\3&0&2&1\end{bmatrix}$$

$$(b{:}14.3)\begin{bmatrix}0&1&0&3\\1&0&2&0\\0&2&-1&2\\3&0&2&1\end{bmatrix}\quad(b{:}15.1)\begin{bmatrix}-1&1&0&3\\1&0&2&0\\0&2&0&3\\3&0&3&1\end{bmatrix}\quad(b{:}15.2)\begin{bmatrix}0&1&0&3\\1&-1&2&0\\0&2&0&3\\3&0&3&1\end{bmatrix}\quad(b{:}15.3)\begin{bmatrix}0&1&0&3\\1&0&2&0\\0&2&-1&3\\3&0&3&1\end{bmatrix}\quad(b{:}16.1)\begin{bmatrix}-1&2&0&1\\2&0&2&0\\0&2&0&1\\1&0&1&1\end{bmatrix}\quad(b{:}16.2)\begin{bmatrix}0&2&0&1\\2&-1&2&0\\0&2&0&1\\1&0&1&1\end{bmatrix}\quad(b{:}17.1)\begin{bmatrix}-1&2&0&1\\2&0&2&0\\0&2&0&2\\1&0&2&1\end{bmatrix}\quad(b{:}17.2)\begin{bmatrix}0&2&0&1\\2&-1&2&0\\0&2&0&2\\1&0&2&1\end{bmatrix}\quad(b{:}17.3)\begin{bmatrix}0&2&0&1\\2&0&2&0\\0&2&-1&2\\1&0&2&1\end{bmatrix}$$

$$(b{:}18.1)\begin{bmatrix}-1&2&0&1\\2&0&2&0\\0&2&0&3\\1&0&3&1\end{bmatrix}\quad(b{:}18.2)\begin{bmatrix}0&2&0&1\\2&-1&2&0\\0&2&0&3\\1&0&3&1\end{bmatrix}\quad(b{:}18.3)\begin{bmatrix}0&2&0&1\\2&0&2&0\\0&2&-1&3\\1&0&3&1\end{bmatrix}\quad(b{:}19.1)\begin{bmatrix}-1&2&0&2\\2&0&2&0\\0&2&0&2\\2&0&2&1\end{bmatrix}\quad(b{:}19.2)\begin{bmatrix}0&2&0&2\\2&0&2&0\\0&2&0&2\\2&0&2&1\end{bmatrix}\quad(b{:}20.1)\begin{bmatrix}-1&2&0&2\\2&0&2&0\\0&2&0&3\\2&0&3&1\end{bmatrix}\quad(b{:}20.2)\begin{bmatrix}0&2&0&2\\2&-1&2&0\\0&2&0&3\\2&0&3&1\end{bmatrix}\quad(b{:}20.3)\begin{bmatrix}0&2&0&2\\2&0&2&0\\0&2&-1&3\\2&0&3&1\end{bmatrix}\quad(b{:}21.1)\begin{bmatrix}-1&2&0&3\\2&0&2&0\\0&2&0&3\\3&0&3&1\end{bmatrix}$$

$$(b{:}21.2)\begin{bmatrix}0&2&0&3\\2&-1&2&0\\0&2&0&3\\3&0&3&1\end{bmatrix}$$

$$(c{:}1.1)\begin{bmatrix}-1&1&0&1\\1&0&1&0\\0&1&1&1\\1&0&1&1\end{bmatrix}\quad(c{:}2.1)\begin{bmatrix}-1&1&0&1\\1&0&1&0\\0&1&1&2\\1&0&2&1\end{bmatrix}\quad(c{:}3.1)\begin{bmatrix}-1&1&0&1\\1&0&1&0\\0&1&1&1\\1&0&3&1\end{bmatrix}\quad(c{:}4.1)\begin{bmatrix}-1&1&0&1\\1&0&2&0\\0&2&1&1\\1&0&1&1\end{bmatrix}\quad(c{:}4.2)\begin{bmatrix}0&1&0&1\\1&-1&2&0\\0&2&1&1\\1&0&1&1\end{bmatrix}\quad(c{:}5.1)\begin{bmatrix}-1&1&0&1\\1&0&2&0\\0&2&1&2\\1&0&2&1\end{bmatrix}\quad(c{:}5.2)\begin{bmatrix}0&1&0&1\\1&-1&2&0\\0&2&1&2\\1&0&2&1\end{bmatrix}\quad(c{:}6.1)\begin{bmatrix}-1&1&0&1\\1&0&2&0\\0&2&1&3\\1&0&3&1\end{bmatrix}\quad(c{:}6.2)\begin{bmatrix}0&1&0&1\\1&-1&2&0\\0&2&1&3\\1&0&3&1\end{bmatrix}$$

$$(c{:}7.1)\begin{bmatrix}-1&1&0&1\\1&0&3&0\\0&3&1&1\\1&0&1&1\end{bmatrix}\quad(c{:}7.2)\begin{bmatrix}0&1&0&1\\1&-1&3&0\\0&3&1&1\\1&0&1&1\end{bmatrix}\quad(c{:}8.1)\begin{bmatrix}-1&1&0&1\\1&0&3&0\\0&3&1&2\\1&0&2&1\end{bmatrix}\quad(c{:}8.2)\begin{bmatrix}0&1&0&1\\1&-1&3&0\\0&3&1&2\\1&0&2&1\end{bmatrix}\quad(c{:}9.1)\begin{bmatrix}-1&1&0&1\\1&0&3&0\\0&3&1&3\\1&0&3&1\end{bmatrix}\quad(c{:}9.2)\begin{bmatrix}0&1&0&1\\1&-1&3&0\\0&3&1&3\\1&0&3&1\end{bmatrix}\quad(c{:}10.1)\begin{bmatrix}-1&1&0&2\\1&0&2&0\\0&2&1&1\\2&0&1&1\end{bmatrix}\quad(c{:}11.1)\begin{bmatrix}-1&1&0&2\\1&0&2&0\\0&2&1&2\\2&0&2&1\end{bmatrix}\quad(c{:}12.1)\begin{bmatrix}-1&1&0&2\\1&0&2&0\\0&2&1&3\\2&0&3&1\end{bmatrix}$$

Figure G.9. Compliant mechanism matrices (CM) for each four-bar compliant mechanism.

Type Synthesis of Compliant Mechanisms

$$\begin{bmatrix} -1 & 1 & 0 & 2 \\ 1 & 0 & 3 & 0 \\ 0 & 3 & 1 & 1 \\ 2 & 0 & 1 & 1 \end{bmatrix}$$
(c:13.1)

$$\begin{bmatrix} 0 & 1 & 0 & 2 \\ 1 & -1 & 3 & 0 \\ 0 & 3 & 1 & 1 \\ 2 & 0 & 1 & 1 \end{bmatrix}$$
(c:13.2)

$$\begin{bmatrix} -1 & 1 & 0 & 2 \\ 1 & 0 & 3 & 0 \\ 0 & 3 & 1 & 2 \\ 2 & 0 & 2 & 1 \end{bmatrix}$$
(c:14.1)

$$\begin{bmatrix} 0 & 1 & 0 & 2 \\ 1 & -1 & 3 & 0 \\ 0 & 3 & 1 & 2 \\ 2 & 0 & 2 & 1 \end{bmatrix}$$
(c:14.2)

$$\begin{bmatrix} -1 & 1 & 0 & 2 \\ 1 & 0 & 3 & 0 \\ 0 & 3 & 1 & 3 \\ 2 & 0 & 3 & 1 \end{bmatrix}$$
(c:15.1)

$$\begin{bmatrix} 0 & 1 & 0 & 2 \\ 1 & -1 & 3 & 0 \\ 0 & 3 & 1 & 3 \\ 2 & 0 & 3 & 1 \end{bmatrix}$$
(c:15.2)

$$\begin{bmatrix} -1 & 1 & 0 & 3 \\ 1 & 0 & 3 & 0 \\ 0 & 3 & 1 & 1 \\ 3 & 0 & 1 & 1 \end{bmatrix}$$
(c:16.1)

$$\begin{bmatrix} -1 & 1 & 0 & 3 \\ 1 & 0 & 3 & 0 \\ 0 & 3 & 1 & 2 \\ 3 & 0 & 2 & 1 \end{bmatrix}$$
(c:17.1)

$$\begin{bmatrix} -1 & 1 & 0 & 3 \\ 1 & 0 & 3 & 0 \\ 0 & 3 & 1 & 3 \\ 3 & 0 & 3 & 1 \end{bmatrix}$$
(c:18.1)

$$\begin{bmatrix} -1 & 2 & 0 & 1 \\ 2 & 0 & 1 & 0 \\ 0 & 1 & 1 & 1 \\ 1 & 0 & 1 & 1 \end{bmatrix}$$
(c:19.1)

$$\begin{bmatrix} -1 & 2 & 0 & 1 \\ 2 & 0 & 1 & 0 \\ 0 & 1 & 1 & 2 \\ 1 & 0 & 2 & 1 \end{bmatrix}$$
(c:20.1)

$$\begin{bmatrix} -1 & 2 & 0 & 1 \\ 2 & 0 & 1 & 0 \\ 0 & 1 & 1 & 3 \\ 1 & 0 & 3 & 1 \end{bmatrix}$$
(c:21.1)

$$\begin{bmatrix} -1 & 2 & 0 & 1 \\ 2 & 0 & 2 & 0 \\ 0 & 2 & 1 & 1 \\ 1 & 0 & 1 & 1 \end{bmatrix}$$
(c:22.1)

$$\begin{bmatrix} 0 & 2 & 0 & 1 \\ 2 & -1 & 2 & 0 \\ 0 & 2 & 1 & 1 \\ 1 & 0 & 1 & 1 \end{bmatrix}$$
(c:22.2)

$$\begin{bmatrix} -1 & 2 & 0 & 1 \\ 2 & 0 & 2 & 0 \\ 0 & 2 & 1 & 2 \\ 1 & 0 & 2 & 1 \end{bmatrix}$$
(c:23.1)

$$\begin{bmatrix} 0 & 2 & 0 & 1 \\ 2 & -1 & 2 & 0 \\ 0 & 2 & 1 & 2 \\ 1 & 0 & 2 & 1 \end{bmatrix}$$
(c:23.2)

$$\begin{bmatrix} -1 & 2 & 0 & 1 \\ 2 & 0 & 2 & 0 \\ 0 & 2 & 1 & 3 \\ 1 & 0 & 3 & 1 \end{bmatrix}$$
(c:24.1)

$$\begin{bmatrix} 0 & 2 & 0 & 1 \\ 2 & -1 & 2 & 0 \\ 0 & 2 & 1 & 3 \\ 1 & 0 & 3 & 1 \end{bmatrix}$$
(c:24.2)

$$\begin{bmatrix} -1 & 2 & 0 & 1 \\ 2 & 0 & 3 & 0 \\ 0 & 3 & 1 & 1 \\ 1 & 0 & 1 & 1 \end{bmatrix}$$
(c:25.1)

$$\begin{bmatrix} 0 & 2 & 0 & 1 \\ 2 & -1 & 3 & 0 \\ 0 & 3 & 1 & 1 \\ 1 & 0 & 1 & 1 \end{bmatrix}$$
(c:25.2)

$$\begin{bmatrix} -1 & 2 & 0 & 1 \\ 2 & 0 & 3 & 0 \\ 0 & 3 & 1 & 2 \\ 1 & 0 & 2 & 1 \end{bmatrix}$$
(c:26.1)

$$\begin{bmatrix} 0 & 2 & 0 & 1 \\ 2 & -1 & 3 & 0 \\ 0 & 3 & 1 & 2 \\ 1 & 0 & 2 & 1 \end{bmatrix}$$
(c:26.2)

$$\begin{bmatrix} -1 & 2 & 0 & 1 \\ 2 & 0 & 3 & 0 \\ 0 & 3 & 1 & 3 \\ 1 & 0 & 3 & 1 \end{bmatrix}$$
(c:27.1)

$$\begin{bmatrix} 0 & 2 & 0 & 1 \\ 2 & -1 & 3 & 0 \\ 0 & 3 & 1 & 3 \\ 1 & 0 & 3 & 1 \end{bmatrix}$$
(c:27.2)

$$\begin{bmatrix} -1 & 2 & 0 & 2 \\ 2 & 0 & 2 & 0 \\ 0 & 2 & 1 & 1 \\ 2 & 0 & 1 & 1 \end{bmatrix}$$
(c:28.1)

$$\begin{bmatrix} -1 & 2 & 0 & 2 \\ 2 & 0 & 2 & 0 \\ 0 & 2 & 1 & 2 \\ 2 & 0 & 2 & 1 \end{bmatrix}$$
(c:29.1)

$$\begin{bmatrix} -1 & 2 & 0 & 2 \\ 2 & 0 & 2 & 0 \\ 0 & 2 & 1 & 3 \\ 2 & 0 & 3 & 1 \end{bmatrix}$$
(c:30.1)

$$\begin{bmatrix} -1 & 2 & 0 & 2 \\ 2 & 0 & 3 & 0 \\ 0 & 3 & 1 & 1 \\ 2 & 0 & 1 & 1 \end{bmatrix}$$
(c:31.1)

$$\begin{bmatrix} 0 & 2 & 0 & 2 \\ 2 & -1 & 3 & 0 \\ 0 & 3 & 1 & 1 \\ 2 & 0 & 1 & 1 \end{bmatrix}$$
(c:31.2)

$$\begin{bmatrix} -1 & 2 & 0 & 2 \\ 2 & 0 & 3 & 0 \\ 0 & 3 & 1 & 2 \\ 2 & 0 & 2 & 1 \end{bmatrix}$$
(c:32.1)

$$\begin{bmatrix} 0 & 2 & 0 & 2 \\ 2 & -1 & 3 & 0 \\ 0 & 3 & 1 & 2 \\ 2 & 0 & 2 & 1 \end{bmatrix}$$
(c:32.2)

$$\begin{bmatrix} -1 & 2 & 0 & 2 \\ 2 & 0 & 3 & 0 \\ 0 & 3 & 1 & 3 \\ 2 & 0 & 3 & 1 \end{bmatrix}$$
(c:33.1)

$$\begin{bmatrix} 0 & 2 & 0 & 2 \\ 2 & -1 & 3 & 0 \\ 0 & 3 & 1 & 3 \\ 2 & 0 & 3 & 1 \end{bmatrix}$$
(c:33.2)

$$\begin{bmatrix} -1 & 2 & 0 & 3 \\ 2 & 0 & 3 & 0 \\ 0 & 3 & 1 & 1 \\ 3 & 0 & 1 & 1 \end{bmatrix}$$
(c:34.1)

$$\begin{bmatrix} -1 & 2 & 0 & 3 \\ 2 & 0 & 3 & 0 \\ 0 & 3 & 1 & 2 \\ 3 & 0 & 2 & 1 \end{bmatrix}$$
(c:35.1)

$$\begin{bmatrix} -1 & 2 & 0 & 3 \\ 2 & 0 & 3 & 0 \\ 0 & 3 & 1 & 3 \\ 3 & 0 & 3 & 1 \end{bmatrix}$$
(c:36.1)

$$\begin{bmatrix} -1 & 1 & 0 & 1 \\ 1 & 1 & 1 & 0 \\ 0 & 1 & 0 & 1 \\ 1 & 0 & 1 & 1 \end{bmatrix}$$
(d:1.1)

$$\begin{bmatrix} -1 & 1 & 0 & 1 \\ 1 & 1 & 1 & 0 \\ 0 & 1 & 0 & 2 \\ 1 & 0 & 2 & 1 \end{bmatrix}$$
(d:2.1)

$$\begin{bmatrix} 0 & 1 & 0 & 1 \\ 1 & 1 & 1 & 0 \\ 0 & 1 & -1 & 2 \\ 1 & 0 & 2 & 1 \end{bmatrix}$$
(d:2.2)

$$\begin{bmatrix} -1 & 1 & 0 & 1 \\ 1 & 1 & 1 & 0 \\ 0 & 1 & 0 & 3 \\ 1 & 0 & 3 & 1 \end{bmatrix}$$
(d:3.1)

$$\begin{bmatrix} 0 & 1 & 0 & 1 \\ 1 & 1 & 1 & 0 \\ 0 & 1 & -1 & 3 \\ 1 & 0 & 3 & 1 \end{bmatrix}$$
(d:3.2)

$$\begin{bmatrix} -1 & 1 & 0 & 1 \\ 1 & 1 & 2 & 0 \\ 0 & 2 & 0 & 2 \\ 1 & 0 & 2 & 1 \end{bmatrix}$$
(d:4.1)

$$\begin{bmatrix} 0 & 1 & 0 & 1 \\ 1 & 1 & 2 & 0 \\ 0 & 2 & -1 & 2 \\ 1 & 0 & 2 & 1 \end{bmatrix}$$
(d:4.2)

$$\begin{bmatrix} -1 & 1 & 0 & 1 \\ 1 & 1 & 2 & 0 \\ 0 & 2 & 0 & 3 \\ 1 & 0 & 3 & 1 \end{bmatrix}$$
(d:5.1)

$$\begin{bmatrix} 0 & 1 & 0 & 1 \\ 1 & 1 & 2 & 0 \\ 0 & 2 & -1 & 3 \\ 1 & 0 & 3 & 1 \end{bmatrix}$$
(d:5.2)

$$\begin{bmatrix} -1 & 1 & 0 & 1 \\ 1 & 1 & 3 & 0 \\ 0 & 3 & 0 & 3 \\ 1 & 0 & 3 & 1 \end{bmatrix}$$
(d:6.1)

$$\begin{bmatrix} 0 & 1 & 0 & 1 \\ 1 & 1 & 3 & 0 \\ 0 & 3 & -1 & 3 \\ 1 & 0 & 3 & 1 \end{bmatrix}$$
(d:6.2)

$$\begin{bmatrix} -1 & 1 & 0 & 2 \\ 1 & 1 & 1 & 0 \\ 0 & 1 & 0 & 2 \\ 2 & 0 & 2 & 1 \end{bmatrix}$$
(d:7.1)

$$\begin{bmatrix} -1 & 1 & 0 & 2 \\ 1 & 1 & 1 & 0 \\ 0 & 1 & 0 & 3 \\ 2 & 0 & 3 & 1 \end{bmatrix}$$
(d:8.1)

$$\begin{bmatrix} 0 & 1 & 0 & 2 \\ 1 & 1 & 1 & 0 \\ 0 & 1 & -1 & 3 \\ 2 & 0 & 3 & 1 \end{bmatrix}$$
(d:8.2)

$$\begin{bmatrix} -1 & 1 & 0 & 2 \\ 1 & 1 & 2 & 0 \\ 0 & 2 & 0 & 1 \\ 2 & 0 & 1 & 1 \end{bmatrix}$$
(d:9.1)

$$\begin{bmatrix} -1 & 1 & 0 & 2 \\ 1 & 1 & 2 & 0 \\ 0 & 2 & 0 & 2 \\ 2 & 0 & 2 & 1 \end{bmatrix}$$
(d:10.1)

$$\begin{bmatrix} 0 & 1 & 0 & 2 \\ 1 & 1 & 2 & 0 \\ 0 & 2 & -1 & 2 \\ 2 & 0 & 2 & 1 \end{bmatrix}$$
(d:10.2)

$$\begin{bmatrix} -1 & 1 & 0 & 2 \\ 1 & 1 & 2 & 0 \\ 0 & 2 & 0 & 3 \\ 2 & 0 & 3 & 1 \end{bmatrix}$$
(d:11.1)

$$\begin{bmatrix} 0 & 1 & 0 & 2 \\ 1 & 1 & 2 & 0 \\ 0 & 2 & -1 & 3 \\ 2 & 0 & 3 & 1 \end{bmatrix}$$
(d:11.2)

$$\begin{bmatrix} -1 & 1 & 0 & 2 \\ 1 & 1 & 3 & 0 \\ 0 & 3 & 0 & 1 \\ 2 & 0 & 1 & 1 \end{bmatrix}$$
(d:12.1)

$$\begin{bmatrix} 0 & 1 & 0 & 2 \\ 1 & 1 & 3 & 0 \\ 0 & 3 & -1 & 1 \\ 2 & 0 & 1 & 1 \end{bmatrix}$$
(d:12.2)

$$\begin{bmatrix} -1 & 1 & 0 & 2 \\ 1 & 1 & 3 & 0 \\ 0 & 3 & 0 & 2 \\ 2 & 0 & 2 & 1 \end{bmatrix}$$
(d:13.1)

$$\begin{bmatrix} 0 & 1 & 0 & 2 \\ 1 & 1 & 3 & 0 \\ 0 & 3 & -1 & 2 \\ 2 & 0 & 2 & 1 \end{bmatrix}$$
(d:13.2)

$$\begin{bmatrix} -1 & 1 & 0 & 2 \\ 1 & 1 & 3 & 0 \\ 0 & 3 & 0 & 3 \\ 2 & 0 & 3 & 1 \end{bmatrix}$$
(d:14.1)

$$\begin{bmatrix} 0 & 1 & 0 & 2 \\ 1 & 1 & 2 & 0 \\ 0 & 2 & -1 & 3 \\ 2 & 0 & 3 & 1 \end{bmatrix}$$
(d:14.2)

$$\begin{bmatrix} -1 & 1 & 0 & 3 \\ 1 & 1 & 1 & 0 \\ 0 & 1 & 0 & 3 \\ 3 & 0 & 3 & 1 \end{bmatrix}$$
(d:15.1)

$$\begin{bmatrix} -1 & 1 & 0 & 3 \\ 1 & 1 & 2 & 0 \\ 0 & 2 & 0 & 2 \\ 3 & 0 & 2 & 1 \end{bmatrix}$$
(d:16.1)

$$\begin{bmatrix} 0 & 1 & 0 & 3 \\ 1 & 1 & 2 & 0 \\ 0 & 2 & -1 & 2 \\ 3 & 0 & 2 & 1 \end{bmatrix}$$
(d:16.2)

$$\begin{bmatrix} -1 & 1 & 0 & 3 \\ 1 & 1 & 2 & 0 \\ 0 & 2 & 0 & 3 \\ 3 & 0 & 3 & 1 \end{bmatrix}$$
(d:17.1)

$$\begin{bmatrix} 0 & 1 & 0 & 3 \\ 1 & 1 & 2 & 0 \\ 0 & 2 & -1 & 3 \\ 3 & 0 & 3 & 1 \end{bmatrix}$$
(d:17.2)

$$\begin{bmatrix} -1 & 1 & 0 & 3 \\ 1 & 1 & 3 & 0 \\ 0 & 3 & 0 & 1 \\ 3 & 0 & 1 & 1 \end{bmatrix}$$
(d:18.1)

$$\begin{bmatrix} -1 & 1 & 0 & 3 \\ 1 & 1 & 3 & 0 \\ 0 & 3 & 0 & 2 \\ 3 & 0 & 2 & 1 \end{bmatrix}$$
(d:19.1)

$$\begin{bmatrix} 0 & 1 & 0 & 3 \\ 1 & 1 & 3 & 0 \\ 0 & 3 & -1 & 2 \\ 3 & 0 & 2 & 1 \end{bmatrix}$$
(d:19.2)

$$\begin{bmatrix} -1 & 1 & 0 & 3 \\ 1 & 1 & 3 & 0 \\ 0 & 3 & 0 & 3 \\ 3 & 0 & 3 & 1 \end{bmatrix}$$
(d:20.1)

$$\begin{bmatrix} 0 & 1 & 0 & 3 \\ 1 & 1 & 3 & 0 \\ 0 & 3 & -1 & 3 \\ 3 & 0 & 3 & 1 \end{bmatrix}$$
(d:20.2)

$$\begin{bmatrix} -1 & 2 & 0 & 2 \\ 2 & 1 & 2 & 0 \\ 0 & 2 & 0 & 2 \\ 2 & 0 & 2 & 1 \end{bmatrix}$$
(d:21.1)

$$\begin{bmatrix} -1 & 2 & 0 & 2 \\ 2 & 1 & 2 & 0 \\ 0 & 2 & 0 & 3 \\ 2 & 0 & 3 & 1 \end{bmatrix}$$
(d:22.1)

$$\begin{bmatrix} 0 & 2 & 0 & 2 \\ 2 & 1 & 2 & 0 \\ 0 & 2 & -1 & 3 \\ 2 & 0 & 3 & 1 \end{bmatrix}$$
(d:22.2)

$$\begin{bmatrix} -1 & 2 & 0 & 2 \\ 2 & 1 & 3 & 0 \\ 0 & 3 & 0 & 3 \\ 2 & 0 & 3 & 1 \end{bmatrix}$$
(d:23.1)

$$\begin{bmatrix} 0 & 2 & 0 & 2 \\ 2 & 1 & 3 & 0 \\ 0 & 3 & -1 & 3 \\ 2 & 0 & 3 & 1 \end{bmatrix}$$
(d:23.2)

$$\begin{bmatrix} -1 & 2 & 0 & 3 \\ 2 & 1 & 2 & 0 \\ 0 & 2 & 0 & 3 \\ 3 & 0 & 3 & 1 \end{bmatrix}$$
(d:24.1)

$$\begin{bmatrix} -1 & 2 & 0 & 3 \\ 2 & 1 & 3 & 0 \\ 0 & 3 & 0 & 2 \\ 3 & 0 & 2 & 1 \end{bmatrix}$$
(d:25.1)

$$\begin{bmatrix} -1 & 2 & 0 & 3 \\ 2 & 1 & 3 & 0 \\ 0 & 3 & 0 & 3 \\ 3 & 0 & 3 & 1 \end{bmatrix}$$
(d:26.1)

$$\begin{bmatrix} 0 & 2 & 0 & 3 \\ 2 & 1 & 3 & 0 \\ 0 & 3 & -1 & 3 \\ 3 & 0 & 3 & 1 \end{bmatrix}$$
(d:26.2)

$$\begin{bmatrix} -1 & 3 & 0 & 3 \\ 3 & 1 & 3 & 0 \\ 0 & 3 & 0 & 3 \\ 3 & 0 & 3 & 1 \end{bmatrix}$$
(d:27.1)

$$\begin{bmatrix} -1 & 1 & 0 & 1 \\ 1 & 1 & 1 & 0 \\ 0 & 1 & 1 & 1 \\ 1 & 0 & 1 & 1 \end{bmatrix}$$
(e:1.1)

$$\begin{bmatrix} -1 & 1 & 0 & 1 \\ 1 & 1 & 1 & 0 \\ 0 & 1 & 1 & 2 \\ 1 & 0 & 2 & 1 \end{bmatrix}$$
(e:2.1)

$$\begin{bmatrix} -1 & 1 & 0 & 1 \\ 1 & 1 & 1 & 0 \\ 0 & 1 & 1 & 3 \\ 1 & 0 & 3 & 1 \end{bmatrix}$$
(e:3.1)

$$\begin{bmatrix} -1 & 1 & 0 & 1 \\ 1 & 1 & 2 & 0 \\ 0 & 2 & 1 & 2 \\ 1 & 0 & 2 & 1 \end{bmatrix}$$
(e:4.1)

$$\begin{bmatrix} -1 & 1 & 0 & 1 \\ 1 & 1 & 2 & 0 \\ 0 & 2 & 1 & 3 \\ 1 & 0 & 3 & 1 \end{bmatrix}$$
(e:5.1)

$$\begin{bmatrix} -1 & 1 & 0 & 1 \\ 1 & 1 & 3 & 0 \\ 0 & 3 & 1 & 3 \\ 1 & 0 & 3 & 1 \end{bmatrix}$$
(e:6.1)

$$\begin{bmatrix} -1 & 1 & 0 & 2 \\ 1 & 1 & 1 & 0 \\ 0 & 1 & 1 & 1 \\ 2 & 0 & 1 & 1 \end{bmatrix}$$
(e:7.1)

$$\begin{bmatrix} -1 & 1 & 0 & 2 \\ 1 & 1 & 1 & 0 \\ 0 & 1 & 1 & 2 \\ 2 & 0 & 2 & 1 \end{bmatrix}$$
(e:8.1)

$$\begin{bmatrix} -1 & 1 & 0 & 2 \\ 1 & 1 & 1 & 0 \\ 0 & 1 & 1 & 3 \\ 2 & 0 & 3 & 1 \end{bmatrix}$$
(e:9.1)

$$\begin{bmatrix} -1 & 1 & 0 & 2 \\ 1 & 1 & 2 & 0 \\ 0 & 2 & 1 & 1 \\ 2 & 0 & 1 & 1 \end{bmatrix}$$
(e:10.1)

$$\begin{bmatrix} -1 & 1 & 0 & 2 \\ 1 & 1 & 2 & 0 \\ 0 & 2 & 1 & 2 \\ 2 & 0 & 2 & 1 \end{bmatrix}$$
(e:11.1)

$$\begin{bmatrix} -1 & 1 & 0 & 2 \\ 1 & 1 & 2 & 0 \\ 0 & 2 & 1 & 3 \\ 2 & 0 & 3 & 1 \end{bmatrix}$$
(e:12.1)

$$\begin{bmatrix} -1 & 1 & 0 & 2 \\ 1 & 1 & 3 & 0 \\ 0 & 3 & 1 & 1 \\ 2 & 0 & 1 & 1 \end{bmatrix}$$
(e:13.1)

$$\begin{bmatrix} -1 & 1 & 0 & 2 \\ 1 & 1 & 3 & 0 \\ 0 & 3 & 1 & 2 \\ 2 & 0 & 2 & 1 \end{bmatrix}$$
(e:14.1)

$$\begin{bmatrix} -1 & 1 & 0 & 2 \\ 1 & 1 & 3 & 0 \\ 0 & 3 & 1 & 3 \\ 2 & 0 & 3 & 1 \end{bmatrix}$$
(e:15.1)

$$\begin{bmatrix} -1 & 1 & 0 & 3 \\ 1 & 1 & 1 & 0 \\ 0 & 1 & 1 & 1 \\ 3 & 0 & 1 & 1 \end{bmatrix}$$
(e:16.1)

$$\begin{bmatrix} -1 & 1 & 0 & 3 \\ 1 & 1 & 1 & 0 \\ 0 & 1 & 1 & 2 \\ 3 & 0 & 2 & 1 \end{bmatrix}$$
(e:17.1)

$$\begin{bmatrix} -1 & 1 & 0 & 3 \\ 1 & 1 & 1 & 0 \\ 0 & 1 & 1 & 3 \\ 3 & 0 & 3 & 1 \end{bmatrix}$$
(e:18.1)

Figure G.9 (continued)

Examples

$$\begin{bmatrix} -1 & 1 & 0 & 3 \\ 1 & 1 & 2 & 0 \\ 0 & 2 & 1 & 1 \\ 3 & 0 & 1 & 1 \end{bmatrix}$$ (e:19.1)
$$\begin{bmatrix} -1 & 1 & 0 & 3 \\ 1 & 1 & 2 & 0 \\ 0 & 2 & 1 & 2 \\ 3 & 0 & 2 & 1 \end{bmatrix}$$ (e:20.1)
$$\begin{bmatrix} -1 & 1 & 0 & 3 \\ 1 & 1 & 2 & 0 \\ 0 & 2 & 1 & 3 \\ 3 & 0 & 3 & 1 \end{bmatrix}$$ (e:21.1)
$$\begin{bmatrix} -1 & 1 & 0 & 3 \\ 1 & 1 & 3 & 0 \\ 0 & 3 & 1 & 1 \\ 3 & 0 & 1 & 1 \end{bmatrix}$$ (e:22.1)
$$\begin{bmatrix} -1 & 1 & 0 & 3 \\ 1 & 1 & 3 & 0 \\ 0 & 3 & 1 & 2 \\ 3 & 0 & 2 & 1 \end{bmatrix}$$ (e:23.1)
$$\begin{bmatrix} -1 & 1 & 0 & 3 \\ 1 & 1 & 3 & 0 \\ 0 & 3 & 1 & 3 \\ 3 & 0 & 3 & 1 \end{bmatrix}$$ (e:24.1)
$$\begin{bmatrix} -1 & 2 & 0 & 2 \\ 2 & 1 & 1 & 0 \\ 0 & 1 & 1 & 1 \\ 2 & 0 & 1 & 1 \end{bmatrix}$$ (e:25.1)
$$\begin{bmatrix} -1 & 2 & 0 & 2 \\ 2 & 1 & 1 & 0 \\ 0 & 1 & 1 & 2 \\ 2 & 0 & 2 & 1 \end{bmatrix}$$ (e:26.1)
$$\begin{bmatrix} -1 & 2 & 0 & 2 \\ 2 & 1 & 1 & 0 \\ 0 & 1 & 1 & 3 \\ 2 & 0 & 3 & 1 \end{bmatrix}$$ (e:27.1)

$$\begin{bmatrix} -1 & 2 & 0 & 2 \\ 2 & 1 & 2 & 0 \\ 0 & 2 & 1 & 2 \\ 2 & 0 & 2 & 1 \end{bmatrix}$$ (e:28.1)
$$\begin{bmatrix} -1 & 2 & 0 & 2 \\ 2 & 1 & 2 & 0 \\ 0 & 2 & 1 & 3 \\ 2 & 0 & 3 & 1 \end{bmatrix}$$ (e:29.1)
$$\begin{bmatrix} -1 & 2 & 0 & 2 \\ 2 & 1 & 3 & 0 \\ 0 & 3 & 1 & 3 \\ 2 & 0 & 3 & 1 \end{bmatrix}$$ (e:30.1)
$$\begin{bmatrix} -1 & 2 & 0 & 3 \\ 2 & 1 & 1 & 0 \\ 0 & 1 & 1 & 1 \\ 3 & 0 & 1 & 1 \end{bmatrix}$$ (e:31.1)
$$\begin{bmatrix} -1 & 2 & 0 & 3 \\ 2 & 1 & 1 & 0 \\ 0 & 1 & 1 & 2 \\ 3 & 0 & 2 & 1 \end{bmatrix}$$ (e:32.1)
$$\begin{bmatrix} -1 & 2 & 0 & 3 \\ 2 & 1 & 1 & 0 \\ 0 & 1 & 1 & 3 \\ 3 & 0 & 3 & 1 \end{bmatrix}$$ (e:33.1)
$$\begin{bmatrix} -1 & 2 & 0 & 3 \\ 2 & 1 & 2 & 0 \\ 0 & 2 & 1 & 1 \\ 3 & 0 & 1 & 1 \end{bmatrix}$$ (e:34.1)
$$\begin{bmatrix} -1 & 2 & 0 & 3 \\ 2 & 1 & 2 & 0 \\ 0 & 2 & 1 & 2 \\ 3 & 0 & 2 & 1 \end{bmatrix}$$ (e:35.1)
$$\begin{bmatrix} -1 & 2 & 0 & 3 \\ 2 & 1 & 2 & 0 \\ 0 & 2 & 1 & 3 \\ 3 & 0 & 3 & 1 \end{bmatrix}$$ (e:36.1)

$$\begin{bmatrix} -1 & 2 & 0 & 3 \\ 2 & 1 & 3 & 0 \\ 0 & 3 & 1 & 1 \\ 3 & 0 & 1 & 1 \end{bmatrix}$$ (e:37.1)
$$\begin{bmatrix} -1 & 2 & 0 & 3 \\ 2 & 1 & 3 & 0 \\ 0 & 3 & 1 & 2 \\ 3 & 0 & 2 & 1 \end{bmatrix}$$ (e:38.1)
$$\begin{bmatrix} -1 & 2 & 0 & 3 \\ 2 & 1 & 3 & 0 \\ 0 & 3 & 1 & 3 \\ 3 & 0 & 3 & 1 \end{bmatrix}$$ (e:39.1)
$$\begin{bmatrix} -1 & 3 & 0 & 3 \\ 3 & 1 & 1 & 0 \\ 0 & 1 & 1 & 1 \\ 3 & 0 & 1 & 1 \end{bmatrix}$$ (e:40.1)
$$\begin{bmatrix} -1 & 3 & 0 & 3 \\ 3 & 1 & 1 & 0 \\ 0 & 1 & 1 & 2 \\ 3 & 0 & 2 & 1 \end{bmatrix}$$ (e:41.1)
$$\begin{bmatrix} -1 & 3 & 0 & 3 \\ 3 & 1 & 1 & 0 \\ 0 & 1 & 1 & 3 \\ 3 & 0 & 3 & 1 \end{bmatrix}$$ (e:42.1)
$$\begin{bmatrix} -1 & 3 & 0 & 3 \\ 3 & 1 & 2 & 0 \\ 0 & 2 & 1 & 2 \\ 3 & 0 & 2 & 1 \end{bmatrix}$$ (e:43.1)
$$\begin{bmatrix} -1 & 3 & 0 & 3 \\ 3 & 1 & 2 & 0 \\ 0 & 2 & 1 & 3 \\ 3 & 0 & 3 & 1 \end{bmatrix}$$ (e:44.1)
$$\begin{bmatrix} -1 & 3 & 0 & 3 \\ 3 & 1 & 3 & 0 \\ 0 & 3 & 1 & 3 \\ 3 & 0 & 3 & 1 \end{bmatrix}$$ (e:45.1)

Figure G.9 (continued)

Design Requirements for Near-Constant-Force Mechanisms. The initial design is a four-bar slider mechanism with one flexural pivot located between the slider and the adjacent rigid connecting-rod segment. The goal of this topological synthesis will be to enumerate the possible four-bar slider mechanisms capable of constant force. Therefore, a topological requirement will be that the mechanism be a four-bar slider mechanism. To maintain this basic topology, acceptable compliant chains must contain two adjacent rigid segments. Furthermore, acceptable mechanisms will contain a rigid segment connected to ground by a type-1 connection (prismatic

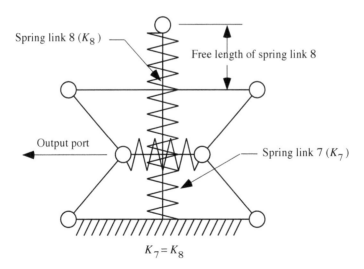

Figure G.10. Exact-constant-force mechanism developed by Jenuwine and Midha [174].

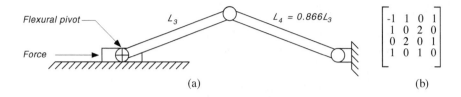

Figure G.11. (a) Near-constant-force mechanism, and (b) its corresponding compliant mechanism matrix (CM).

pair). To further limit the extent of the enumeration process, flexural pivots will not be allowed as connections to compliant segments. The input to the mechanism will remain the single force (or deflection) input at the slider. Finally, to generate a resistive force due to a single input, the rigid-body degrees of freedom for the resulting mechanisms should be less than 1 (i.e., $F_r < 1$).

Topological Synthesis of Near-Constant-Force Mechanisms. Because we are investigating four-bar compliant mechanisms, the topological synthesis work is largely complete. As can be seen by examining the compliant element matrices in Figure G.4, only three of the five compliant chains enumerated in Section G.6.1 contain two adjacent rigid segments. After combining the possible connections between segments with the three acceptable segment combinations, applying the restriction on clamped connections between rigid segments, and removing isomorphic topologies, 63 compliant chains result. After forming kinematic inversions (forming mechanisms), removing isomorphic topologies, and eliminating mechanisms that do not contain a type-1 connection between the ground segment and an adjacent rigid segment, 71 mechanisms result. As noted earlier, to limit the extent of the enumeration process, mechanisms containing flexural pivots connected to compliant segments will be removed from further consideration. As a result of this restriction, the number of mechanisms under consideration is reduced to 32. The compliant mechanism matrices for these mechanisms are presented in Figure G.12.

By examining the rigid-body degrees of freedom for these mechanisms, four more mechanisms are removed from further consideration. These mechanisms contain only kinematic pairs as connections between segments and their subsequent rigid-body degrees of freedom is 1, ($F_r = 1$). For all the other mechanisms the rigid-body degrees of freedom are less than 1, $F_r < 1$. In this example, the input has been specified and a topological analysis is not required. The 28 compliant mechanisms, with the input specified, that meet the design requirements are presented in Figure G.13. It should be noted that in addition to the mechanisms shown in Figure G.13, flexural pivots connected to compliant segments were disclosed in the U.S. patent received for this design [177]. In general, when a design is to be protected as

Examples

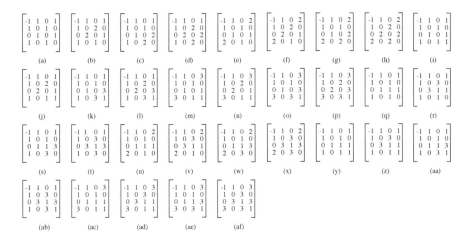

Figure G.12. Compliant mechanism matrices for the 32 compliant mechanisms that satisfy the topological requirements for the near-constant-force mechanism.

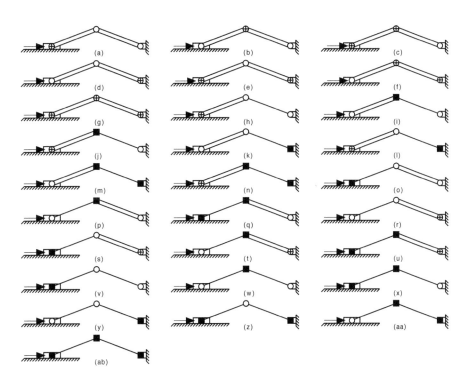

Figure G.13. Twenty-eight near-constant-force four-bar slider mechanisms with the input shown.

an intellectual property, all the possible topologies should be disclosed to enhance patent coverage and prevent competitors from designing around the patent. Conversely, type synthesis may be used as an effective tool to work around competitors' designs or investigate the uniqueness of a new design.

Example: Parallel-Guiding Mechanisms. As noted in Section 10.2, a parallel-guiding mechanism is a mechanism with two opposing segments that remain parallel throughout the mechanism's motion. The basic mechanism is a rigid-body four-bar in which opposing links have the same length to form a parallelogram. Compliant parallel-guiding mechanisms have gained wide usage in optical equipment and micromachined sensors. Because they form the basis for so many mechanisms with industrial applications, the enumeration of compliant four-bar parallel guiding mechanisms is of interest.

Design Requirements for Parallel-Guiding Mechanisms. The goal of this enumeration process will be to determine the topologies of compliant four-bar mechanisms capable of parallel-guiding motion. To ensure this motion, the mechanism must contain a rigid segment opposite the ground segment (nonadjacent). To further limit the extent of the enumeration process, flexural pivots will not be allowed as connections to compliant segments. Since the relative motion is the only design requirement, a topological analysis is not required and the degrees of freedom for acceptable mechanisms are not specified. The original mechanism and its corresponding compliant mechanism matrix are provided in Figure G.14.

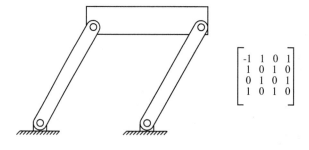

Figure G.14. Rigid-link parallel-guiding mechanism.

Examples

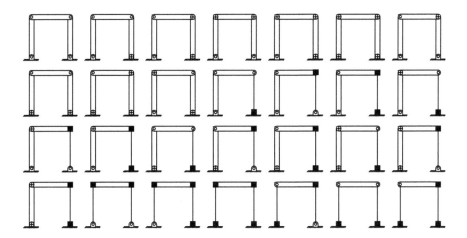

Figure G.15. Twenty-eight configurations of the parallel-guiding mechanism resulting from type synthesis.

Topological Synthesis of Parallel-Guiding Mechanisms. For this example a direct route is taken to the final list of acceptable mechanisms, by applying the design constraints directly to the mechanisms whose compliant mechanism matrices are listed in Figure G.9. These requirements can be applied to the matrices automatically. The requirement that a rigid segment is opposite the ground segment can be stated: If $c(i, i) = -1$ and $c(i, j) = 0$, then $c(j, j) = 0$. The limitation excluding flexural pivots connected to compliant segments can be stated: If $c(i, i) = 1$, then $c(i, j) \neq 2$. By applying these requirements, the 28 mechanism configurations depicted in Figure G.15 are determined. Of these mechanisms, three are fully compliant—they contain no pin joints (kinematic pairs). If a further goal of the enumeration process were to eliminate kinematic pairs, these three mechanisms would be the only acceptable designs.

G.7.1 Discussion

The examples serve to illustrate the importance of design requirements to limit the number of designs that must be considered. Moreover, the use of the matrix formulation for mechanism topology has the additional benefit that these requirements can be checked in an automated fashion. This means that the entire topological synthesis procedure can be automated to enumerate possible solutions, remove isomorphic topologies, and apply design requirements.

INDEX

A
Acceleration analysis 123–125
Active forces 140–141
Adjoint method 326
Advantages of compliant
 mechanisms 2–6
Alternating stress 93
Aluminum alloys
 mechanical properties 403
Analytical gradients 326
Ananthasuresh, G.K. xvi, 301
Angular acceleration 124
Angular deflection approximation
 fixed-pinned segments 152
Angular velocity 123
Arithmetico-geometrical means 421
Artificial material density 321
Axial stiffening 41
Axial stiffness 22
Axial stress 26, 61

B
Beam deflections
 cantilever, linear 34–36
 linear elastic 407–410
 maximum elastic 28
 moment load 43
 see also Large deflections
 see also Pseudo-rigid-body model
Bending stress 26, 61

Bernoulli-Euler equation 34, 42, 45
Bicycle brakes 6222–225, 243–244, 349
Bistable mechanisms 355–384
 buckled beams 382
 cam mechanisms 382–383
 classes 359
 contact force 365
 critical force 365
 critical torque 364
 double slider 377–382
 examples 206–207
 four-bar mechanisms 359–372
 fully compliant example 200
 microbistable mechanisms 202–205
 slider crank 372–382
 stability 355–357
 theorems 362
 Young mechanisms 367–372
Bistable switch 206–207, 365–367
 fatigue 101
Bow 3, 8
Brigham Young University xvi, 103, 367
Brittle materials 26, 32
 microelectromechanical systems 26
 static failure 73–77
 stress concentrations 67, 86
 see also Coulomb-Mohr theory
 see also Modified-Mohr theory
Buckled beams 382
Burmester theory 297–298

451

C

Calculus of variations 315
Cam, bistable 382–383
Cam, passive 183
Catapult 8
Category 14
CGS units 18
Chain algorithm 55261–274
  flexibility matrix 266–268
  shooting method 268–271
Chains, kinematic 112–113, 425
Change point 113360
Characteristic pivot
  fixed-pinned segments 145
  pinned-pinned segments 173
  small-length flexural pivots 136
Characteristic polynomial 432
Characteristic radius
  fixed-pinned segments 145
  initially curved segments 167
Characteristic radius factor 145
  end-moment loading 165
  fixed-guided segments 163
  fixed-pinned segments 148–150
  initially curved segments 167
  pinned-pinned segments 173
  rule of thumb 151–152
Clutch, overrunning 2, 3, 184
Commercial software 205–209
Completely reversed stress 88–92
Complex number method 118–123
  acceleration analysis 123–125
  loop closure 119
  multiloop mechanisms 122–123
  slider crank 120–122
  synthesis 126–131
  velocity analysis 123–125
Complex numbers 118
Compliant mechanisms
  advantages 2–6
  challenges 6–8
  definition 2
  fully compliant 13
  historical background 8–10
  in nature 10–11
  nomenclature 11–15
  partially compliant 13
Compliers 5, 188
Composite materials 32
Connection-type enumeration 437–441
Connection-type vector 432
Conservative load 22
Constant force 337–346
  configurations 340
  dimensional synthesis 339–341
  examples 343–344
  force equation 339342–343
  position analysis 338–339
  robot end effector 4345–346
  stress 344–345
  type synthesis 442–448
Constrained optimization 306
Constraints 306–308
Contact force 365
Continuum models 303–304
Convex linearization 324
Corrosion 87
  fretting 87
Coulomb-Mohr theory 73–77
Coupled equations 288–290
  strongly coupled 289
  uncoupled 289
  weakly coupled 289
Crank rocker 113, 114
  see also Slider crank
Crank slider, *see* Slider crank
Creep 7, 2132–34, 99
  see also Stress relaxation
Critical force 365
Critical torque 364
Cross-axis flexural pivots 8, 9, 180189–190
Curvature 34, 42, 50
Curved segments, *see* Initially curved segments

D

Deflection equations
  fixed-pinned segments 150
  initially curved segments 168
  pinned-pinned segments 173
  small-length flexural pivots 136–138
Degrees of freedom 111–112
  Gruebler's equation 112
Delphi Automotive Systems xvi, 425
Deltoid joints 187–189
Design domain 301
Design parameterization 312
Design variables 306–308
Diagrams, free-body 220–225
Diagrams, skeleton 11, 15
Dimensional synthesis
  definition 127
Direct method 326
Displacement inverter 334
Displacement loads 26–28
Displacement vector 260
Distortion energy theory 68–73

Index

Distributed compliance 303
Double crank 113, 114
Double rocker 113, 114
Double slider mechanisms
   bistable mechanisms 377–382
Ductile materials 69
   static failure 68–73
   stress concentrations 67, 86
   see also Distortion energy theory
   see also Maximum shear stress theory
Ductility 26
Dyad 129
Dyne 18

E

Effective stress 68
Efficiency
   mechanical 310
Elastostatic analysis 304–305
Elliptic integrals 45–55
   beam, combined loads 53–55
   beam, end force 47–53
   complete 46
   evaluation of 421–423
   first kind 46
   incomplete 46
   second kind 46
   third kind 46
End effector
   constant force 4345–346
End-moment loading, prbm
   characteristic radius factor 165
   parametric angle coefficient 165
   pseudo-rigid-body model 165–166
   spring constants 166
   stiffness coefficient 165
   stress 166
   summary 416
Endurance limit 78, 80, 81
   estimation 82
   modification factors 83–87
Energy storage 38–40
Equivalent diameter 85

F

Fatigue 7
   completely reversed stress 88–92
   corrosion 87
   cycles to failure 82
   endurance limit 78, 80, 81, 82
   failure analysis 77–104
   fatigue strength 80, 82
   fluctuating stresses 93–98
   high-cycle 78, 80, 81
   infinite life 78, 80
   low-cycle 78, 80
   modification factors 83–87
   modified Goodman 93
   plastics 98–102
   residual stresses 87
   S-N diagram 78, 80
   stress concentrations 86
   stress life model 79
   temperature effects 87
   testing 102
Fatigue strength 80
   estimation 82
   modification factors 83–87
Fatigue stress-concentration factor 67
Fatigue testing 102
Feasibility 324
Finite difference method 325
Finite element analysis 55260–261,
   304–305
Fixed-fixed segments 175–180
   case I 175–176, 178
   case II 176–177, 178
   case III 177, 179
   examples 243–244
   linear 410
   simplified model 179420
Fixed-guided segments 162–165
   characteristic radius factor 163
   examples 197–200, 207–209, 222–225
   linear 410
   parameterization limit 163
   parametric angle coefficient 164
   spring constants 164
   stress 164
   summary 415
Fixed-pinned segments 145–161
   angle of applied force 146
   angular deflection approximation 152
   characteristic pivot 145
   characteristic radius 145
   characteristic radius factor 145148–150
   deflection equations 150
   examples 157–160, 195–202
   linear 407
   parameterization limit 148, 150
   parametric angle coefficient 152
   practical implementation 160–161, 200
   pseudo-rigid-body angle 145
   rule of thumb 151–152
   spring constants 157

Fixed-pinned segments (Continued)
  stiffness coefficient 152–156
  stiffness coefficient, average 156
  stress 157
  summary 412–414
Flexibility 23–26
  measures of 308–309
Flexibility matrix 266–268
Flexural rigidity 22
Fluctuating stresses 93–98
  alternating stress 93
  mean stress 93
Force at Free End
  see Fixed-pinned segments
Force vector 260
Force-deflection relationships 219–256
Forces
  active 140–141
  nonfollower 140
  passive 140–141
Four-bar mechanism
  acceleration 123–125
  bistable mechanisms 359–372
  change point 113
  crank rocker 113, 114
  double crank 113, 114
  double rocker 113, 114
  Grashofian 113360
  kinematic coefficients 125–126
  mechanical advantage 114
  position analysis 116–117
  pseudo-rigid-body model 239–248
  triple rocker 113
  velocity 123–125
Frecker, Mary I. xvi, 301
Free choices 128, 130
Free variables, see Free choices
Free-body diagrams 220–225
Free-flex pivot
  see Cross-axis flexural pivot
Fretting corrosion 87
Freudenstein's equations 120
Fully compliant mechanism
  definition 13
Fully stressed design 327
Function generation 127–129, 281–283, 290–292
Functional 315
Functional type 13
Functionally Binary Fixed-Pinned
  see Fixed-pinned segments
Functionally binary pinned-pinned
  see Pinned-pinned segments

G

Gauss' scale 421
Generalized convex approximation 324
Generalized coordinates 225–226
Geometric nonlinearities 21
  axial stiffening 41
  large strains 22
  see also Large deflections
  see also Stress stiffening
Gradient 324
Grashof's law 113360
Grashofian mechanism 113360
Gripper mechanism 201–202, 235–237, 320–321
Ground 112
Ground structure 320
Ground structure parameterization 319–321
Gruebler's equation 112

H

Hessian 324
High-cycle fatigue 78, 80, 81
Hoeken straight-line mechanism 276–277
Homogenization method 322
Hyperelasticity 21

I

Ice-cream scoop 37
Ideal mechanical system 230
Infinite life 78, 80
Infinitesimal displacements 298
Initially curved segments, prbm 166–170
  characteristic radius 167
  characteristic radius factor 167
  deflection equations 168
  spring constant 169
  stiffness coefficient 169
  stress 170
  summary 417–418
Inversion, kinematic 112–113
Isomorphism determination 429–433

J

Jensen, Brian xvi
Joints
  higher pairs 112
  kinematic pairs 13, 111
  lower pairs 111, 112
  passive 183–185
  pin 12
  prismatic 13

Index

Joints (Continued)
  pseudo 13
  quadrilateral, or Q
  revolute 12, 111
  sliding 13, 111

K

Karush-Kuhn-Tucker conditions 307
Kind 14
Kinematic chains 112–113
Kinematic coefficients 125–126, 242
  four-bar mechanism 125–126
  slider crank 126
Kinematic inversion 112–113
Kinematic pairs, *see* Joints
Kinematic synthesis
  see Rigid-body replacement
Kinematics 111
Kinetics 111
Kinetostatic synthesis
  see Synthesis with compliance

L

Lagrange multipliers 307
Landen's scale 421
Large deflections 21, 22
  analysis 42–55
  elliptic integrals 45–55
  moment load 43
  numerical methods 259–274
  see also Pseudo-rigid-body model
Large strains 22
LIGA 17
Linear elastic deflections 34–38, 407–410
  cantilever 34–36
  versus nonlinear 21–22
Linear motion mechanism 207–209
Link
  binary 13
  characteristics 14–15
  compound 15
  definition 12
  family 15
  functionally binary 13
  functionally ternary 13
  ground 112
  homogeneous 15
  nonhomogeneous 15
  quaternary 13
  simple 15
  structurally binary 13
  structurally ternary 13
  ternary 13

Living hinges 9, 100 144–145, 161 181–183
  example 200
  geometry 182
  molding 183
Load factor 85
  axial 85
  bending 85
  shear 85
  torsion 85
Loop closure 119 280–281
Low-cycle fatigue 78, 80
Lubrication 2
Lumped compliance 303

M

Maintenance, reduced 2
Marin correction factors 83–87
Material density parameterization 321–323
Material nonlinearities 21
  hyperelasticity 21
  nonlinear elasticity 21
  plasticity 21
  see also Creep
Materials 28–32, 401–405
  brittle 26
  composites 32
  metals 30
  physical properties 401
  plastics 31, 402
  properties 401–405
  ratio of strength to modulus 29–30
  wood 32
Matrix representation 425–429
Maximum shear stress theory 68–73
Mean compliance 308
Mean stress 93
Mechanical advantage 113–115
Mechanical efficiency 310
Mechanism
  crimping 2, 184
  definition 1
  rigid body 1 111–131
  synthesis 126–131
MEMS, *See* Microelectromechanical systems
Metals 30
Michell continua 319
Micro mirror device 192–193
Microbistable mechanisms 202–205
Microelectromechanical systems 15–18
  electrostatic actuator 76, 91
  materials 26

Microelectromechanical systems (Continued)
  micro mirror device 192–193
  microbistable mechanisms 202–205
  static failure 73
  surface micromachining 17
  Young bistable mechanisms 367–372
Midha, Ashok xvi
Mobility 111–112
  Gruebler's equation 112
Modified Goodman 93
Modified-Mohr theory 73–77
Modulus of rigidity 191
Moore rotating-beam device 102
Motion generation 130, 284–286, 294–297
Moving asymptotes 324
Multicriteria formulations 310–312
Multi-degree-of-freedom 254–255
Multiloop mechanism analysis 122–123
Murphy, Morgan D. xvi, 425
Mutual potential energy density 328
Mutual strain energy 309

N
National Science Foundation xvii
Nature 10–11
Nodes 304
Noise, reduction 3
Nomenclature 11–15
Nominal stress 67
Nonconservative load 22
Nonfollower force 140
Nonlinear elastic deflections
  versus linear 21–22
  see also Elliptic integrals
  see also Pseudo-rigid-body model 21
Nonlinear elasticity 21
Nonlinearities
  structural 21
  see also Geometric nonlinearities
  see also Material nonlinearities
Notch sensitivity 67
Number synthesis
  definition 127
Numerical methods 55, 259–274
  chain algorithm 261–274
  finite element analysis 260–261
  see also Chain algorithm
  see also Finite element analysis

O
Objective function 306–308
Optimal synthesis 301–332
  algorithms 324–325
  constrained optimization 306
  constraints 306–308
  continuum models 303–304
  design domain 301
  design parameterization 312
  design variables 306–308
  elastostatic analysis 304–305
  feasibility 324
  homogenization method 322
  Karush-Kuhn-Tucker conditions 307
  Lagrangian multipliers 307
  multicriteria formulations 310–312
  objective function 306–308
  optimality criteria methods 327–331
  sensitivity analysis 324325–326
  shape optimization 312, 319
  size optimization 312–318
  structural optimization 305–306
  topology optimization 312, 319–323
  unconstrained optimization 306
Optimality criteria methods 327–331
  fully stressed design 327
  uniform strain energy density 327
Ortho-planar spring
  infinite life 96
  maximum deflection 70
  principal stresses 64

P
Parallel guiding mechanisms 197–200, 346–353
  applications 347–350
  force analysis 350–352
  position equations 347350
  type synthesis 448–449
  with cross-axis pivots 190
Parallel motion 346–353
  bicycle brakes 6, 349
  see Parallel guiding mechanisms
Parallelogram joints 186–187
Parameterization limit 148, 150, 163
Parametric angle coefficient
  end-moment loading 165
  fixed-guided segments 164
  fixed-pinned segments 152
Part-count 2
Partially compliant mechanism
  definition 13
Passive cam 183
Passive forces 140–141
Passive joints 180183–185
Path generation 127129–130, 283–284, 293–294

Index

Pennsylvania State University xvi, 301
Pennsylvania, University of xvi, 301
Physical properties 401
Pinned-pinned segments, prbm
   characteristic pivot 173
   characteristic radius factor 173
   deflection equations 173
   examples 202–205
   spring constants 174
   stiffness coefficient 174
   stress 174
   summary 418–419
Pivots, cross-axis flexural
   see Cross-axis flexural pivot
Plasticity 21
Plastics 31
   fatigue 98–102
   mechanical properties 402
   see also Polypropylene
Plating, effect on fatigue 87
Pocket knife 383
Polypropylene 30, 31, 69, 181, 402
Polysilicon 26, 30, 73, 76
Position analysis 115–123
   complex number method 118–123
   four-bar mechanism 116–117
   loop closure 119
   multiloop mechanisms 122–123
   slider crank 117 120–122
Precision 3, 17
Precision points 127, 288
Prescribed timing 129, 130
Principal stresses 62–66
Properties of sections 399–400
Pseudo joints 13
Pseudo-rigid-body angle
   fixed-pinned segments 145
   small-length flexural pivots 136
Pseudo-rigid-body models 135–209
   commercial software 205–209
   modeling of mechanisms 194–209
   optimization 298
   see also Initially curved
      segments 166–170
   see also End-moment loading 165–166
   see also Fixed-fixed segments 175–180
   see also Fixed-guided segments 162–165
   see also Fixed-pinned segments 145–161
   see also Pinned-pinned
      segments 170–174
   see also Small-length flexural
      pivots 136–145

Q

Q-joints 180 185–189
   deltoid joints 187–189
   parallelogram joints 186–187
Quadrilateral joint, see Q-joint

R

Ratio of strength to modulus 29–30
References 385–397
Reliability factor 86
Residual stresses 87
Rigid-body replacement 275–286
Rigidity 22
   axial 22
   flexural 22
   see also Stiffness
Rigidity, modulus of 191
Roberts straight-line mechanism 299

S

Sandia National Laboratories 17, 18
Scotch-yoke mechanism 133
Section properties 399–400
Segment, compliant 14–15
   category 14
   characteristics 14–15
   kind 14
Segment-type enumeration 436–441
Sensitivity analysis 324 325–326
   analytical gradients 326
   finite difference method 325
Sequential linear programming 324
Sequential quadratic programming 324
Shape functions 304
Shape optimization 312, 319
Shear stress 62
   torsion 62
   torsion, rectangular 62, 192
Shooting method 268–271
Simply supported beam 408–409
Size factor 84
   axial loading 85
   bending, steel 84
   equivalent diameter 85
   torsion, steel 84
Size optimization 312–318
Skeleton diagrams 11, 15
Slider crank
   bistable 372–382
   complex number method 120–122

Slider crank (Continued)
   compliant 195–197, 220–222, 338–346
   kinematic coefficients 126
   position analysis 117
   pseudo-rigid-body model 248–254
Slider rocker, *see* Slider crank
Small-length flexural pivots 136–145
   active and passive forces 140–141
   characteristic pivot 136
   definition 136
   deflection equations 136–138
   examples 142–144, 201–202, 244–248
   pseudo-rigid-body angle 136
   spring constants 139
   stress 141–142
   summary 411–412
   see also Living hinges
S-N diagram 78, 80
Socket 183
Special-purpose mechanisms 337–353
Split-tube flexures 193–194
   stress 194
Spring constants
   end-moment loading 166
   fixed-guided segments 164
   fixed-pinned segments 157
   initially curved segments 169
   pinned-pinned segments 174
   small-length flexural pivots 139
Stability 355–357
Stainless steels
   mechanical properties 405
Standard form 129
Static failure 67–77
   brittle materials 73–77
   ductile materials 68–73
   see also Coulomb-Mohr theory
   see also Distortion energy theory
   see also Maximum shear stress theory
   see also Modified-Mohr theory
Steels
   mechanical properties 404–405
Steels, stainless
   mechanical properties 405
Stiffness
   definition 10, 22
   measures of 308–309
   versus strength 22–23
Stiffness coefficient
   end-moment loading 165
   fixed-pinned segments 152–156
   intially curved segments 169
   pinned-pinned segments 174
Stiffness matrix 260, 267, 304
Straight-line mechanism
   Hoeken 276–277
   Roberts 299
Strain energy 308
Strain energy density 308, 328
Strength
   definition 10, 22
   versus stiffness 22–23
Stress
   axial 26, 61
   bending 26, 61
   constant force mechanisms 344–345
   end-moment loading 166
   fixed-guided segments 164
   fixed-pinned segments 157
   fluctuating 93–98
   initially curved segments 170
   nominal 67
   pinned-pinned segment 174
   principal 62–66
   residual 87
   shear 62
   shear from torsion 62
   shear from torsion, rectangular 62, 192
   small-length flexural pivots 141–142
   split-tube flexures 194
   von Mises 68, 94
   Young mechanisms 370
Stress concentration factor 67
Stress concentrations 67, 69, 86
   brittle materials 67, 86
   ductile materials 67, 86
Stress life model 79
Stress raisers, *see* Stress concentrations
Stress relaxation 732–34, 99
   see also Creep
Stress stiffening 2241–42
Structural nonlinearities 21
   see also Geometric nonlinearities
   see also Material nonlinearities
Structural optimization 305–306
Structural type 13
Structural universe 320
Structures 27
   compliant 12
   definition 1
Super structure 320
Surface factor 84
Surface micromachining 17
Synthesis 126–131, 275–299
   Burmester theory 297–298

Synthesis (Continued)
  dyad 129
  free choices 128, 130
  function generation 127–129, 281–283, 290–292
  motion generation 130, 284–286, 294–297
  optimization 301–332
  path generation 127129–130, 283–284, 293–294
  rigid-body replacement 275–286
  standard form 129
  with compliance 286–297
  see also Optimal synthesis 301–332
  see also Type synthesis 425–449
Synthesis with compliance 286–297
  coupled equations 288–290

T
Temperature effects 87
Theoretical stress-concentration factor 67
Titanium alloys
  mechanical properties 403
Toggle position 115
Topological synthesis 435–442
  connection-type enumeration 437–441
  segment-type enumeration 436–441
Topology optimization 312, 319–323
  ground structure parameterization 319–321
  homogenization method 322
  material density parameterization 321–323
Torsion 62
  rectangular cross section 62
Torsional hinges 190–193
  angular displacement 191
  rectangular cross section 192
  shear stress 191
Torsional spring constant, *see* Spring constant 169
Tresca theory, *see* Maximum shear stress theory
Triple rocker 113
Type synthesis 425–449
  characteristic polynomial 432
  compliant element matrix 436
  compliant mechanism matrix 426–429
  connection-type designation 428
  connection-type enumeration 437–441
  connection-type vector 432
  definition 126
  isomorphism determination 429–433
  matrix representation 425–429
  segment-type designation 428
  segment-type enumeration 436–441
  topological synthesis 435–442
  vertex-vertex adjacency matrix 426

U
Unconstrained optimization 306
Uniform strain energy density 327
Uniformly distributed load 408, 410
Utah Center of Excellence Program xvii

V
Variational calculus equation 315
Velocity analysis 123–125
Vertex-vertex adjacency matrix 426
Vibration, reduction 3
Virtual displacements 228–230
Virtual work, principle of 219230–256
  definition 230
  examples 235–237, 243–248
  four-bar mechanism 239–248
  generalized coordinates 225–226
  multi-degree-of-freedom 254–255
  virtual displacements 228–230
  virtual work 228–230
von Mises stress 68, 94
von Mises-Hencky, *see* Distortion energy theory

W
Wear, reduced 2, 17
Weight, reduced 2, 4
Wood 32
Work required for deflection 40

Y
Young bistable mechanisms 367–372
  definition 367
  design of 369–370
  micromechanisms 367–372
  stress 370

CPSIA information can be obtained
at www.ICGtesting.com
Printed in the USA
BVHW042107070619
550155BV00028B/236/P